KU-210-900

CREATION'S
TINY MYSTERY

CREATION'S TINY MYSTERY

Robert V. Gentry

Earth Science Associates
Knoxville, Tennessee

Front Cover: Photomicrograph of single ^{218}Po halo
(Magnification × 1840)

Back Cover: Photomicrograph of cluster of ^{218}Po halos
(Magnification × 490)

First printing, October 1986
© 1986 by
Robert V. Gentry

All rights reserved. No part of this book may be reproduced in any form or by any means, electronic or mechanical, including photocopying, recording, or by any information storage and retrieval system, without permission in writing from the author.

Published by Earth Science Associates, Box 12067
Knoxville, Tennessee 37912-0067
Printed by The Eusey Press, Leominster, Massachusetts

Library of Congress Catalog Card Number: 86-081671

ISBN 0-9616753-1-4

DEDICATION

To Patricia
Patti Lynn
Michael
and
David

CONTENTS

FOREWORD

An open letter to the readers of *Creation's Tiny Mystery:*

If I were to follow the unwritten, but commonly understood, guidelines laid down by my fellow evolutionists, many of whom are agnostics like myself, when presented with a book written by a fundamentalist Christian on the topic of "creation," I would ignore the work. Of course, I might kick over the traces a bit, skim through the thing quickly—one *must* be fair, you know—and then give the document a decent quiet burial in the nearest wastebasket. After all, those among us who have brains in our head instead of rocks—presumably put there by the dead hands of ancient superstition—know that (1) science and religion are immiscible, (2) true scientists cannot be creationists, (3) creationists cannot be scientific, let alone scientists, (4) the last factor is doubled and redoubled—in spades—for fundamentalists, (5) as the good nongray Judge Overton has decreed: there is no science in "creation-science," in fact, (6) those poor—but well-heeled by the radical right—fumblers don't even know what science is. The preceeding six commandments—others may be confidently added as time goes on—may be referred to as the A&S Doctrine, in honor of the guiding cosmic luminaries, Isaac Asimov and Carl Sagan.

Fortunately, my scientific education came from teachers who fostered an impertinent curiosity alloyed with a tolerant skepticism. I have news for my evolutionary colleagues: "there are more things in heaven and earth, than are dreamt of in..." the A&S Doctrine. Quite apart from the matter of constitutional justice, which has been decisively treated in the works of Cord and Bird, the question of "origins" remains a challenge not only to the human intellect, but also to the human spirit. *Creation's Tiny Mystery* is a fine documentation of the research of a tenacious, courageous scientist. Robert V. Gentry writes lucidly of his meticulous experimentation with radioactive halos in ancient minerals. Many scientists with international reputations, such as Truman P. Kohman, Edward Anders, Emilio Segre, G.N. Flerov, Paul Ramdohr, Eugene Wigner. E. H. Taylor, etc., have commented favorably in regard to Gentry's integrity and the professional quality of his data. A non-Darwinian evolutionist like me is struck by how often creationists and evolutionists look at the same information, e.g., the fossil record, and extract from it mutually exclusive interpretations.

It is generally believed that science must remain essentially conservative, even "fundamentally conservative"—no pun intended—if its domain is to progress in a nice orderly fashion. This intellectual strategy can lead to an institutionalized bureaucracy of mind, theory, and investigation, that would require a Carroll Quigley to unravel. What are we to think of the chairman of the physics department who urged Gentry to follow a "more conventional thesis problem" that would not lead to an "embarrassment" to the university? Should Svante Arrhenius have played it safely also? Galileo? How many scientists, today, would give up their doctoral work in adherence to a principle? In writing of his struggle to do his own work, to publish his own interpretations that were consistent with his data, Gentry is fighting for academic freedom and intellectual decency for all scientists who defy the established opinion of the day. The investigation of anomalies can be critical to the structure of scientific revolutions, as Thomas Kuhn has suggested.

Creation's Tiny Mystery can be profitably read by all scientists, regardless of their specific discipline, by evolutionists and nonevolutionists alike. Also, it is a challenge to students of government and philosophical thought. Gentry has called into question the practice of science in the institutionalized public arena. Environmental scientists will find Gentry's "young earth model" especially interesting in regard to the problem of nuclear waste confinement. I wonder if his information is being buried somewhere at the bottom of our "tower of Babel" on this problem? Perhaps it is intellectually inconvenient to recognize the potential merit of Gentry's measurements...? In this era of burgeoning governmental waste, it should be encouraging to learn of steps to reduce expenses, even in the research area, but I find it discomforting that "Oak Ridge National Laboratory's budget required marked cutbacks..." such as Gentry's $1.00/year subcontract. Methinks this smacks of evolutionary hubris, especially after Gentry's testimony at Little Rock. Hoyle put it rather well in *Ossian's Ride:* "In science and mathematics, the important thing is what is being said, not who is saying it." Robert V. Gentry is a scientist in the tradition of Galileo. He, his work, and his Weltanschauung do not deserve the premature obituary that my evolutionary colleagues are preparing for it.

W. Scot Morrow, Ph.D.
Associate Professor of Chemistry
Wofford College

PREFACE

Several years ago the TV mini-series *Roots* catapulted to fame the black American author whose book had traced in captivating words the record of the hardships of his forebears. In a sense this book too is about roots, for ultimately it deals with the "roots" of our planet and how and when it came into existence. My method of tracing those roots has been through probing the historical "records" of Earth's crustal rocks.

This quest for truth about origins unfolds a personal odyssey about my experiences in exploring the microscopic world enclosed within the foundation rocks of the earth. The central thesis of this book is that the Creator left decisive evidence enabling us to identify the Genesis rocks of our planet. But genuine evidence for creation falsifies the evolution model of origins, irrespective of how many pieces of the evolutionary puzzle seem to fit together.

By most popular accounts, scientists are thought to be fair, open-minded, and honest, always ready to accept new evidence, even if it conflicts with cherished theories. This book presents the other side of the story—the twenty-year-long effort to publish my discoveries supporting creation. Some people have perceived these results as a threat to their beliefs and have reacted quite negatively. Many others, however, have exhibited an outstanding spirit of fairness to a minority viewpoint. To them I wish to express my sincere appreciation. Without such persons among the ranks of editors, scientists, colleagues, and administrators, my controversial results would not have been published and, equally as important, I would not have gained access to the modern scientific instrumentation so necessary for the progress of my research. Even though this book takes exceptions to the evolutionary viewpoint, many of my good friends and associates espouse this view. In dealing with philosophical preferences, it is necessary to separate personal friendships from one's inherent right to differ on a particular issue.

Throughout the years my brother Joe, my wife's brother Dan, and many others, including Eugene Anderson, John Boyle, Don Jones, George Sharpe, Calvin and Agatha Thrash, and the late Ralph Crawford and Meta Schneider, contributed greatly in many ways to this research project. My mother remained a special inspiration until her demise while this book was in preparation.

Also, I am most appreciative for Dick and Sheila Frezza's timely financial support in getting this book started. The following individuals offered valuable suggestions and/or editorial assistance while this book was in various stages of preparation: Randy Barnes, Art Battson, Jack Blanco, Cal Beisner, Gene Chaffin, Steven Connor, John Eisele, Roland Fanselau, Lee Greer II, Don Jones, Russell Humphreys, Ed Karlow, Lane Lester, David McQueen, Richard Niessen, Jim Melnick, Phil Morrison, Ethel Nelson, Gretchen Passantino, Nancy Pearcy, Ralph Scorpio, Nancy Stake, Stephen Talbott, Calvin Thrash, and Richard Utt.

Finally, I have been exceptionally fortunate that my wife, Patricia, and our children, Patti Lynn, Michael, and David, have been such loyal supporters throughout the many long years when it seemed that nothing would ever result from my research project. This book could not have been written without my wife's unfailing assistance. She collaborated on many of the chapters and oversaw all editorial changes during the numerous manuscript revisions. Patti Lynn and David also used their journalistic abilities to provide special editorial input as the book was in its final stages of preparation, and Michael materially assisted in getting the Radiohalo Catalogue in its final form.

OVERVIEW

Debate over the origin of man is as much alive in the eighties as it was during the famous Scopes trial of 1925. A 1982 Gallup Poll found the public about evenly divided between belief that God created man within the last 10,000 years and belief in some form of evolution. At the very heart of the question of the origin of man is the matter of the origin of the earth.

How did the earth arrive at its present condition? Was it through slow, random, evolutionary changes? Or is there evidence the earth was called into existence by an infinite Creator who is above and beyond His creation? This book deals with these questions as I tell of my efforts to unlock the secrets of nature hidden within the Precambrian granites—the foundation rocks of the earth.

According to modern evolutionary theory, our planet originated from the accumulation of hot, gaseous material ejected from the sun, and the Precambrian granites were among the first rocks to form during the cooling process. University science courses convinced me that the evolution of the earth was just a part of the cosmic evolution of the universe. As a result I became a theistic evolutionist. Years later I began to re-examine the scientific basis for that decision. My thoughts turned to the age of the earth and the Precambrian granites. Were they really billions of years old? The supposed proof of their great age involved certain concentric ring patterns found in the granites. Under the microscope a tiny radioactive particle could be seen at the center of the rings, like the bull's eye at the center of an archery target. These microscopic-sized ring patterns became known as radioactive halos because of their radioactive origin and their halo-like appearance.

Adventure in Science

My enthusiasm for pursuing research on radioactive halos began over two decades ago while I was teaching and working toward a doctorate in physics at the Georgia Institute of Technology in Atlanta. I was informed, however, that the age of the earth had already been scientifically determined, and it was not something the physics department wanted to have reinvestigated. Concerns were expressed that I might find something which would conflict with the accepted evolutionary time scale, and this could be a cause of considerable embarrassment to Georgia Tech. Since the outlook for my research on radiohalos was unfavorable, my plans for completing the doctorate program were forfeited.

Working at home, I used a microscope to search for radiohalos in thin, translucent sections of granite-type rocks. One spring day in 1965 I was pondering over some special types of halos; there seemed to be conflicting requirements as to their origin. According to evolutionary geology, the granites now containing these special halos had originally formed as hot magma slowly cooled over long ages. On the other hand, the radioactivity responsible for these special halos had such a fleeting existence that it would have disappeared long before the magma had time to cool and form the granite rocks. I wondered how this baffling problem would be resolved.

As I peered into the microscope to view these tiny halos again, some profound questions flashed through my mind: Was it possible that the Precambrian granites were not the end product of slowly cooling magma, but instead were the rocks God created when He spoke this planet into existence? Were the special halos evidence of an instantaneous creation? Were they the Creator's fingerprints in Earth's primordial rocks? Was creation a matter of science as well as faith? I determined to explore these questions.

My goal, then, was clear: to pursue an investigation of these halos with the aim of publishing definitive results in well-known scientific journals. I felt the scientific community needed to examine my work prior to presenting it to nonscientists as evidence of creation. My investigations would require expensive research equipment, and the prospects of gaining access to such equipment seemed dim. There was no laboratory space save that carved from a small room in my house and no equipment but a borrowed microscope. Even the granite-type rocks used in my studies had been borrowed from a university in Nova Scotia. Personal funds were almost nonexistent. At the time I could not visualize where this meager beginning would lead in the future.

Though I was an unknown in the scientific community when my research began, a few years later a way opened for me to affiliate for one year as a guest scientist at one of America's national research laboratories. Exceptionally cordial relations were established, and my stay was extended for thirteen years until June 30, 1982. During that time the laboratory's facilities were accessible for all phases of my research, including work on the special radiohalos.

The story behind these investigations, some of which provide evidence for a worldwide flood and young earth, is related in the pages of this book. It provides a behind-the-scenes account of the events surrounding the publication of over twenty reports in notable scientific journals. And it reveals how the scientific establishment reacts when one of its superstatus theories is threatened.

Creation on Trial

The book also details the last year of my guest appointment at the national laboratory, when I was faced with one of the most difficult decisions of my life: whether or not to testify as an expert witness in the 1981 Arkansas creation/evolution trial. The friendship and good will I had established with other scientists over the years were at stake, as was the opportunity to continue my research at this laboratory. As the trial drew near, a number of prominent evolutionists persisted in declaring that scientific evidence for creation was nonexistent.

It seemed the time had come for this claim to be publicly examined. I decided to confront the issue by testifying for creation at the Arkansas trial. There my work would be scrutinized by renowned scientists. They would have an opportunity to expose any flaws. If the special halos in Precambrian granites were not evidence for creation, they should be able to provide an alternative explanation—one which could be scientifically verified. But if the evidence for creation could withstand the scrutiny of some of the world's leading evolutionists and remain untarnished, this scientific truth should not remain hidden from the public.

At the trial, the American Civil Liberties Union (ACLU) argued against the Arkansas law requiring balanced teaching of evolution and creation science. They contended that creation science is religion in disguise because there is no scientific evidence for creation. All their science witnesses, including a world authority in geology, agreed to this view before the court. Under cross-examination the Deputy Attorney General asked this geologist whether he could explain the special halos in the granites. He responded that I had found a "tiny mystery" which scientists would someday solve.

This was a moment I had long waited for—a moment of truth. By postponing the day of reckoning to the indefinite future, one of the world's foremost geologists had deftly sidestepped a major confrontation with the evidence for creation. Yet press reports carried virtually no mention of this event. Moreover, after widely publicizing the evolutionary witnesses' testimony during the first week of the trial, some of the nation's leading newspapers let my testimony fade into oblivion as the trial drew to a close. When my testimony began, some of the media representatives actually left the courtroom.

In other instances the media reports, especially those in various scientific magazines, dealt a fatal blow to my hopes of continuing research at the national laboratory. One prestigious science journal denied me the right to correct a misleading account of my testimony—an action that had far-reaching effects on my research endeavors.

The aftermath of the Arkansas trial was a difficult period, one of those times marked by apparent failure. The ACLU had convinced the judge that my results were irrelevant to the creation/evolution issue. I went to the trial to settle the question of whether valid scientific evidence exists for creation. Yet my presence there had produced only an admission that I had found "a tiny mystery." The scientific press generally cooperated with the ACLU and their expert witnesses in writing my scientific obituary. My search for truth wasn't over, but my contributions to science seemed destined to remain entombed in obscurity.

Then some other thoughts occurred to me. The trial had been the crucial test of the scientific evidences for creation. Indeed, those evidences had stood unrefuted after the most critical examination. Like nothing else could have done, the trial had shown that creation does have a scientific basis. I began to realize that the secrets locked within the granite rocks—the secrets until now hidden within earth's invisible realm—provided the key which unlocked the scientific truth about the origin of the earth and humankind as well. *I sensed this information might be of considerable import to the millions of individuals on this planet who are ardently searching for truth about their roots and their destinies.* Thus the impetus for this book was born out of the ashes of my apparent defeat at the trial.

Creation Science: A Cause for Investigation?

During the Arkansas trial I listened carefully for any new, irrefutable evidence for evolution—such as the synthesis of life from inert matter. During a news conference there I remarked, if this were accomplished, evolution would again be acceptable to me. My intent is not to disparage the evolutionists who advised and testified for the ACLU, but I do question whether their mind sets even allowed them to consider that they might be scientifically wrong.

As Americans, the ACLU and others have the right to oppose the teaching of creation science in the public schools; likewise, I have the right to believe that *if* the public schools are going to teach about origins, students should have the option of studying either the evolution or creation model of origins. If there is unambiguous scientific evidence that one view is true, this should not be kept secret. Under our form of government, citizens may advocate whatever position they choose as far as the Constitution and the courts allow. But is it ethical for the scientific organization which is mandated to advise the Federal Government to unfairly represent the case for creation science in order to maintain preferential treatment of evolution in the public schools? I refer to the most esteemed scientific organization in America—the National Academy of Sciences.

In the spring of 1984 the Academy released a booklet *"Science and Creationism: A View from the National Academy of Sciences"* (National Academy of Sciences 1984). On page two the booklet describes the Academy as a private, self-supporting organization of distinguished scientists which was chartered over a century ago by the U.S. Congress to advise the Federal Government in matters of science and technology. In its official role, the Academy had a double responsibility to act in the highest traditions of science and objectively examine the scientific merits of all evidences for creation. But a prerequisite for this undertaking required that the Academy be open-minded on this issue. The booklet contains a declaration which unmistakably reveals its position:

> . . . The hypothesis of special creation has, over nearly two centuries, been repeatedly and sympathetically considered and rejected on evidential grounds by qualified observers and experimentalists. In the forms given in the first two chapters of Genesis it is now an invalidated hypothesis. . . .
> Confronted by this challenge to the integrity and effectiveness of our national education system and to the hard-won evidence-based foundations of science, the National Academy of Sciences cannot remain silent.

To do so would be a dereliction of our responsibility to academic and intellectual freedom and to the fundamental principles of scientific thought. As a historic representative of the scientific profession and designated advisor to the Federal Government in matters of science, the Academy states unequivocally that the tenets of 'creation science' are not supported by scientific evidence, [and] that creationism has no place in a *science* curriculum at any level...(National Academy of Sciences 1984, 7)

Under the guise of defending intellectual freedom and the integrity of the national education system, the Academy has clearly impugned the scientific integrity of the Bible. If special creation, as described in Genesis, has truly been "rejected on evidential grounds" and "invalidated," as the Academy says, then the Academy should provide the basis for these claims, or else tell where such evidence can be found. But the Academy's booklet fails on both of these counts. Instead, it arbitrarily promotes the view that certain scientific results confirm the evolutionary model, without mentioning all the uncertainties connected with those results. Throughout the booklet plausibility arguments based on questionable assumptions are used to support the evolutionary scenario. In its official capacity as the designated adviser to the Government in matters of science, the Academy has done its utmost to promote evolution as truth. Doubtless there are many who believe that meritorious recognition should be given for this action. History may even record that the timely publication of their booklet was one of the Academy's greatest achievements.

The other possibility is that the Academy will gain lasting fame in history for having opened its own Pandora's box. From the economic standpoint, if genuine scientific evidence for creation has been published in leading scientific journals *and if* the Academy has ignored this evidence while extolling evolution as the only truly scientific theory of origins, should not there be an investigation of this matter? The potential cost for negligence in advising the Government of this information could be enormous. For example, millions of dollars are granted annually by government agencies to fund a variety of evolution-oriented research projects. One well-funded effort concerns attempts to synthesize life from nonliving matter. All such research is based on the fundamental evolutionary assumption that in the distant past life began spontaneously, by chance. However, valid scientific evidence that the earth was created shows the evolutionary scenario to be wrong, and the belief that life began by chance crumbles. Taxpayers have a stake in learning whether the Academy has tried to maintain the status quo of evolution by remaining silent about evidences for creation. And

Americans have more at stake in this issue than their money, almost none of which is used to investigate the scientific basis for creation.

The National Academy of Sciences and Academic Freedom

The format of the Academy's booklet—by excluding a fair presentation of the evidences for creation—suggests the Academy wished to secure the condemnation of creation science on the basis of the eminent reputations of Academy members. Using private funds, the booklet was distributed gratis to numerous public school officials and legislators across America (36,000 to high school superintendents and science department heads, and 9,000 to U.S. Congressmen, governors, and other influential Americans). Clearly the Academy has assumed a leadership role in the growing movement to maintain the exclusive teaching of evolution in public school science courses.

Americans need to be aware of what this action of the Academy means in terms of one of their most cherished heritages. By employing authoritarian measures to promote evolution as truth and creation science as error, the Academy seems to have directly contradicted itself on intellectual freedom. How did this happen?

On April 27, 1976, eight years before its booklet on creation science was published, the Academy adopted a magnificent resolution, quoted below, which aptly represents what America stands for—the freedom to express minority views without fear of repression:

AN AFFIRMATION OF FREEDOM OF INQUIRY AND EXPRESSION
I hereby affirm my dedication to the following principles:
. . . That the search for knowledge and understanding of the physical universe and of the living things that inhabit it should be conducted under conditions of intellectual freedom, without religious, political or ideological restriction.
. . . That all discoveries and ideas should be disseminated and may be challenged without such restriction.
. . . That freedom of inquiry and dissemination of ideas require that those so engaged be free to search where their inquiry leads, free to travel and free to publish their findings without political censorship and without fear of retribution in consequence of unpopularity of their conclusions. Those who challenge existing theory must be protected from retaliatory reactions.

. . . That freedom of inquiry and expression is fostered by personal freedom of those who inquire and challenge, seek and discover.

. . . That the preservation and extension of personal freedom are dependent on all of us, individually and collectively, supporting and working for application of the principles enunciated in the United Nations Universal Declaration of Human Rights and upholding a universal belief in the worth and dignity of each human being.

This *Affirmation* is a marvelous statement of conscience. It focuses attention on the plight of many dissident foreign scientists who might otherwise have been forgotten. We would expect that influential scientists, especially Academy members, would be foremost in adhering to its principles. It is tragic that this prestigious organization, which espoused such high ideals in defense of dissidents, would subsequently advocate a plan that could adversely affect the lives of many school-aged Americans.

In its *Affirmation* the Academy urges that those who search for truth should do so under our right of freedom of inquiry and expression. Does this include public school students in America? Does the Academy believe they have the right to ask, to probe, or to critically inquire about creation science without fear of recrimination from their teachers? After their teachers inform them that *"the Academy states unequivocally that the tenets of 'creation science' are not supported by scientific evidence,"* how many students will ask about it? The few who might venture to do so will now run the risk of being ridiculed because of the invidious comparison which Dr. Frank Press, President of the Academy, makes in his Preface to the booklet:

. . . Teaching creationism is like asking our children to believe on faith, without recourse to time-tested evidence, that the dimensions of the world are the same as those depicted in maps drawn in the days before Columbus set sail with his three small ships, when we *know* from factual observations that they are really quite different. (National Academy of Sciences 1984, 5)

The thrust of Press's innuendo is clear. He insinuates that creationism, equated in the booklet with the first two chapters of Genesis, is a deception which ignores demonstrable scientific evidence. Thus, Press's judgment comes close to insulting those Americans who accept the scientific validity of the Genesis account of creation. It is difficult to conceive of a more effective method of intimidation than for a teacher to quote the above statement in answer to any question about the scientific merits of creation.

Later in his Preface, Press confirms his unalterable faith in evolution:

The theory of evolution has successfully withstood the tests of science many, many times. Thousands of geologists, paleontologists, biologists, chemists, and physicists have gathered evidence in support of evolution as a fundamental process of nature. Our understanding of evolution has been refined over the years, and indeed its details are still undergoing testing and evaluation. For example, some scientists currently debate competing ideas about the rate at which evolution occurred. One group believes that evolution proceeded in small, progressive stages evenly spread throughout the billions of years of geological time; another group believes that there were alternating periods of relatively rapid and slow changes throughout time.

Creationists cite this debate as evidence for disagreement about evolution among scientists; some even suggest that scientists who advocate the latter hypothesis are actually supporting a process similar to that of creationism. What these creationists fail to understand, however, is that neither scientific school of evolutionary thought questions the scientific evidence that evolution took place over billions of years. Rather, the debate centers on only the finer details of *how* it took place. (National Academy of Sciences 1984, 6)

If, as Press claims, the debate centers only on how evolution took place, rather than whether it occurred, in effect the Academy has decreed that creation must be false. Therefore, students have no choice but to accept evolution in their science curricula. Is this suppression of inquiry consistent with the principles of academic freedom for students? Or is it an example of how those in authority can repress an unpopular belief? Some may think that teachers in free America would never attempt to intimidate students for questioning evolution. Unfortunately, this environment existed thirty years ago when I was pursuing my university studies, and as this book reveals, it still exists. The widespread distribution of the Academy's booklet, reflecting the views of confirmed evolutionists, can only be expected to make it worse for conscientious, inquiring students who will not be cowed by proclamations issued by the Academy.

What causes those in the National Academy of Sciences and others, who are confirmed in their evolutionary convictions, to be so entrenched in their views? Perhaps the reason can be found in the following considerations:

Staunch evolutionists are convinced that their theory must be essentially correct because numerous pieces of scientific data from astronomy, geology, and biology seem to mesh naturally to form the beautiful mosaic of evolution. What is often overlooked is that the evolutionary mosaic is actually held together by a glue known as the *uniformitarian principle*. In reality this *principle* is only an assumption that the cosmos, including the earth and life thereon, evolved to its present state through the action of known physical laws. If the *uniformitarian principle* is wrong, then all the pieces in the evolutionary scenario become unglued, and the mosaic disintegrates. Consequently, this *principle* is crucial to the overall concept of evolution.

But valid, scientific evidence for creation would contradict the *uniformitarian principle*. The billions of years postulated for the earth to evolve from some nebulous mass would evaporate when confronted by evidence of an instantaneous creation. The age-dating techniques thought to establish a great age of the earth would be invalidated. The essential time element needed for the geological evolution of the earth and the biological evolution of life on earth would vanish. Thus, unambiguous evidence for creation would devastate the entire evolutionary scenario.

At the Arkansas trial, creation and evolution met in a direct confrontation. The ACLU had the grand opportunity to discredit the evidence for creation. They failed to do this. Instead they minimized the significance of the special halos by having them labeled a "tiny mystery." This ploy was so successful that the judge mimicked the ACLU position—using the term "minor mystery"—when he rendered a verdict favorable to evolution.

As effective as this strategy was in winning the court battle at Little Rock, the court of world opinion has yet to give its verdict on the creation/evolution controversy. This verdict will be rendered in part by those who read this book. In arriving at a decision the reader might reflect on another facet of the label "tiny mystery," not considered by the ACLU. In itself each of the special halos is very tiny; smaller still is a single atom. But enough atoms combined can make a mountain. Likewise, the trillions of "tiny mysteries," embedded in basement rocks all over this planet, together form *Creation's Tiny Mystery*—a Gibraltar of evidence for creation.

By the end of this book the reader should have in hand sufficient information to decide whether the National Academy of Sciences is correct in claiming that special creation is an invalidated hypothesis—or whether the Creator chose to leave positive evidence of creation, thus showing that it is the evolutionary hypothesis which is invalid.

1

Radiohalos and the Age of the Earth

Like most students attending state universities in the fifties I was immersed in the theory of evolution in a first-year biology class. The professor argued persuasively in favor of biological evolution of life over immense periods of time. He presented evolution as the inevitable outworking of the natural laws of the universe, a theory that could be explained in terms of mechanisms observable today. It was the only explanation of origins presented in that class and throughout the remainder of my university curriculum.

I was one of the many Americans brought up in a conservative religious environment which conflicted with the evolutionary concepts taught at the university. However, my convictions were not strong enough to raise questions about the inconsistencies between Genesis and evolution. Students who did were not always treated with respect. As an aspiring scientist, the wisest course was to shun anything controversial. Just as millions of Americans nightly trust their favorite television news commentators to be objective and truthful, my classmates and I trusted that our education was giving us the whole story. Scientific evidence for Genesis was never mentioned; we assumed none existed.

Evolution as a Total Framework

The biological arguments for evolution were not sufficiently convincing for me to become an evolutionist. The final persuasion came several years later when I enrolled in a graduate physics course in cosmology, a field which deals with the origin and development of the universe. The course focused on the Big Bang model, so named because it pictures the universe as having its beginning in a gigantic primeval explosion.

In some respects this theory appealed to me philosophically. It was fascinating to think that science could probe the ultimate beginning of the universe, and this tended to overshadow many uncertainties in the theory. Yet a major question remained: A basic tenet of physics is that matter and energy can neither be created nor destroyed. *But the standard Big Bang theory supposes that absolutely nothing existed before it occurred billions of years ago — neither matter, nor space, nor even time itself.* Logically, then, if the Big Bang had occurred at all, it had to involve the creation of matter. According to the laws of physics this was an impossibility. Here was a fundamental contradiction that I was unable to resolve. Was it realistic to believe the universe had evolved from an event for which there was no scientific explanation?

One day in class the discussion focused on these issues. Sensing an uneasiness developing about the entire concept, the professor mentioned that decades earlier a Catholic theologian, Georges Lemaître, had postulated a possible solution. Lemaître, who was also a cosmologist, suggested that God might have initiated the Big Bang. Why not, I thought. After all, God can do anything: He *could* have started the Big Bang. The final exam for the course was to calculate when the Big Bang had occurred. My result was 5.7 billion years ago, which was considered the right answer at that time. (In the last three decades this figure has escalated to about 17 billion.)

I kept that final exam as a reminder of how much my views about origins had changed during my collegiate days. My university education had transformed me into a theistic evolutionist, one who believed that God intended the Genesis account of creation to be an allegory picturing the total evolution of the cosmos. The pieces of the puzzle now seemed to fall into place—the six days of creation were just six vast, indefinite periods of time. The biological evolution of life on earth was intertwined with the geological evolution of our planet, and everything was traceable to the mystical Big Bang. Science and God were really together after all, and I could still believe in a God who always told the truth.

After receiving my M.S. in physics from the University of Florida in 1956, I became involved in military applications of nuclear weapons effects at Convair-Fort Worth (now General Dynamics). Two years later I continued the same work at what is now the Martin-Marietta Corporation in Orlando, meanwhile zealously defending evolution whenever the occasion arose.

Then someone confronted me with a major obstacle to my belief in a God of truth and my allegorical acceptance of Genesis. He pointed out that God had rewritten the Genesis record of creation in one of the Ten Commandments.

For in six days the Lord made heaven and earth, the sea, and all that in them is, and rested the seventh day. . . (Exodus 20:11)

The context of this passage seemed to indicate that the "days" were literal, not figurative. If this were true, I could no longer associate the six days of creation with six long geological periods of the earth's development, and my basis for believing in theistic evolution would be negated. This was disturbing. Were the Commandments allegorical as well? Where did it all stop? Was anything that God said reliable? Was He really a God of truth? Did He even exist? My package plan uniting God and science seemed to have collapsed. Somehow I had to find time to reinvestigate the scientific evidences for evolution. This long-term goal caused me to re-evaluate my work in the defense industry. For the next two years I taught at the University of Florida and pondered the question of origins while my wife completed her degree in mathematics.

The Question of Origins Reopened

Again I examined the evidence, trying to determine which factors were most important in leading me to accept evolution. It seemed ironic that I had accepted a theological solution (God initiating the Big Bang) to remedy a crucial defect in a supposedly scientific theory (matter and energy from nothing). This brought to mind the supposition that the earliest stars had accumulated from matter synthesized in the Big Bang. The problem was that fragments of an ordinary explosion don't reaccumulate. Then why would matter formed in the greatest of all possible explosions ever reunite to form stars? My doubts about this were later confirmed when I learned an astronomer had said, "If stars did not exist, it would be easy to prove that this is what we expect" (Aller and McLaughlin 1965, 577). And what caused trillions of stars to cluster into the highly ordered systems observed in different galaxies? Could all this have resulted by chance from such a vast homogeneous expansion of matter?

Coming closer to home, how reasonable was it to believe that the origin of our planet was just the last phase in the evolutionary development of the universe? The Big Bang is presumed to have produced just hydrogen and helium, only two of the ninety-two elements of the earth's crust. Where, then, did the remaining ninety elements supposedly originate? Theoretically they came from thermonuclear fusion reactions that occurred billions of years ago deep inside certain stars. In this scenario, space became lightly sprinkled with these other elements when those stars later exploded (supernovae). Assuming all this, how did supernova remnants from throughout the vast reaches of interstellar space reaccumulate to become the raw matter for the solar system? My cosmology course never explained this any more than it explained how stars could develop from the Big Bang. And just how valid was the idea that the planets had their origin in an enormous ring of gases surrounding the sun? What produced the gaseous ring? And what justification was there for believing the earth had its beginning when part of that ring coalesced into a hot, molten sphere—the proto-earth?

Yet, one piece of scientific evidence lent credibility to the entire scenario. My training in physics had led me to place unquestioning confidence in the radiometrically determined age of the earth. These data apparently provided a direct link between the earth's geological evolution and the presumed evolutionary development of the universe. According to radiometric dating techniques, the oldest rocks on earth formed several billion years ago when a hot, molten proto-earth began to cool. Timewise this fitted plausibly into the Big Bang framework. My earlier acceptance of the Big Bang scenario, including biological evolution and the geological evolution of the earth, hinged on my belief that radiometric dating techniques established an ancient age for the earth. But was my belief well founded? It was time to do some critical thinking about the assumptions used in these techniques.

Radioactivity and the Age of the Rocks

Radiometric (or radioactive) dating of rocks involves the decay of some "parent" element into its stable (nonradioactive) end product. As an example, uranium is a parent element which decays to its end product, radiogenic lead. (It is called radiogenic lead to distinguish it from other lead which is not derived from radioactive decay.) By measuring (1) how much of a parent element in a rock has decayed into its end product, and (2) the current rate of this decay, most geologists believe they can assess the date when the parent was incorporated into the rock, or equivalently, the period of time that has elapsed since the rock formed.

My attention turned to the question of whether the decay rates of different radioactive elements have always been what they are at present. A uniform decay rate would mean, for example, that the amount of uranium in a rock would constantly diminish while the end product, radiogenic lead, would constantly increase. In this instance the ratio of uranium to radiogenic lead would be a measure of the time since the rock solidified. If, however, the decay rate was much higher sometime in the past, then the radiogenic lead would have rapidly accumulated in the rock—what normally would have taken eons would have been accomplished in a short period.

On the basis of the uniform decay rate assumption, the rock would be falsely judged to be quite ancient, not because the data (meaning the ratio of uranium to radiogenic lead) was wrong, but because of an erroneous premise. How important, then, it was to know the truth about this matter. My university physics courses had taught me to believe the assumption of uniform decay was beyond question, but no proof was given. Did such proof actually exist? If so, I needed to find it, for some weighty matters about the evolutionary scenario hung in the balance.

The assumption of constant decay rates is an integral part of the evolutionary premise that all physical laws have remained unchanged throughout the history of the universe. This is the *uniformitarian principle,* the glue that holds all the pieces in the evolutionary mosaic together. If it is wrong, all the pieces become unglued and evolution disintegrates. Understandably, scientists who are convinced that evolution is beyond question might have difficulty in considering variable decay rates. To do this would be equivalent to admitting that the *uniformitarian principle* might be in error, which would be tantamount to agreeing that evolution could be wrong. My acceptance of evolution had been quite firm; yet I always remained willing to consider new evidence. Thus I didn't feel any inhibitions about continuing my inquiry into radiometric dating and the crucial question about decay rates.

In the summer of 1962 I was awarded a National Science Foundation Fellowship for three months to attend the Oak Ridge Institute of Nuclear Studies Summer Institute in Oak Ridge, Tennessee. My free time was devoted to studying about radioactivity and the age of the earth. The following fall I taught physics full-time and concurrently pursued graduate studies in physics at the Georgia Institute of Technology in Atlanta. The investigation of radioactive dating techniques was sandwiched between teaching duties and course work. My attention was increasingly drawn to a tiny radioactive phenomenon found in certain rocks because it was thought to be the evidence for the constancy of radioactive decay rates throughout earth history. It occurred to me that a reinvestigation of this phenomenon might serve as an appropriate thesis topic for the doctoral degree. Before approaching the physics department chairman with this suggestion, I perused most of the important scientific reports on the subject. The next three sections are a summary of my initial findings.

The Puzzle of the Rings in the Rocks

The scientific literature revealed a fascinating story that began to unfold in the late 1800's, when improved microscopes became available. Mineralogists realized the microscope could be a powerful tool to examine many features of rocks and minerals, hidden from normal view. They especially wanted to see through pieces of rock to learn how the different minerals were interlaced. To accomplish this they learned to prepare thin, translucent slices of minerals. Mineral specimens that appeared clear of defects with the unaided eye were now often seen to contain tiny grains of other minerals. Most of these tiny grains aroused little interest; mineralogists just assumed they were embedded in the host mineral when it crystallized.

Some of the tiny grains attracted attention, not because of their own appearance, but because of what appeared around them. Mineralogists saw that these grains were surrounded by a series of beautifully colored, concentric rings. Under the microscope the tiny ring patterns resembled a miniature archery target, with the grain at the center as the bull's eye. Because of their halo-like appearance and because they exhibited color variations known as pleochroism in certain minerals, these concentric ring patterns came to be known as pleochroic halos.

Upon further study mineralogists found that what appeared as a series of flat, concentric rings under the microscope was actually a cross section of a group of spherical shells. To illustrate: If an onion is thinly sliced from top to bottom, the onion rings with the largest diameter will be in the slice through the center. The off-center onion slices will still show the ring pattern, but the diameters of the rings will be smaller. This is similar to what mineralogists found when they examined adjacent slices of a mineral containing a pleochroic halo. Thin slices immediately above and below the center grain showed reduced ring sizes when compared to the slice through the center. This proved that the two-dimensional pleochroic halo seen under the microscope was actually a slice of a group of tiny, concentric microspheres.

The presence of the tiny grain in the center was thought to hold the key to the origin of the halos. Some mineralogists speculated that an organic pigment might have been trapped in the halo center when the mineral formed, only to diffuse out later to form tiny colored spheres. However, no one could identify the pigment or satisfactorily explain how diffusion could produce multiple spheres. Pleochroic halos defied explanation until, about the turn of this century, uranium and some other elements were discovered to be radioactive.

The Radioactive Nature of the Halos

In 1907 the solution to the halo puzzle came into focus in the geology laboratory of Professor John Joly of Trinity College in Dublin. Joly was quite familiar with halos, especially those in biotite, a dark mica that is easily split into thin slices. Joly realized the diffusion hypothesis could not explain either the well-defined edges of the halo rings, or their regular sizes. He began to consider a radioactive origin for the halos.

By that time scientists knew that uranium is the parent of a radioactive decay chain, with the successive daughter products being called the members of the chain. This decay chain is shown in Figure 1.1 along with other relevant information. Joly was also aware that uranium and its radioactive daughter products decayed in one of two ways: (1) By ejecting a very light fragment (the beta particle), which causes little damage as it passes through matter, or (2) by ejecting a much heavier nuclear fragment (the alpha particle), which interacts rather strongly as it passes through a substance. Because of its light weight the beta particle is easily bounced around and thereby takes a rather unpredictable, zigzag path before it finally comes to rest in matter. The alpha particle, on the other hand, is heavy enough to plow almost straight ahead before it stops.

As Joly thought about which of these particles might be responsible for the halos, doubtless he quickly realized that the light beta particles would be unlikely to produce coloration changes in the mica, and that their zigzag paths could not yield sharp boundaries. The heavier alpha particles seemed much more promising candidates. Studies had shown that most alpha emitters in the uranium decay chain had different energies, with the isotope ^{238}U being the lowest. (See Fig. 1.1 for further explanation.)

Was there a connection between the different energies of those alpha particles and the different sizes of the rings in the halos Joly had observed? Alpha particles having different energies would travel slightly different distances in a mineral. What if some uranium was in the tiny halo center? Could alpha particles from uranium and its daughters cause enough damage to the surrounding mineral to produce the pleochroic halos?

In a mineral, alpha particles lose their energy quite rapidly through collisions with other atoms. A single alpha particle will ionize about 100,000 atoms along its path of travel, leaving in its wake a short damage trail which remains as a permanent scar. On an atomic scale the damage to the mineral is so small that by itself each tiny scar is invisible. Any mineral, such as mica, which contains trace amounts of uranium, will also contain some

Figure 1.1 Glossary of Technical Terms

Radioactive atoms are capable of spontaneously changing, or decaying, to atoms of a different type. A *parent* radioactive atom *decays* into a *daughter* atom in various ways, one of which is by the emission of an *alpha* (α) particle. Numerous types of radioactive atoms occur in nature, but only three are the initiators of a *decay chain.* For this book the one beginning with uranium-238 (^{238}U in scientific notation) is most important.

The numerical superscript denotes the number of protons and neutrons in the nucleus and signifies how *heavy* the element is. *Isotopes* of the same element have different masses but nearly identical chemical behavior—as for example (^{238}U and ^{235}U). An alpha (α) particle has a mass of 4.

Uranium-238 initiates a chain of steps which ends in the element lead (chemical symbol Pb). The ^{238}U decay chain, as shown below, has some daughters which decay by emitting a beta (β) particle, which is nearly 2000 times lighter than the more massive alpha (α) particle. The type of decay is shown by the symbols α and β.

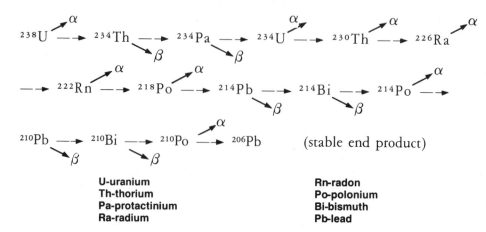

U-uranium   Rn-radon
Th-thorium   Po-polonium
Pa-protactinium   Bi-bismuth
Ra-radium   Pb-lead

The *half-life* of a radioactive isotope is the time required for half the atoms in any collection to decay. If 1000 atoms exist at a certain time, then only 500 will remain after one half-life, after two half-lives only 250 atoms of the original collection will remain, and so forth. *Half-life* and *decay rate* are closely related quantities. Isotopes that decay quickly have short half-lives; those that decay more slowly have longer half-lives. At present ^{238}U is decaying very slowly with a half-life of 4.5 billion years.

alpha-damage trails from the uranium atoms that have already decayed. Generally, however, the uranium atoms are uniformly dispersed throughout the mineral so that the damage trails do not overlap. Thus, a mineral may be filled with invisible alpha-damage trails. Even in the instances when several uranium atoms, or even several hundred, are close enough to produce overlapping trails, this amount of overlap is still insufficient to produce noticeable color changes in the mineral.

In contrast, imagine billions of uranium atoms clustered in the tiny grain at a halo center. Alpha particles ejected from this grain can be compared to the appearance of a vast array of needles stuck into a point. To Joly it seemed quite plausible that the overlapping damage effects of this sunburst pattern of alpha particles might just be sufficient to produce the coloration seen in a halo. Figure 1.2 illustrates this effect.

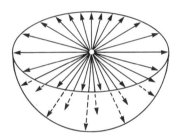

Figure 1.2 Sunburst Effect of Alpha-Damage Trails
Sunburst pattern of alpha-damage trails produces a spherically colored shell around the halo center. Each arrow represents 5 million alpha particles emitted from the center. Halo coloration initially develops after 100 million alpha decays, becomes darker after 500 million, and very dark after 1 billion.

Only one main question now remained: Did the sizes of the halo rings correspond to the path lengths of the uranium series alpha particles in mica? Measurements had shown alpha particles from the uranium decay chain traveled from about three to seven centimeters in air before coming to rest. Joly calculated that in mica alpha particles travel only 1/2000 as far as in air. Reducing the measured air-path lengths of the uranium series alpha particles by this factor gave values which did correspond to the ring sizes of one halo type he had found. The pieces of the puzzle had fallen into place. Joly proposed that alpha emission from the tiny halo center could account for both the sphericity and the size of the different shells comprising the halos. Moreover, the fact that alpha particles do most damage near the end of their paths would explain why the outer edges of halo rings could be darker than the interior regions. Thus Joly specifically identified uranium and a companion element, thorium, as radioactive elements that could produce pleochroic halos. Quite appropriately, they later became known as radioactive halos, or radiohalos.

Figure 1.3 graphically illustrates the idealized three-dimensional cross section of a uranium halo. Color photos of uranium halos appear in the Radiohalo Catalogue. Those photos show five rings of the uranium halo; these can be accounted for by the eight alpha emitters in the uranium decay chain as shown in Fig. 1.3. Figure 1.1 shows there are also six beta emitters in this chain, but, as just discussed, their interaction with mica is insufficient to produce halo rings.

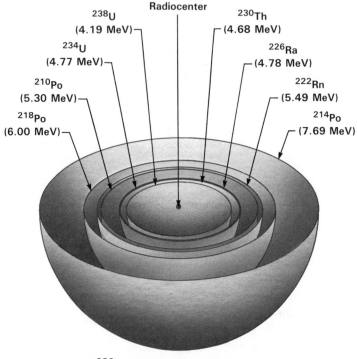

Radiocenter

^{238}U
(4.19 MeV)

^{234}U
(4.77 MeV)

^{210}Po
(5.30 MeV)

^{218}Po
(6.00 MeV)

^{230}Th
(4.68 MeV)

^{226}Ra
(4.78 MeV)

^{222}Rn
(5.49 MeV)

^{214}Po
(7.69 MeV)

^{238}U HALO CROSS SECTION
(^{238}U half-life = 4.5 billion years)

Figure 1.3 Uranium Halo Cross Section
Idealized three-dimensional illustration of a uranium halo obtained by slicing the halo through the center. Each halo ring is identified by the appropriate isotope and its alpha energy in MeV (Million electron Volts).

Radioactive Halos and the Decay Rate Question

Radiohalos, small though they were, soon commanded the attention of many scientists who were interested in questions about the age and origin of the earth. Physicists speculated that halos might provide the data needed to settle the question of whether the decay rate had always been constant. Geologists were vitally interested in this topic because they wanted to use radioactivity as a means of age determination. The question of the age of the earth was still vigorously debated in some geological circles, and this fact generated considerable interest in Joly's results on the measurement of uranium and thorium halo sizes. (For simplicity, the discussion in the rest of this chapter will focus only on the uranium halo.)

The reason for this interest was significant: physicists theorized that halo sizes were directly related to past radioactive decay rates. It was believed that faster decay rates would produce more energetic alpha particles, and hence larger-sized halo rings. Thus, standard-sized rings were thought to prove a constant decay rate whereas a deviation in size was thought to indicate a change in the decay rate sometime during earth history. For many years Joly studied the ring sizes of halos in rocks believed to represent some of the oldest geological ages. In 1923 Joly published a report asserting that uranium halos had ring sizes that varied with age (Joly 1923, 682). The implication was that the radioactive decay rate had varied with time. Of course this result called into question all the radioactive methods of dating rocks. However, the few researchers who studied halos later on disagreed with Joly's conclusions. And they seemed to believe that their own research had nearly settled all remaining questions about the matter. But was this true? Did they have adequate and comprehensive data? More importantly, were halo ring sizes actually a measure of past decay rates?

Microscopic Chances

By the end of 1962, the close of my first graduate quarter at Georgia Tech, I concluded that radioactive halos definitely needed further investigation. I discussed the results of my preliminary study with the physics department chairman and suggested my work could be expanded into a thesis for my doctoral degree. His initial reaction was not very favorable. He felt radioactive dating techniques were almost beyond question and believed my chances of finding anything new about pleochroic halos were "microscopic."

Moreover, he was unwilling to give me that chance of finding anything new. His stated concern was what might happen if perchance I did succeed. Would the end result of my research be an embarrassment to Georgia Tech and many of its faculty? He strongly advised me to give up my interest in radioactive halos and the age of the earth and pursue my doctoral program with a more conventional thesis topic, if I wanted to continue my graduate program at Georgia Tech.

Fortunately, a year of grace was granted for me to make a decision. To do that I needed to investigate the halos themselves, rather than just read about what other investigators had found. In lieu of teaching in the summer of 1963 at Georgia Tech, I borrowed funds for a research trip to Dalhousie University in Halifax, Nova Scotia, where the late physicist G. H. Henderson had conducted a decade-long series of halo investigations during the 1930's.

This trip proved to be a launching point for an intensive study of radioactive halos and their startling revelation about the earth's origin.

2

The Genesis Rocks

The halo photographs in Henderson's scientific reports showed much more clearly defined rings than those reported by Joly. Both investigators had used the dark mica, biotite, in their halo searches. Henderson, however, had used thinner slices, and this had given the sharper rings. Henderson's uranium halos were the very ones needed to make my own measurements. Was his collection of thin sections still available? Correspondence with the geology and physics departments at Dalhousie was not encouraging. Henderson had died many years earlier, and most of his halo collection had been lost. It seemed that a trip to Nova Scotia was the quickest way to obtain more information about the intriguing halos.

The trip was an experience in frugal living, and for a week it appeared little would come of it. Then, early the next week, the head of the physics department returned from a brief trip and managed to locate the few remaining thin sections from Henderson's original halo collection. A few days later my funds had almost run out, but my studies of the thin sections had barely begun. The trip was made a success when it was agreed the halo specimens could be loaned to me for further study. In addition, the geology department gave me many fresh specimens of mica from their museum collection. I returned to Atlanta, borrowed a microscope, and set up a makeshift laboratory in my home.

Unfortunately, Henderson's remaining thin sections did not contain the best uranium halos pictured in his reports. Some with better ring definition had to be found, and this activity began to consume much of my time outside of my teaching duties. The mica specimens given to me at Dalhousie became the source material for my own search. Halos with large centers were common in these specimens, but such halos did not exhibit the delicate ring structure produced by those with point-like centers. Perfect uranium halos with clearly defined rings were needed to settle the question of variable-sized halo rings that Joly had reported. I spent long, tedious hours scanning different pieces of mica, but the perfect uranium halos remained elusive.

The end of my second year at Georgia Tech was approaching, and the time had come to decide about my graduate program. My interest in learning the scientific truth about the age of the earth was stronger than ever. And so was my conviction that radioactive halos might be the key to unlock that truth. At the same time, the physics department chairman remained firm that research on radioactive halos was not an acceptable thesis topic for my doctoral degree; so I left Georgia Tech at the end of that academic year and spent the summer of 1964 doing independent research on halos, using my own funds. (Fortunately, my wife was in total agreement with this decision.) Savings and borrowed funds do run out, though, and that fall I became a substitute high school math teacher in the Atlanta area.

The A, B, C, and D Halos

In addition to the uranium (and thorium) halos, Henderson had reported four other types which he designated as simply A, B, C, and D halos. Along with searching for perfect uranium halos, my attention focused on the D halos. Under the microscope this halo type appeared as a uniformly colored disk with a somewhat fuzzy periphery. It was only about half the size of a fully developed uranium halo; yet it much resembled a uranium halo in an early stage of development when only the inner rings are visible. I became curious about Henderson's tentative association of this halo type with an isotope of radium having a half-life of about 1,600 years. (Figures 1.1 and 1.3 show where this isotope, ^{226}Ra, fits into the uranium decay chain.) The micas in which the D halos had been found were thought to be so old that all the original radium should have died away; only the stable end product was thought to remain in the centers. Henderson claimed the radioactivity in the D halo centers halos should be dead, or "extinct." However, no one had shown this was true, and I decided it was worth investigating. Who knew? Perhaps some new information about the age of the earth would present itself in the process.

The small number of radioactive atoms in the halo centers meant a low rate of alpha-particle emission—only a few particles per month were expected from the uranium-halo centers. Autoradiography was the only technique that could show exactly where an alpha particle originated; hence it was the only technique which could determine whether the D halo centers were still radioactive. The autoradiographic experiments required the use of a special photographic emulsion capable of recording the passage of a single alpha particle. The first step was to split the mica specimen so that the D halo centers were either exposed on the surface or else very close to it. (The specimens chosen sometimes contained uranium halos and one or more of the A, B, or C halos as well.) Step two consisted of pouring a thin layer of this special emulsion over the exposed surface. Under these conditions, nearly half of all the alpha particles ejected from the various halo centers would pass up into the alpha-sensitive emulsion; there they would leave very short trails of ionized atoms. These short trails would remain invisible until the emulsion was developed; after development they appeared as short black tracks when viewed under the microscope. The emulsion-covered halo specimens were placed in a freezer to insure that the tiny trails didn't fade away during the several-week or more storage time.

In the early experiments the emulsion often slid over the sample during the development process. This slippage destroyed the exact registration between the emulsion and the halo centers and made it impossible to know which, if any, of the alpha tracks were actually from the halo centers. A change in procedure remedied this difficulty, and soon I had a technique for maintaining registration throughout the experiments.

After the emulsion was developed, I sometimes observed a few short alpha tracks radiating from both the uranium and the D halo centers. I expected the tracks from the uranium halo centers, but the tracks from the D halos were a surprise. Something long held to be a fact was not true: the D halo centers were not extinct after all. (Later experiments have strongly suggested that the D halos are just uranium halos in an early stage of development, not really a complete surprise considering their almost identical appearance.) It had taken a lot of effort to come to this conclusion, but in the world of science it wasn't much of a discovery. And it didn't seem to have anything to do with my main interest in the age of the earth.

Unspectacular though they were, I decided to present the results of these initial investigations at the January 1965 annual meeting of the American Association of Physics Teachers in New York City. My wife encouraged me to take this trip, even though it depleted the last of our financial reserves. Some new acquaintances, Drs. C.L. and A.M. Thrash, learned of this venture and soon after became the primary sponsors of my research for the next year and a half. This was a difficult time for us, and my research would surely have ended without their help.

Extinct Halos Intrude on the Scene

For a while it seemed the experiments on the D halos were done for no good purpose. In retrospect, it appears they were the most important experiments I could have done at the time. They successfully focused my attention on the A, B, and C halos. Without that focus it is quite possible that my research would soon have ceased. For over a year I had dismissed the A, B, and C halos as being unimportant, not worthy of investigation. Outwardly it seemed that the autoradiographic experiments hadn't shown anything startling at all. In contrast to the uranium and D halos, there was, with one possible exception, a complete absence of alpha tracks from the A, B, and C halos after the emulsion was developed. But it was this general nothingness that finally attracted my attention; it occurred to me that the radioactivity that produced these halos really was extinct! I remembered that Henderson had described these halos in considerable detail and had discussed extinct radioactivity in connection with them. I now went back and carefully reviewed his evaluation.

My measurements of the various halo ring sizes confirmed his tentative conclusion that the A, B, and C halos had originated with alpha radioactivity from three isotopes of the element polonium. These three isotopes— polonium-210, polonium-214 and polonium-218 (in scientific notation ^{210}Po, ^{214}Po and ^{218}Po)—are all members of the uranium decay chain. This didn't necessarily mean that the ^{210}Po, ^{214}Po, and ^{218}Po halos were generated by polonium atoms derived from uranium, but for reasons to be discussed shortly, Henderson postulated that this was the case. He theorized that, sometime in the past, solutions containing uranium and all its daughters must have flowed through tiny cracks, cleavages or conduits in the mica. Under these special conditions he proposed that the isotopes necessary to produce the different polonium halos would gradually accumulate at certain points along the path of the solution. Supposedly, after a certain time, a sufficient number of atoms would be collected for a polonium halo to form.

Earlier this explanation had seemed so plausible that I promptly accepted it and almost lost interest in the A, B, and C halos. However, since the emulsion experiments had shown that their radioactivity was extinct, I became quite interested in why they were extinct and began to think more critically about Henderson's proposed mode of origin. Figures 2.1-2.3 show idealized three-dimensional views of the ^{210}Po, ^{214}Po and ^{218}Po halos. (Color photos of these halos appear in the Radiohalo Catalogue.)

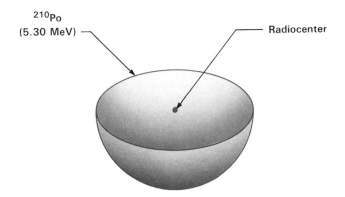

²¹⁰Po
(5.30 MeV)

Radiocenter

²¹⁰Po HALO CROSS SECTION

(²¹⁰Po half-life = 138.4 days)

(²¹⁰Pb half-life = 22 years)

Figure 2.1   ²¹⁰Po Halo Cross Section
Idealized three-dimensional illustration of a ²¹⁰Po halo obtained by slicing the halo through the center. Each halo ring is identified by the appropriate isotope and its alpha energy in MeV (Million electron Volts).

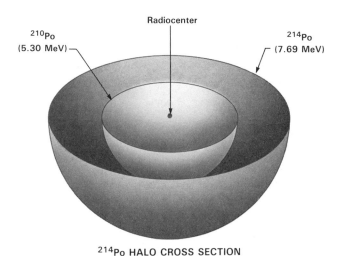

Radiocenter

²¹⁰Po
(5.30 MeV)

²¹⁴Po
(7.69 MeV)

²¹⁴Po HALO CROSS SECTION

(²¹⁴Po half-life = 164 microseconds)

(²¹⁴Pb half-life = 26.8 minutes)

Figure 2.2   ²¹⁴Po Halo Cross Section
Idealized three-dimensional illustration of a ²¹⁴Po halo obtained by slicing the halo through the center. Each halo ring is identified by the appropriate isotope and its alpha energy in MeV (Million electron Volts).

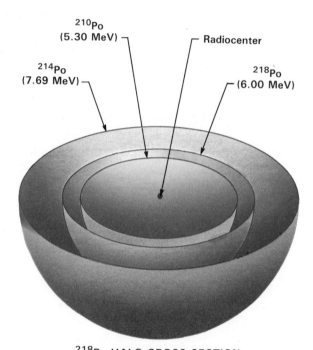

^{210}Po
(5.30 MeV)

Radiocenter

^{214}Po
(7.69 MeV)

^{218}Po
(6.00 MeV)

^{218}Po HALO CROSS SECTION

(^{218}Po half-life = 3 minutes)

Figure 2.3 ^{218}Po **Halo Cross Section**
Idealized three-dimensional illustration of a ^{218}Po halo obtained by slicing the halo through the center. Each halo ring is identified by the appropriate isotope and its alpha energy in MeV (Million electron Volts).

Could Henderson's hypothesis for the secondary origin of polonium halos be tested? He had suggested this should be done. His entrance into Canadian defense work during World War II and his death soon afterward prevented him from doing the tests himself. I began to examine polonium halos closely and paid special attention to why Henderson felt it was necessary to explain polonium halos by some sort of secondary mechanism. Of course! The reason was the vast difference in the decay rate, or average life span, between the uranium atoms and the polonium atoms. Any hypothesis proposed for the origin of the polonium halos had to take this difference into account. On the average, uranium atoms are now decaying so slowly that it would take 4.5 billion years for half of them to undergo radioactive decay. In contrast, the three isotopes responsible for the origin of the polonium halos, namely ^{210}Po, ^{214}Po and ^{218}Po, decay far more rapidly. Their brief life spans present some unique problems in formulating a satisfactory hypothesis for the origin of the respective halos.

The following synopsis, showing what types of radioactivity fit into evolutionary model of the origin of our planet, will enable the reader to more readily grasp the significance of these problems.

Modern Cosmology and Extinct Natural Radioactivity

According to the evolutionary Big Bang scenario, our planet began as a hot molten sphere several billion years ago. Cosmologists admit that the Big Bang, if it occurred, could have produced only hydrogen (H) and helium (He) and that the earliest stars were composed of just these two lightest elements. They assume the heavier chemical elements, of which the earth is mostly comprised, originated in thermonuclear reactions (nucleosynthesis) in the hot interiors of various stars. Supposedly these elements were ejected into space when these stars later exploded as supernovae. They further believe that the newly synthesized elements from one or more supernovae eventually reaccumulated to form vast interstellar clouds of gas. It is assumed that one of these clouds later condensed to form the primeval sun and then the embryonic planets of our solar system. Cosmologists believe a long time elapsed between nucleosynthesis and the formation of the primordial earth. And they also think a certain kind of radioactivity can reveal the approximate length of this period.

In particular, they envision that some radioactive elements formed at nucleosynthesis decayed so slowly that significant fractions of the original amounts have survived to the present— uranium and thorium are examples. They also believe, however, there were other elements whose decay was slow enough for them to be initially incorporated into the primordial earth, but which have almost completely decayed away during the last several billion years. Extinct natural radioactivity is the term used for this special category of radioactive elements. Scientists have diligently sought for extinct natural radioactivity in various rocks because they think it could provide an upper limit on the time interval between nucleosynthesis and the formation of the primordial earth. Because they believe this interval was tens or hundreds of millions of years long, they have searched for certain long half-life (tens of millions of years) radioisotopes in Earth's crustal rocks. One isotope of plutonium (not to be confused with polonium) with a half-life of 83 million years has been found and accepted as extinct natural radioactivity because it fits in with the Big Bang scenario. (Modern cosmologists would consider it useless to look for decay products of relatively short half-life radioactivity, for in their view it would be impossible for there to be such evidence of extinct natural radioactivity.)

The Enigma of the Polonium Halos

The polonium halos in granites present a unique challenge to the evolutionary view of earth history because their origin can be traced directly to certain known isotopes, none of which have long half-lives. Figure 1.1 shows that ^{210}Pb and ^{210}Bi, whose respective half-lives are 22 years and 5 days, successively beta decay to ^{210}Po, the alpha emitter whose half-life is 138 days. Because beta decays do not cause coloration, this means a ^{210}Po halo radiocenter could have initially contained any one of these three isotopes, and a ^{210}Po halo would still have resulted. Likewise, Figure 1.1 shows that a ^{214}Po halo could have initially contained the beta emitters ^{214}Pb, or ^{214}Bi, whose respective half-lives are about 27 minutes and 20 minutes, or the alpha emitter ^{214}Po, whose half-life is 164 microseconds. There is no beta progenitor for ^{218}Po; so the ^{218}Po halo must have originated with this isotope, whose half-life is just three minutes.

Clearly, any of these isotopes which might have formed in a far distant supernova would quickly have decayed away. Never by any stretch of the imagination could they have survived the eons that supposedly elapsed before the hot primeval earth formed. Even in the hypothetical situation where polonium isotopes are imagined to initially exist on the primeval earth, they would never survive the hundreds of millions of years presumably required for its surface to cool down and finally crystallize into granite-type rocks. Thus conventional geological theory considers it impossible for polonium to be a primordial constituent of Earth's granite rocks.

This impossibility is what motivated Henderson to propose a secondary origin of polonium from uranium. Henderson classified polonium halos as extinct only in the sense that the polonium in the halo centers had already decayed away. Never did he hint that polonium halos might represent extinct natural radioactivity, and for over a year and a half neither did this possibility once enter my mind. I simply assumed Henderson's idea for a secondary origin for them was correct—there seemed to be no alternative. Nevertheless, I was puzzled by the fact that in most cases there was no visual evidence of a concentration of uranium near the polonium halos.

Even more puzzling was how the various polonium isotopes would be expected to separate to form the different halo types. Technologically, separation of isotopes is quite difficult because they have almost identical chemical properties. And something else bothered me: Henderson's theory of polonium halo formation primarily involved uranium solutions flowing along tiny conduits or cleavages in the mica. I found, however, polonium halos were also visible in clear areas that were free from those defects. The coloration that I expected to see if uranium had flowed through those areas was generally absent. It was a curious situation. Was it possible that uranium flowed through the mica without leaving a trail of coloration to mark its passage?

About this time a special acid etching technique was discovered that was capable of locating very small amounts of uranium in mica. Application of this technique to regions of mica near polonium halos showed only evidences of trace amounts of uranium (a few parts per million) that exist throughout all mica specimens—there was no concentration of uranium in or near the halo centers in the clear areas. All my attempts to confirm Henderson's hypothesis for a secondary origin of polonium halos had failed. It seemed that polonium halos had not originated with radioactivity derived from uranium. But what other possibility was there? It was most perplexing, like having the solution to a problem but not knowing exactly what the problem was.

Polonium Halos: A Revolutionary New Interpretation

One spring afternoon in 1965 I was examining some thin, transparent sections of mica under the microscope, a task which had been my main research occupation for over a year. Winter had begun to fade, and on that particular day I had moved the microscope to the living room. The afternoon sun beaming through the front windows provided a more conducive atmosphere for contemplation than the shadowy back room that normally served as my laboratory. Again I puzzled over the origin of some beautifully colored polonium halos. The conflicting requirements concerning their origin continued to mystify me. According to evolutionary geology, the Precambrian granites containing these special halos had crystallized gradually as hot magma slowly cooled over long ages. On the other hand, the radioactivity which produced these special radiohalos had such a fleeting existence that it would have disappeared long before the hot magma had time to cool sufficiently to form a solid rock. It was a true enigma. Would I ever resolve it?

Looking up from the microscope I became aware that our home was quiet—our three boisterous young children were asleep. I wondered what they would think if they were old enough to understand what my research was all about.

Back to work. Again I peered through the microscope and could vividly see polonium halos in the thin sections of mica. At that moment the following verses in the Bible flashed through my mind—and immediately triggered some awesome questions:

> *By the word of the Lord were the heavens made; and all the host of them by the breath of his mouth.*
> *For he spake, and it was done; he commanded, and it stood fast. (Psalm 33:6,9)*

Was it possible that the granites had not crystallized out of a slowly cooling magma? Was it possible that the earth had not begun as a molten sphere? Was it even possible that the chemical elements of our planet were not the result of nucleosynthesis in some distant supernova at all—but instead were created instantly when the Creator spoke this planet into existence?

Were the polonium halos mute evidence of extinct natural radioactivity? Was, then, the half-life of ^{218}Po—just three brief minutes—the measure of time that elapsed from the creation of the chemical elements to the time that God formed the granites?

In my search for the truth about the age of the earth, had I discovered evidence for its instantaneous creation?

Were the tiny polonium halos God's fingerprints in Earth's primordial rocks? Could it be that the Precambrian granites were the Genesis rocks of our planet?

I was stunned by these thoughts. Doubtless there were trillions of polonium halos scattered throughout the Precambrian granites around the world. If each one was evidence for creation, it was staggering to think how vast and pervasive this evidence really was! What would its effect be on radiometric and geologic calculations of the age of the earth? How might it affect the way that scientists viewed evolution? Gradually I realized the tremendous implications.

The Impact of Creation on Evolution and the Age of the Earth

Confirmed evolutionists believe that by objective scientific investigations they have been able to fit together numerous pieces of scientific data from astronomy, geology, and biology to construct the beautiful mosaic of evolution. The glue which holds this evolutionary mosaic together is the *uniformitarian principle*. In reality this *principle* is only an assumption that the cosmos, including the earth and life on it, evolved to its present state through the unvarying action of known physical laws. It is the foundation of all radiometric and geological dating methods. Without it there is no basis for assuming that radioactive decay rates have been constant and thus no basis for believing the earth is billions of years old.

Nor is there any basis for *geological uniformitarianism*—the assumption that present rates of accumulation, decomposition and erosion have been constant throughout earth history. After all, geological processes are governed by physical laws. Since valid scientific evidence for an instantaneous creation contradicts the *unifomitarian principle,* it must also contradict *geological uniformitarianism*. Thus the adhesive for all the interlocking pieces in the evolutionary scenario dissolves, and the mosaic falls apart.

Nowhere is this disintegration more apparent than in the area of time. Unambiguous evidence for creation falsifies all aspects of the theory of evolution because it invalidates the basis for the radioactive dating techniques thought to support a great age of the earth. In particular, an instantaneous creation of the granites collapses several billion years of earth history to almost nothing. Comparison of Figure 2.4 (a) and (b) shows how evidence for creation results in a reassignment or elimination of some of the major events in the evolutionary scenario and a drastic telescoping of the time intervals. The billions of years believed necessary for the earth to evolve from some nebulous mass simply evaporate when confronted by such evidence. The essential time element needed for evolution to occur just vanishes.

Primordial and Secondary Rocks

If most of evolutionary time had vanished, there had to be another framework of earth history. Using different premises, could the Genesis account of creation and a worldwide flood provide such a framework? The basement rocks of the continents, the Precambrian granites, would be classed among the primordial Genesis rocks of our planet. And what about the vast rock formations that had been laid down by the action of water—those that contain plant, animal, and marine fossils? Evolutionary theory held that it took hundreds of millions of years for all these sedimentary rocks to accumulate and millions more for erosion to carve out scenic wonders such as the Grand Canyon. But these conclusions were hinged on *geological uniformitarianism*.

EVOLUTION MODEL

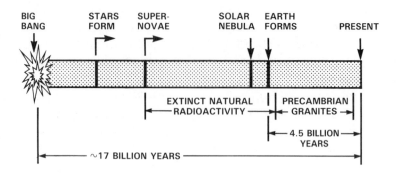

CREATION MODEL

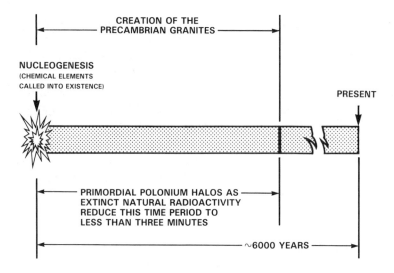

Figure 2.4 Models of Origins
(a) Based on Evolution (b) Based on Creation

If that assumption was incorrect, then I had to ask: Had the major sedimentary formations on the earth's crust resulted from singular, catastrophic events rather than uniformitarian processes? If different premises were used, was it possible that the raw data from geology could also fit into a creation framework of earth history that included catastrophism?

Ideas like these would not have occurred to me ten years earlier. Admittedly, my interest in pursuing this research was sparked by some philosophical questions concerning the Genesis account of earth history, but I determined to be faithful to the scientific evidence no matter where that led me. These new ideas concerning polonium halos would have to meet scientific standards. The only sure guarantee that bias was not creeping into my work would be to study this phenomenon as objectively as possible and present the results in well-known scientific journals. The scientific community attempts to guard itself against bias by publishing experimental results in its refereed literature. Such a forum would enable my data to be scrutinized carefully by researchers from many disciplines, and any errors in methodology or principle would be discerned.

If polonium halos in the granites were part of the evolutionary development of the earth from the Big Bang, they must be explainable on the basis of established physical laws; their origin would have to be traceable to the effects of known chemical elements. I reasoned that, even if I failed to uncover evidence for a conventional explanation, my suggestions of a rapid crystallization of the Precambrian granites would afford other researchers an opportunity to respond with contrary evidence, if such existed.

To obtain some informed feedback on my ideas as soon as possible, I decided to write up the essential details and thereby obtain private critical reviews. Dr. Robert Page, then Director of the Naval Research Laboratory in Washington, DC, agreed to have some of his staff examine the manuscript. The consensus of their opinion was that, if these ideas were published in the open scientific literature, they ''should certainly create comment and some hard analysis . . . which is all to the good.'' I was encouraged that the mystery of the origin of the polonium halos might yet turn into a real adventure in science.

Precambrian Granites—The Genesis Rocks

Tentatively, I identified the Precambrian granites as primordial (or Genesis) rocks because they (1) contain the polonium halos, (2) are the foundation rocks of the continents, and (3) are devoid of the fossils seen in sedimentary rocks. Such granites are coarsely crystalline rocks composed primarily of the light-colored minerals quartz and feldspar, and smaller amounts of biotite and hornblende. I would need to be careful when referring to granites because geologists often used this term to cover a variety of rocks, some of which are not at all similar to the Precambrian granite shown in Figure 2.5.

Figure 2.5 A Precambrian Granite from Finland

It was interesting to learn that the origin of the Precambrian granites (hereafter referred to as simply granites) had been a controversial topic in geology for many decades. One school of geologists speculated that granites, especially the massive formations known as plutons, had crystallized at great depths from slow-cooling magma. The opposite school held that the granites had resulted from recrystallization of pre-existing, deeply buried sedimentary rocks. Eventually both views had become accepted as possible explanations for different types of granites. Yet there was no experimental "standard" by which to judge the relative merits of the two views. There was no direct proof of either hypothesis because massive granitic plutons had never been been observed to form. Neither had sedimentary rocks such as limestones or sandstones been observed to transform into a granite. So, in practice there was no compelling experimental evidence that proved either view was correct.

I reasoned that if the polonium halos in the granites were primordial, it logically followed that the granites must also be primordial—they must be Earth's Genesis rocks. It seemed that a crucial test of this idea hinged on determining whether the polonium halos in the granites were derived secondarily from uranium. If more exhaustive experimentation failed to reveal a secondary origin of those halos, then the primordial hypothesis would remain intact. The research I had in mind would require very expensive modern laboratory facilities. My long-term goal was to conduct in-depth research and disseminate the results through publication in the world's leading scientific journals. Possibly this might be a difficult task because of the strong evolutionary stance of these journals. In the summer of 1965 my short-term goal was to generate the necessary interest for funding that further work.

Were the polonium halos the fingerprints God left to identify the Genesis rocks of our planet? This question provided the driving, motivating force behind all my research.

3

Polonium Halos Go to Press

By late 1965 my investigations of polonium halos had yielded some results that could be submitted for publication. It seemed prudent, however, to begin with another phase of my research which concerned some puzzling, abnormally large halos. I submitted a report to *Applied Physics Letters,* a journal known for rapid publication of new and interesting results in physics. It successfully passed peer review (the screening process used to decide the suitability for publication), and was published early in 1966 (see Gentry 1966a in the References).

Misfits in the Evolutionary Mosaic

Soon afterward I submitted my experimental results on polonium halos for publication to the same journal. Near the end of the manuscript I included the following suggestion about the origin of polonium halos:

> . . . It is difficult to reconcile these results with current cosmological theories which envision long time periods between nucleosynthesis and [the earth's] crustal formation. It is suggested these [polonium] halos are more nearly in accord with a cosmological model which would envision an instantaneous fiat creation of the earth.

I had been naive enough to think that something this straightforward might pass peer review. It didn't. The editor sent the referee's comments, quoted below in their entirety, to me. The "x x x x"'s were substituted by the editor in lieu of certain remarks made by this referee.

> The author appears to be a perfectly competent technician who does not understand or employ the scientific method. He has observed certain

phenomena (halos with anomalous radii) and has considered certain explanations and rejected them. To illustrate his logic, I quote from the next-to-last paragraph of his cover letter, " . . . many of these variant halos cannot be accounted for on the basis of a hydrothermal mode of formation . . . and hence they do represent extinct natural radioactivity from the cosmological standpoint." Failing to think of any other possible explanations, he concludes that the earth was formed by instantaneous fiat. In one blow he implicitly rejects all the carefully accumulated evidence of decades which is in complete conflict with his remarkable conclusion.

He is undoubtedly well aware of the findings of the modern science of geochronology. The scientific approach would be to use all these results to his advantage and try to find a compatible explanation. Without going into a long harangue about "pseudoscience," let me simply say that x x x x, and I regard the reasoning displayed in this manuscript in its present form as unworthy of publication. The experimental observations, minus any wild speculation, might be appropriately reported in a journal such as *Nature.*

Uncomplimentary comments aside, there was one positive note. The reviewer did concede that my investigations might merit publication in the well-respected British scientific journal *Nature,* if the "wild speculation," i.e., the implications for creation, were omitted from the manuscript. This experience taught me a valuable lesson: I was going to have to be more cautious about expressing the implications of the polonium halos if my results were to be published.

A New Affiliation and Better Research Opportunities

Clearly my manuscript would have to be revised before sending it to *Nature.* Possibly more experimental work needed to be done. In the meantime I decided to present my results on polonium halos at the 1966 annual spring meeting of the American Geophysical Union in Washington, DC. This was a national meeting attended by thousands of earth scientists. Only a small number heard my presentation; nevertheless this occasion served to bring my results before the scientific community in a limited way. Perhaps more importantly, at that time at least, this presentation became known to the science faculty at Columbia Union College in nearby Takoma Park, Maryland. They expressed interest in my affiliating with Columbia Union College to continue my research. This new affiliation was effective in July 1966. It was a most welcome change. Acquisition of a quality research microscope and freedom to use the standard laboratory facilities available there made it considerably easier to pursue my investigations. The supportive attitude of all the science faculty, especially Dr. Don Jones, was a source of great encouragement.

Additional experimental results were soon obtained. These were incorporated into a revised manuscript and sent to *Nature*. By leaving out any direct reference to implications for creation, this manuscript successfully passed peer review and was published in early 1967 (Gentry 1967). Using the same strategy I submitted another manuscript on halos to *Earth and Planetary Science Letters*, an international earth science journal published in Amsterdam; this manuscript was also accepted and subsequently appeared in this journal late in 1966 (Gentry 1966b).

Although research on halos occupied most of my time, my general interest in the age of the earth had led me to preliminary investigations of carbon-14 fossil dating. In fact, as early as 1965 my attention was attracted to a report in *Nature* concerning the possible carbon-14 build-up in the atmosphere resulting from the 1908 Tunguska meteor explosion in Russia. My investigations of this topic were summarized in a manuscript I submitted for publication to *Nature*. The manuscript successfully passed peer review and was published in September of 1966 (Gentry 1966c).

Extended Peer Review and Controversy

I realized my 1967 report in *Nature* on polonium halos had not settled the question of their origin in the eyes of my scientific colleagues. To more accurately test whether polonium halos were of secondary origin, I needed a method that would allow me to determine whether uranium solutions had ever passed through a specimen of mica. A newly discovered technique made this evaluation possible. It was based on the fact that an atom decaying by alpha emission leaves a very tiny damage pit when the nucleus of the atom recoils into the mica. By etching the mica with acid, these tiny pits could be enlarged sufficiently to become visible under the microscope. Thus, any uranium solution which might have supplied radioactivity for polonium halos in a piece of mica, must also have produced numerous *additional* recoil pits from those radioactive atoms which decayed in transit. (All mica specimens have a background density of damage pits from trace amounts of uranium.) On this basis the mica specimens containing polonium halos should have a higher damage-pit density than the adjacent areas which are devoid of polonium halos. However, a long series of experiments showed no difference in the density of damage pits between the two specimens.

These experiments provided evidence against the secondary origin of polonium halos in mica.

I wrote up these new results and submitted them for publication in *Science,* a journal with an outstanding reputation among all scientific disciplines. My first draft, submitted in May 1967, concentrated on the experimental results and contained only oblique reference to any implications. As usual, two anonymous referees were chosen to review the manuscript. Referee A approved the manuscript. Referee B wanted more explanation about how polonium halos in granites had originated. My revised manuscript was somewhat more explicit, for I suggested that

> the experimental evidence indicates the inclusions of the polonium halos contained the specific alpha emitters responsible for the halos (or possibly in certain cases beta decaying lead precursors) at the time when the mica crystallized, and as such these particular halos represent extinct radioactivity.

Reviewer B objected to this statement, claiming that I had proposed a "very weak and contradictory argument," and said the manuscript should not be accepted. However, since this referee had not criticized the experimental data, I had the opportunity to ask for further consideration. After some discussion with the editorial office, it was agreed that the manuscript could be revised and that different referees (C and D) would be selected.

My next revision avoided direct references to the contradiction which polonium halos in granites pose to the conventional view of earth history. Instead the implications were phrased in the form of a series of questions. After some delay, I learned referee C had approved this revised manuscript. My hopes were high that referee D would do likewise.

Soon I received another letter from the editorial office, stating that referee D had raised some serious questions which had to be answered before the article could be published. Reviewer D had made some penetrating observations about the possible meaning of my results: Did they suggest a radically different model for the origin of the earth? Part of his review reads as follows:

> Gentry proposes in this and previous papers that "extinct radioactivity" is responsible for halos whose "parents" are polonium and/or lead isotopes with half-lives ranging from 3 minutes to 21 years, and it is clear that he means "extinct natural radioactivity" by his statements that "the inclusions of the polonium halos contained the specific alpha emitters responsible for the halos (or possibly in certain cases beta decaying lead precursors) at the time when the mica crystallized," and "it is not clear just how the existence of short half-life radioactivity may be reconciled with current

cosmological theories which envision long time spans between nucleosynthesis and crustal formation.'' Does he mean to imply that current cosmological (and geological) theories are possibly so wrong that all of the events leading from galactic, or even protosolar, nucleosynthesis to the formation of crystalline rock minerals could have taken place in a few minutes?

Of course the answer was yes! It was gratifying to see the experimental data spoke so loudly that the implications of polonium halos as extinct natural radioactivity could not be overlooked. Figure 2.4(a) illustrates the evolutionary meaning of extinct natural radioactivity and Figure 2.4(b) illustrates the creation implications of polonium halos as extinct natural radioactivity. Despite evidence to the contrary, referee D concluded that Henderson's model of secondary polonium halo formation must somehow be correct. The tenor of his comments made it seem futile to request further consideration of my manuscript. Yet one aspect of his response compelled me to persevere.

A seldom violated rule of the peer review process is that the scientists who act as referees remain anonymous to the authors of the submitted manuscripts. But this reviewer actually requested the editorial office to make his name and address known to me. On the reviewer's statement form he even invited me to contact him directly. Encouraged by his frankness, I telephoned him immediately.

At the very outset of this first conversation he asked my opinion of the implications of polonium halos in granites. Such a direct question deserved an equally direct response. I replied that they seemed to be evidence for creation. Surprisingly enough, he didn't hang up! Instead, this world-renowned authority in radiometric dating continued to ply me with incisive questions over the next hour. At the end of the conversation he was sufficiently impressed with the evidence to suggest that certain other experiments be conducted to enable him to further evaluate the implications of my work. These additional experiments required research equipment not available at Columbia Union College.

Initial Experiments at Oak Ridge

A search for the necessary facilities led me to inquire at the Oak Ridge National Laboratory in Oak Ridge, Tennessee. Years before, while I was still in Atlanta, a staff scientist, Roger V. Neidigh, had kindly assisted in getting some experiments done at this outstanding research complex. I was again extremely fortunate that another Laboratory scientist, John W. Boyle, took a personal interest in arranging for the additional experiments then needed. Without his cordial and very able cooperation they could not have been done.

With the results of these new experiments and a newly revised manuscript in hand, I visited referee D at his own laboratory. This fair-minded colleague made an exhaustive study of the new results and concluded that polonium halos in granites were more perplexing than he had first thought to be the case. The lack of evidence to support the hypothesis that they originated from some secondary source of uranium puzzled him. He indicated a willingness to consider this revised manuscript for publication provided there was no mention of the possibility that polonium halos may have originated with primordial polonium. This report, "Fossil Alpha-Recoil Analysis of Variant Radioactive Halos," was subsequently published in the June 14, 1968, issue of *Science* (Gentry 1968; Appendix—this notation indicates the report cited is also reprinted in the Appendix).

An Invitation to Join a National Laboratory

In addition to my research on polonium halos, I had continued to study some unusal halo types known as the dwarf and giant halos. Their rarity and uncommon sizes suggested they might have originated with an unknown type of radioactivity. In late 1968 the U.S. Atomic Energy Commission (AEC) first became aware of my research on dwarf and giant halos through a contact I initiated with the scientist who was then Chairman of the AEC. Subsequently, arrangements were made for me to give a seminar on my research at the Lawrence Radiation Laboratory (now the Lawrence Berkeley Laboratory) and the Oak Ridge National Laboratory (ORNL). Both laboratories were among several around the world which were then initiating a search for superheavy elements—chemical elements with atomic weights heavier than any previously discovered in nature. Because the dwarf and giant halos seemed to be evidence of unknown radioactivity, I was invited to affiliate with ORNL as a guest scientist and join them in their search for superheavy elements. This one-year opportunity, which stretched to thirteen years, greatly accelerated my research.

Before joining the Oak Ridge National Laboratory the AEC wrote letters of introduction enabling me to visit two well-known Soviet scientists who were involved in the search for superheavy elements. My trip to the Soviet Union in the spring of 1969 included stops in Moscow and Dubna, where the Soviet nuclear laboratory equivalent of ORNL is located.

My move to ORNL occurred in July 1969. By 1970 I had completed a series of new experiments on giant halos using the advanced scientific instrumentation available there. A manuscript detailing those results was prepared for publication. After it passed the standard internal review process at ORNL, it was submitted to *Science*. With minor revisions this report was published in August 1970 as *"Giant Radioactive Halos: Indicators of Unknown Alpha-Radioactivity?"* (Gentry 1970; Appendix). Eight possible explanations for the origin of the giant halos were examined, but at that time none, including superheavy elements, could be identified as the final solution. The origin of the giant halos remained an enigma, and this attracted attention to my research.

Search for Halos in Lunar Rocks

Soon after joining ORNL as a guest scientist, I submitted a proposal to the National Aeronautics and Space Administration (NASA) to search for halos in rocks returned from the Apollo 11 mission to the moon. This proposal was accepted by NASA, and a search was made of the thin sections of the lunar rocks then available. No halos were found. This is not surprising when one considers that the minerals which most often contain halos (in earth rocks) are generally absent from the lunar rocks returned on the Apollo missions. In addition, most of those lunar rocks had recrystallized from molten material produced by meteorite impact. Any halos, if they had existed, would have been destroyed in this process. My summary report on these investigations was published in the *Proceedings of the Second Lunar Science Conference* (Gentry 1971a).

Polonium Halo Analysis

The same advanced analytical techniques employed to study the giant halos were also adaptable to the study of polonium halos. Most of my earlier research on polonium halos had involved the optical microscope, in combination with chemical etching and neutron irradiation techniques. These procedures were quite useful in showing that uranium was generally absent around the polonium halos, but they could not reveal the composition of the halo centers. With the equipment available at ORNL, I analyzed the centers of the halos, the tiny specks where the radioactive atoms themselves were originally encased. Using advanced mass spectrometry techniques I discovered that polonium halo radiocenters contained a composition of the chemical element lead which was different from any previously known. This new type of lead, greatly enriched in the isotope ^{206}Pb, could not be accounted for by uranium decay; yet it was exactly that expected on the basis of the decay of polonium in the halo center. These experimental results, along with others obtained on the puzzling dwarf halos, formed the basis of another report published in *Science* in 1971 (Gentry 1971b).

I expected the discovery of this new type of lead in polonium halo radiocenters to attract more attention to my work on polonium halos than my previous reports. Evidence that this had happened came in 1972 when I received an invitation to contribute a review article on radioactive halos for the *Annual Review of Nuclear Science (ARNS)*. My review article was published in the 1973 edition (Gentry 1973). My *ARNS* article briefly discussed (1) limitations in the original arguments used to establish a uniform radioactive decay rate over geological time, (2) characteristics of a number of unusual types of radioactive halos (dwarf and giant halos) whose origin was still under investigation, and (3) evidence for the existence of primordial polonium halos featuring the results of my most recent experiments at Oak Ridge. In that article I again drew attention to the implications associated with their existence:

> Now the reason for the various attempts to account for Po halos by some sort of secondary process is quite simple; the half-lives of the respective Po isotopes are far too short to be reconciled with slow magmatic cooling rates for Po-bearing rocks such as granites ($t_{1/2}$ = 3 min for ^{218}Po). (Gentry 1973, 356).

A Novel Theory of Polonium Halo Origin

About the time that I was preparing the *ARNS* review article, a colleague who had become interested in my work privately suggested an alternative explanation of polonium halos. He speculated that an uncommon (isomeric) form of radioactivity might have been the source of the polonium. Some colleagues and I used mass spectrometry techniques to investigate this possibility but found no experimental evidence to support it. (Chapter 5 cites the results of a renowned nuclear physicist who later excluded the isomer hypothesis on the basis of his theoretical studies.) Our results were published in *Nature* in August of 1973 (Gentry et al. 1973; Appendix). The following quote from that report shows how attention was again focused on the implications of the polonium halos in Precambrian granites:

...A straightforward attempt to account for the origin of these Po haloes by assuming that Po was incorporated into the halo inclusion at the time of host mineral crystallization meets with severe geological problems: the half-lives of the polonium isotopes ($t_{1/2}$ = 3 min for ^{218}Po) *are too short to permit anything but a rapid mineral crystallization, contrary to accepted theories of magmatic cooling rates.* (Gentry et al. 1973, 282—italics mine)

Suggesting a rapid synthesis of the earth's basement rocks was like raising a red flag before some of my colleagues. Such statements invited scientists to refute my results if it could be done.

Objections Refuted

Indeed, even as the experimental work for this report in *Nature* was underway, three scientists were preparing to contest my results on polonium halos in granites. Their report appeared in the June 22, 1973, issue of *Science* (Moazed et al. 1973). The following quote shows the nature of their objections:

We now report the results of a series of measurements made on polonium-type halos. Our measurements do not support the polonium halo hypothesis. We cannot definitely rule out the existence of polonium halos, but it appears that there is no evidence requiring, or even firmly suggesting, their existence. *It was realized very early that their existence would cause apparently insuperable geological problems since the relevant polonium half-life is of the order of minutes.* Polonium halos would require that the polonium atoms become part of the inclusion within minutes of the formation of the polonium and that in this very short time the polonium must be so far removed from the parent uranium mass that its presence or location is no longer evident. (Moazed et al. 1973, 1272—italics mine).

The issues had begun to focus. These scientists saw that polonium halos in granites presented "apparently insuperable geological problems" to the conventional view of earth history. To protect this view they suggested that polonium halos might not even exist, claiming instead they might just be uranium halos.

A later review of my work, "Mystery of the Radiohalos," *Research Communications Network*, aptly noted the futility of their effort to eliminate polonium halos from the granites:

To date there has been only one effort to dispute Gentry's *identification* of polonium halos. As it turned out, that effort might better never have been written, the authors having been impelled more by the worry that polonium halos "would cause apparently insuperable geological problems," than by a thorough grasp of the evidences. . . . (Talbott 1977, 6—emphasis his: Appendix)

In preparing my reply to the Moazed et al. report I spent months studying uranium and polonium halos, both in mica and in another mineral, fluorite. The Radiohalo Catalogue (see Contents) shows photographs of a variety of those halos. Fluorite sometimes occurs along with mica in the so-called granitic pegmatites—regions within granites where crystals of different minerals can be quite large (several feet long in certain instances). The polonium halos in fluorite are virtually identical to their counterparts in mica. Sometimes they occur along tiny cracks and fissures and sometimes in regions free from mineral defects. Polonium halos in fluorite in defect-free regions are significant because this mineral does not exhibit the perfect cleavage property of mica. Since no cleavages exist for uranium solutions to have flowed in a laminar fashion through fluorite crystals, this excludes the possibility that polonium halos in defect regions could have originated secondarily from uranium daughter radioactivity. This is the same conclusion reached earlier in this chapter when the origin of polonium halos in mica were investigated using alpha-recoil techniques.

A number of new experimental techniques were incorporated into my response to the 1973 report of Moazed et al. A variety of experimental results, obtained with particle accelerators and a scanning electron microscope equipped with x-ray fluorescence capabilities, formed the basis for unambiguously identifying three different types of polonium halos in granites. I elaborated on a new standard for halo-size measurements to show conclusively that polonium halos are easily distinguished from uranium halos by their ring structure. Electron-induced, x-ray fluorescence analyses of selected uranium and polonium halo centers confirmed this difference: the uranium halo centers showed considerable amounts of uranium and only a small amount of lead, whereas the Po halo centers showed only the lead.

I submitted the manuscript to *Science* detailing the results of these experiments. After some revision it was published in April 1974 (Gentry 1974; Appendix). It contains the following statements about an alternative framework of earth history:

. . . It is also apparent that Po halos do pose contradictions to currently held views of Earth history.

. . . A further necessary consequence, that such Po halos could have formed only if the host rocks underwent a rapid crystallization, renders exceedingly difficult, in my estimation, the prospect of explaining these halos by physical laws as presently understood. . . . (Gentry 1974, 62)

. . . The question is, Can they [Po halos] be explained by presently accepted cosmological and geological concepts relating to the origin and development of Earth? (Gentry 1974, 66)

I stated these implications plainly, thus inviting my scientific colleagues to challenge the evidence; however, no one responded to this report.

The Spectacle Halo

During a routine examination of a mica specimen from the Silver Crater Mine near Faraday Township in Ontario, Canada, I discovered a most unusual pattern of ^{210}Po halos. In the more than 100,000 halos which I had examined under the microscope, none had even faintly resembled the connecting circular patterns observed in this "spectacle halo" (a photograph of which is shown in Figure 3.1 and in the Radiohalo Catalogue). Incidentally, the shape of this special halo is completely different from any of the known crystallization patterns, all of which yield minerals with straight edges. No mineral crystallizes in circles; yet for some reason the radiocenters of the "spectacle halo" did. From its appearance it was the crown jewel of halos. If single or small groups of polonium halos had defied explanation by conventional scientific principles, it was certain that the intricate array of polonium halos in the "spectacle halo" could only further compound the problems of explanation. Because of its special value, a variety of analytical techniques were used in some exhaustive studies of this special halo pattern.

The experimental results on this unique halo, obtained in collaboration with several of my colleagues, were first submitted for publication to *Geophysical Research Letters* in the spring of 1974. In this manuscript I made some explicit remarks about the constraints which polonium halos place on cosmological theories. One reviewer recommended that the manuscript be rejected, while the other recommended that it should be published. The latter made the rather astonishing comment that the experimental results were "*...indeed impossible to understand in terms of known nuclear physics and geochemistry.*" In spite of this remark the editor rejected both this manuscript and the revised version.

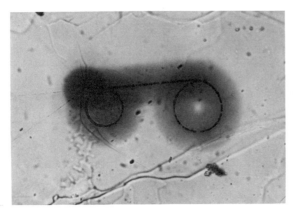

**Figure 3.1 Spectacle Halo in Mica from the Silver Crater Mine
(Magnification ×300)**

It seemed futile to press the issue further with this journal; so I revised the manuscript again, with the cosmological implications of polonium halos toned down, and submitted it to *Nature*. This time it passed peer review and was published in the December 13, 1974, issue of that journal. The statements below show how the report focused attention on the implications of the polonium halos relative to a rapid synthesis of Precambrian rocks:

> Polonium radiohaloes occur widely and not infrequently (total about 10^{15}-10^{20}) in Precambrian rocks, but their existence has so far defied satisfactory explanation based on accepted nucleocosmogeochemical theories. Do Po haloes imply that unknown processes were operative during the formative period of the earth? Is it possible that Po haloes in Precambrian rocks represent extinct natural radioactivity and are therefore of cosmological significance? (Gentry et al. 1974, 564; Appendix)

The last chapter emphasized that when I associate polonium halos in granites with extinct natural radioactivity, scientists understand this to imply only a few minutes elapsed from nucleosynthesis to the formation of a solid earth. As Figure 2.4 (b) illustrates, the only "nucleosynthesis" that could accomplish this feat is the "nucleogenesis" initiated by the Creator—that of a virtually instantaneous creation of the earth.

This report did not go unnoticed. In a letter to *Nature*, Professor J. H. Fremlin, a leading radiophysicist in England, resurrected the idea that polonium halos in granites were secondarily derived from uranium, but provided no new data to support his suggestion (Fremlin 1975). Moreover, he tended to overlook much of my published evidence showing polonium halos in granites had originated independently of uranium. Years earlier it occurred to me that this type of thing might continue indefinitely unless I could find polonium halos which were definitely of secondary origin and show how they differed from polonium halos in granites. My search was successful, and the results were so relevant to the question of polonium halo origin in granites that I briefly mentioned them in my response (Gentry 1975) to Fremlin's letter.

Unfortunately, some colleagues overlooked these new data the next year when they too proposed a secondary origin of polonium halos in granites (Meier and Hecker 1976). Their oversight was more understandable than the case of others (Hashemi-Nezhad et al. 1979) who later overlooked my complete report on the new data published in 1976 (Gentry et al. 1976a; Appendix). As the next chapter shows, that 1976 report describes where secondary polonium halos were discovered and how they were found to be intrinsically different from the polonium halos in granites. The evidence in this report directly contradicted the idea of a secondary origin for polonium halos in granites. But we shall see later that some scientists would still find it difficult to accept this conclusion.

4

Secondary Polonium Halos Fuel the Controversy

During the early seventies I began to contemplate where secondary polonium halos might have formed, realizing that the first requirement was an abundance of uranium to supply the secondary polonium atoms. Whatever the host substance, it must have allowed rapid movement of those atoms; otherwise, because of the short half-lives, all of them would have decayed before they could be captured. Of course, even in a matrix where polonium atoms could move freely, there must also exist microscopic sites where polonium would be collected in order for the halos to form. Summarizing, I was seeking geological specimens that were (1) high in uranium, (2) capable of having allowed rapid movement of secondary polonium atoms, and (3) possessing microscopic-sized capture sites for those polonium atoms.

Uranium in Coalified Wood

These special requirements brought to mind a reference to radioactivity in wood about which I learned several years earlier (Jedwab 1966). Further checking revealed that pieces of wood, partially turned to coal, some as large as logs, had been found in certain uranium mines in western states. The mines were located in the uranium-rich sedimentary deposits in the region geologically known as the Colorado Plateau. Previous microscopic studies of thin slices of these specimens showed halos, having formed around uranium-rich sites. The evidence suggested the wood had been in a water-soaked, gel-like condition at some earlier period in earth history. At that time solutions rich in uranium had passed through the wood, thus permitting the accumulation of uranium at certain sites with an affinity for that element. Secondary halos had then formed around those uranium centers.

These earlier studies were intriguing. If these coalified wood specimens contained microscopic sites which had captured uranium, possibly other sites might have captured polonium. Coalified wood specimens had been found in a number of uranium mines, but they were an uncommon occurrence. Moreover, some of those mines were now closed. Ordinarily it would have been a long and arduous task to collect such specimens. Fortunately, though, I obtained a variety of coalified wood pieces from a colleague who had earlier collected samples from the mines for his own investigations (Breger 1974).

It occurred to me that, irrespective of whether or not these specimens contained secondary polonium halos, they might contain important clues relative to the age of the earth and occurrence of a worldwide flood. To understand my thoughts at that time requires a brief description of some different types of rocks and their histories.

The Origin of Sedimentary Rocks

Scientists generally agree that sedimentary rocks are initially the result of transport and deposition by ice, wind, or water. Many sedimentary (or secondary) rocks, such as shale, sandstone, and limestone, often contain the fossil remains of plants and animals from both terrestrial and marine environments. The Precambrian granites, which are one type of crystalline rocks, do not contain fossils.

While there is general agreement on what sedimentary rocks are, views differ regarding *how rapidly* and *under what conditions* they actually formed. The evolutionary view, based on *geological uniformitarianism,* is that they ordinarily formed slowly over hundreds of thousands or millions of years by geological processes operating at the same rates as observed at present. Interestingly, some geologists now admit that some individual layers could have formed rapidly under "storm" conditions (Ager 1981).

One immediate problem with the uniformitarian viewpoint is the difficulty in finding a location where sedimentary rock formations are in the process of developing at present. River and ocean sediments are forming today, but it is questionable whether any of these will ever turn into the massive limestone and sandstone formations seen in various parts of the world. Nevertheless, evolutionary geologists usually assume that the different sedimentary formations accumulated from the build-up of marine deposits left from the ebb and flow of inland seas over millions of years.

The alternate view of how most sedimentary rocks formed is based on the occurrence of supernaturally induced, catastrophic events associated with a worldwide flood. The scriptural record indicates that the entire earth was covered with water for over a hundred days. Sedimentary material could have been deposited both during the time when the waters were rising and again when they were receding. The scriptural statement, "fountains of the great deep were broken up," suggests that parts of the earth's crust were broken open, implying that the flood was a period characterized by intense volcanic activity. Volcanic eruptions in the ocean basins would have triggered tidal action, resulting in the burial of animal, marine, and plant remains into freshly deposited sediments. The existence of well-preserved fossils in sedimentary rocks is often cited as evidence of a very rapid burial, in agreement with the above scenario.

A rapid deposition of different sediments would also be expected to produce only occasional erosion between successive layers. A prime example of uniform layering of successive formations can be seen in the Grand Canyon. If the horizontal sedimentary layers seen there were really separated by vast periods of time, one would expect to find deep irregular cuts and other signs of erosion *within* the different layers. Instead, such features are the exception rather than the rule.

Radiometric Dating of the Colorado Plateau Deposits

Many geologists pay little attention to these arguments for the flood scenario, perhaps because they believe radiometric dating confirms their views of an ancient age of the sedimentary formations. In particular, radiometric dates of 55 million to 80 million years have been assigned (Stieff et al. 1953) to some of the Colorado Plateau formations where the coalified wood specimens are found. On the basis of the flood model these formations were deposited within a few months of each other only a few thousand years ago. Which was correct? Did radiometric dating justify an ancient age of the coalified wood, or had misplaced confidence in the *uniformitarian principle,* and hence a constant decay rate, led my colleagues to misinterpret the data? Perhaps too, some of the data had escaped their attention.

A singular thought occurred to me. The coalified wood specimens I was soon to receive might have been parts of trees that were growing immediately prior to the flood. My anticipation began to build. When the secrets of the granites were unlocked, they appeared as rocks that were created—the Genesis rocks. Likewise, did these coalified wood specimens contain secrets that would link them to another part of the Genesis record—the account of a recent worldwide flood?

Secondary Polonium Halos: Another Discovery

My observations of the coalified wood specimens from the Colorado Plateau agreed with the major conclusions of other investigators. Evidence indicated that sometime in the past, prior to coalification, a uranium solution had infiltrated the wood when it was in a water-soaked, gel-like condition. As earlier noted, other investigators had reported the halos around uranium-rich centers. These I saw as well, often in abundance. This encouraged me to continue the search for secondary polonium halos in these specimens.

In this case persistence paid off—the long awaited day arrived. In a number of the coalified wood thin sections I discovered secondary polonium halos in greater numbers than the secondary uranium halos. Amazingly enough, sometimes there were over a hundred of them in just one square inch of a coalified wood thin section! Curiously, I found that the polonium halos in these specimens were of only one type—those that had formed from the accumulation of ^{210}Po. None of the other two polonium halo types that occur in granites were seen. The reason for the absence of the ^{214}Po and ^{218}Po halos became clear after I reflected on the difference in the half-lives of the three isotopes.

In brief, the ^{210}Po atoms lived long enough (half-life of 138 days) for them to be captured from the infiltrating uranium solution before they decayed away. In contrast, the other two polonium isotopes, with half-lives of minutes or less, decayed away before they could accumulate at the tiny polonium capture sites. Nature had provided the most favorable conditions for producing secondary polonium halos, namely, an abundant uranium supply coupled with high mobility. Yet even under these optimum conditions *only one type of polonium halo had formed.*

These experimental data presented an insurmountable obstacle to the idea of a secondary origin of polonium halos in granites. That is, if only one polonium halo type could form secondarily under the best natural conditions, what was the scientific basis for theorizing that all three types could form secondarily in the granites? In these rocks both the high uranium content and rapid transport capability were missing.

And this was not all. Most of the secondary ^{210}Po halos in coalified wood exhibited elliptical rather than the circular cross-sections typical of halos in minerals. How were these unusual halos produced? The simplest reconstruction of events pictures uranium solutions infiltrating water-soaked wood that was freshly emplaced in the Colorado Plateau deposits. Halo radiocenters, composed of lead and selenium, accumulated atoms of ^{210}Po out of that solution. In less than a year, secondary ^{210}Po halos developed from the alpha decay of those atoms. Naturally, these halos first formed as spheres and hence initially had a circular outline, just as the halos in minerals. However, as pressure from overlying sediments increased, the gel-like wood was easily compressed, thus leading to the development of the elliptical halos as shown in Figure 4.1 (a) and the Radiohalo Catalogue. Their occurrence in three geological formations suggests they all originated at about the same time, in agreement with the flood-related scenario.

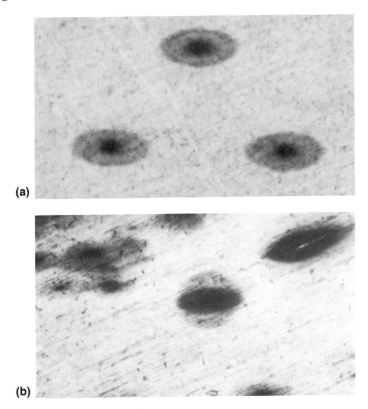

Figure 4.1 Polonium Halos in Coalified Wood
(a) shows elliptical ^{210}Po halos in coalified wood typical of those occurring in the Triassic, Jurassic, and Eocene formations in the Colorado Plateau, and (b) shows the dual ^{210}Po halo (magnification ×250)

It could be argued, however, that secondary polonium halos might have formed in three widely spaced but almost identical geological scenarios instead of the one scenario related to the flood. To be fair, we must carefully examine this possibility.

Here we must realize that the formation of secondary polonium halos required an extraordinarily complex, interrelated series of geological events. The basic ingredients were: (1) water, (2) uprooted trees as the source of the logs and smaller wood fragments, (3) a rich uranium concentration near the wood, and (4) a compression event occurring after the uranium solution invaded the wood, but prior to its becoming coalified. The gel-like condition of the wood suggests only a short time had elapsed since the trees had been uprooted. At the very time the wood was in this special condition, it had to be infiltrated by a solution that had recently dissolved uranium from a nearby deposit. Note that, if the water had contacted the uranium deposit *after* infiltrating the wood, there would have been no radioactivity in solution and hence no possibility of forming secondary halos. The same is true if the wood had already turned to coal before contact with the uranium solutions.

The evolutionary scenario requires that the complex sequence of events described above must have been repeated more than ten million years later in the same geographical location. That this scenario would occur a third time, again in the same area about fifty million years later, seems improbable. Yet the issue must not be decided merely on the basis of improbability. Instead, we must determine whether this interpretation is in harmony with *all* the scientific data. Previously, geologists drew conclusions about the history of the Colorado Plateau formations based on data then available to them. We must now focus special attention on the new data presented by the halos in coalified wood to see if these earlier conclusions are still justified.

New Data Support the Global Flood Model

It is quite significant that the elliptical polonium halos appear in coalified wood specimens from three different geological formations in the Colorado Plateau deposits. The importance of this observation can hardly be overestimated. In the evolutionary scenario those formations represent three geological periods: Triassic, 180 to 230 million years ago; Jurassic,135 to 180 million years ago; and Eocene, 35 to 60 million years ago. The occurrence of the elliptical secondary ^{210}Po halos in specimens from all of these formations is evidence par excellence that the wood in all of them was in the *same gel-like condition* when infiltrated by a uranium solution. These data fit the flood model perfectly.

Another vital piece of scientific data relates to the question of how much time elapsed from the formation of the circular polonium halos to the time of compression. The length of this period would have remained uncertain had it not been for the discovery of "dual" polonium halos such as shown in Figure 4.1 (b) and the Radiohalo Catalogue. These "dual" ^{210}Po halos, which I have seen thus far in Triassic and Jurassic specimens, exhibit both a circular and elliptical outline. (The search for dual halos in "Eocene" wood has been hindered by lack of material.) Initially, I was puzzled as to how two differently shaped halos could develop around the same center. Then I realized that the halo centers, composed of lead and selenium, could also have captured another uranium daughter, ^{210}Pb. Since this isotope of lead decays with a half-life of about 22 years to ^{210}Po, a second ^{210}Po halo could develop within about 20 years after the first one had formed. If there was no deformation of the wood, then both halos would remain circular and they would exactly overlap. Or if the wood was deformed after about 20 years, then both halos would be compressed into an elliptical shape and they still would overlap.

However, if deformation of the wood occurred within just a few years after the introduction of the uranium, then only one ^{210}Po halo could have been compressed because only one (from ^{210}Po) had then formed. Several years later another circular halo could develop (as ^{210}Pb decayed to ^{210}Po) and superimpose on the elliptical halo. Provided there was no further deformation, these two halos would retain their respective shapes and now appear as the "dual" halo shown in Figure 4.1 (b). From this sequence a very relevant conclusion emerges: only a few years elapsed from the introduction of the uranium to the time when the wood was compressed. These data very specifically support the flood model, which includes considerable readjustment and deformation of freshly deposited sedimentary rocks in the years after the flood waters receded.

Additional data on the coalified wood specimens were obtained in collaboration with some colleagues. We studied radiohalos in coalified wood using the same type of advanced scientific instruments that had been used on halos in granites. A report describing the outcome of these collaborative studies was published in the October 15, 1976, issue of *Science* (Gentry et al. 1976a; Appendix). The evidence obtained in these experiments suggested a common source for the uranium in all the coalified wood specimens. These data implied only one uranium solution had infiltrated the different wood specimens.

This result, coupled with the observations just described, permits some rather firm conclusions to be drawn. In particular, a single uranium solution means the uranium infiltration occurred nearly simultaneously in *all* the wood specimens. And since the elliptical polonium halos show the wood specimens taken from the Jurassic, Triassic, and Eocene formations were all in the same gel-like condition at the time of infiltration, it inevitably follows that these geological formations were all deposited at about the same time. Likewise, the presence of dual polonium halos in wood specimens taken from both Jurassic and Triassic deposits provides strong evidence that the event which compressed the wood occurred simultaneously in both cases. This is exactly what would be expected on the basis of a near simultaneous deposition of all the wood at the time of the flood.

On the other hand, the data just discussed directly contradict the view that the Jurassic, Triassic, and Eocene formations in the Colorado Plateau were laid down tens of millions of years apart. If the evolutionary scenario were correct, the wood in the Triassic (oldest) formation would have turned into coalified wood millions of years before the Eocene layer was deposited. In this case compressed halos could not have formed. The above evidences contradict the evolutionary view that a hundred million years or more separate certain formations in the Colorado Plateau, supporting instead a rapid deposition of them all.

Earlier in this chapter I noted that well-preserved fossils in various geological formations around the world are often cited as evidence of a rapid burial. This raises a significant question: Is there any similar physical evidence, apart from the compressed halos, which would suggest that the wood pieces now in the Colorado Plateau formations were encased in sediments somewhat rapidly (that is, before decay set in)? Such evidence, if it exists, would be most clearly impressed on the investigator who actually collected the coalified wood specimens from the uranium mines which were then operating in Colorado, New Mexico, Utah, and Wyoming. That scientist, who worked for the U.S. Geological Survey, subsequently published a report on his studies (and later kindly provided me with many coalified wood specimens). One sentence in the following excerpt from his report succinctly describes the condition of the wood pieces as he first saw them:

> The coalified wood in these sediments ranges in size from finely divided intergranular fragments visible with a hand lens to entire tree trunks many feet long and still having attached branches and roots. The larger pieces of coalified wood are compressed or uncompressed, black or brown in color, and may or may not contain siliceous, calcitic, or dolomitic fillings replacing the original pithy cores. *Some coalified fragments are still flexible when first collected but become brittle when dried.* Black and brown fragments are occasionally superimposed upon each other; the former have the

appearance of lignite, whereas the latter outwardly resemble vitrain. . . . (Breger 1974, 100—italics mine)

I suggest the flexibility of some freshly collected wood fragments is strong evidence of a rapid deposition.

Returning to the subject of my own studies of the coalified wood specimens, I now summarize some other implications of the investigations published in the 1976 *Science* report:

(1) Uranium to lead ratios were found suggesting that the various Colorado Plateau formations are only several thousand years old instead of the 60 to 200-million-year age required by the evolutionary time scale. Timewise this evidence agrees with the scriptural chronology concerning the time (ca 2300 b.c.) when the worldwide flood occurred. Thus, the entire radiometric age-dating scheme developed over the past eighty years is called into question.

(2) The coalification process—whereby organic material such as plant vegetation or wood turns into coal—can occur in a year or less. This result contradicts the presumed tens of thousands of years (or more) thought necessary for the coalification process. Interestingly, I have found references to experimental data suggesting that, under certain laboratory conditions, the process of coalification can occur over just a few days (Stutzer 1940, 105-106; Larsen 1985). Such data are consistent with my results.

A Professor Notes the Silent Response

The 1976 *Science* report (Gentry et al. 1976a; Appendix) on halos in coalified wood questions both the conventional geologic age-dating schemes and the uniformitarian interpretation of the entire geologic column. It provides data that distinguish the multiple, primordial polonium-halo types in granites from the single, secondary polonium-halo type in coalified wood. These results challenge all aspects of evolutionary geology, and they did not go unnoticed. A few months after the report was published I received the following letters:

(January 27, 1977)

Dear Dr. Gentry:

I have been patiently scanning the "letters" section of *Science* since the publication by you and your colleagues of your findings on radiohalos.

The silence is deafening—I think it can be interpreted as "stunned silence," coming as closely as it did on the "neutrino crisis" stemming from a paper published in January 1976 on the absence of the expected neutrino flux from the sun.

Your results will not greatly trouble the engineer, whether he is a mining engineer, a geophysical engineer, or a ground-water engineer. But the impact on the science of geology, in possibly changing the accepted views as to the duration of geologic time, will be felt for many years.

We are indebted to you and your colleagues for your painstaking observation, the careful wording of your paper, and the courage you have manifested in presenting evidence that contravenes the conventional wisdom of the geological profession. I might add that the findings have direct application in the search for a semi-permanent containment for radwastes.

Again, my commendations for a difficult job, extremely well done.

Very truly yours,

/s/ Raphael G. Kazmann

Raphael G. Kazmann
Professor of Civil Engineering
Louisiana State University

(March 9, 1977)

Dear Dr. Gentry:

Thank you for the reprints. It is apparent that you and your coworkers are unearthing fundamental information which will be difficult, if not impossible, to include in the accepted, uniformitarian-evolutionary, scheme.

Here at LSU we are considering organizing a one or two day conference on geologic time including the age of the sun. There will probably be a number of invited papers and I will suggest to the conference organizer that you be invited, once the decision has been made. If you have any thoughts on possible speakers, please let me know.

Best wishes,

/s/ Raphael G. Kazmann

Raphael G. Kazmann
Professor of Civil Engineering
Louisiana State University

Professor Kazmann correctly perceived that the data have called the evolutionary scheme into question. He also understood that if conventional dating techniques have been in error, as the data suggested, this might raise questions about the procedures currently used to select nuclear waste storage sites. To explore these matters further, he organized a symposium addressing the problems and methods used in measuring geologic time.

Debating the Time Scale

The symposium, "Time: In Full Measure," was held at Louisiana State University in April 1978. There were five invited speakers, including me. The symposium dealt primarily with the various aspects of time measurement and the age of the geological formations. Professor Kazmann, as the convener, published an account of those proceedings in the September 1978 issue of *Geotimes* (Kazmann 1978), a monthly publication of the American Geological Institute, and in the January 9, 1979, issue of *EOS* (Kazmann 1979), a weekly publication of the American Geophysical Union. His summary (Kazmann 1979) of my presentation at the symposium is as follows:

> . . . His [Gentry's] specialty is the study of minute halos in mica and biotite crystals and, more recently, in coalified wood from uranium-bearing sands in the Colorado Plateau and the Chattanooga Shale. The halos are created by alpha-particles of differing energies emitted by such substances as uranium, thorium, polonium, and other radioactives. He presented microphotographs of an assortment of radiohalos in biotite, fluorite, and cordierite and then a diagram whereby the lines produced by the various alpha emitters can be identified. Among the eight emitters are two isotopes of uranium and three of polonium. [Gentry 1974; Gentry et al. 1974]
>
> The polonium halos, especially those produced by ^{218}Po, are the center of a mystery. The half-life of the isotope is only 3 min. Yet the halos have been found in granitic rocks . . . in all parts of the world, including Scandinavia, India, Canada, and the United States. The difficulty arises from the observation that there is no identifiable precursor to the polonium; it appears to be primordial polonium. If so, how did the surrounding rocks crystallize rapidly enough so that there were crystals available ready to be imprinted with radiohalos by alpha particles from ^{218}Po? This would imply almost instantaneous cooling and crystallization of these granitic minerals, and we know of no mechanisms that will remove heat so rapidly; the rocks are supposed to have cooled over millennia, if not tens of millennia.
>
> His studies of halos in coalified wood [Gentry et al. 1976a; 315] bear directly on the meeting's topic: geochronology. There he and his co-workers were able to define the tiny uranium centers and to distinguish the various halos produced by different alpha emitters.

However, since the deposits from which the coalified wood was obtained are considered to be of Cretaceous age, and possibly of Jurassic or Triassic age, the ratio between ^{238}U and ^{206}Pb should be low. Instead a number of such halos have been found with uranium-lead ratios ranging from about 2200 to over 64,000. If isotope ratios are to be used as a basis for geologic dating, then presently accepted ages may be too high by a factor of 10,000, admitting the possibility that the ages of the formation are to be measured in millennia. *Thus ages of the entire stratigraphic column may contain epochs less than 0.01% the duration of those now accepted and found in the literature.* . . . (Kazmann 1979, 19—italics mine)

The publication of this clearly stated evaluation of my results was an important event in my research. Kazmann's account of the LSU symposium in both *Geotimes* and *EOS,* two nationally circulated geological news magazines, brought my work to the attention of a much larger segment of the geological community. It was difficult to believe that my contribution to the LSU symposium would go unchallenged.

5

Reverberations from Scientists

Soon after Kazmann's summary appeared in *EOS* (Kazmann 1979), I received a copy of a letter from the eminent geochronologist, Professor Paul Damon, University of Arizona, Tucson, to Dr. A. F. Spilhaus, Editor of *EOS*. Even though Damon's letter was critical, I was elated because his comments focused squarely on the implications of my work. He intended for his criticisms to be published in *EOS*. If this was done, I hoped to have the privilege of responding to his assertions by presenting a further explanation of my position. This would be an opportunity to clarify the issues. The problem was that, as the following letter shows, initially Dr. Spilhaus only asked my opinion of Damon's criticism without offering me the opportunity to respond:

(February 8, 1979)

Dear Mr. Gentry:

I would appreciate your comments on whether the remarks of Paul Damon in his letter of January 23, a copy of which was sent to you, are scientifically sound. If they are, it will be my inclination to publish them as soon as possible in *EOS*. Please let me know your opinion by return mail if possible. I will also consider further commentary on this article as it becomes available.

Sincerely yours,

/s/ A. F. Spilhaus

A. F. Spilhaus, Jr.
Executive Director, American Geophysical Union

I called and then wrote to Dr. Spilhaus, requesting that he allow me to respond to Damon's letter. A few weeks later the following letter was received:

(April 3, 1979)

Dear Professor Gentry:

As you well know and expressed in your letter, the conclusions you reached from the interpretation of your halo data are considered untenable except by a very tiny minority of the earth sciences community. Nevertheless, my reviewers feel that you are due an opportunity to respond to Damon's comments, but you must make that response short. . . .

Sincerely yours,

/s/ Fred Spilhaus

I was quite pleased to receive this letter, for there were some important matters at stake. Damon's letter left no doubt he understood that, *if* polonium halos in granites were primordial, this meant the earth had indeed formed very rapidly, thus calling into question the entire science of geochronology (radiometric age-dating). The first sentence in his letter, later published in *EOS*, is quoted below:

I was dismayed by Raphael G. Kazmann's conclusion in his review of a symposium on "Cosmochronology, geochronology, and the neutrino crisis" (Time: In Full Measure, *Eos Trans. AGU*, 60 (2), pp. 21-22, January 9, 1979) that essentially casts in doubt the entire science of geochronology, on the basis of an absurd interpretation of the origin of "polonium" halos in minerals observed by Robert Gentry. . . . (Damon 1979, 474)

The "absurd interpretation" referred to here is my claim that primordial polonium halos exist in granites. Primordial polonium halos invalidate the assumption of uniform decay over endless time. Without this premise there is no factual basis for a radiometrically derived 4.5-billion-year age of the earth. The last paragraph of his letter concludes:

The history of science includes many examples of valid observations that have been given unacceptable interpretations. One need not doubt the validity of Gentry's observations of the existence of halos with certain characteristics in order to reject his interpretation as reported by Kazmann. However, I certainly hope that Kazmann and his fellow engineers do not design structures such as nuclear reactor sites based upon the short time scale suggested by a misinterpretation of Gentry's apparently valid observations! (Damon 1979, 474)

Damon agrees that my observations on polonium halos are "apparently valid," but he rejects the possibility that they are of primordial origin without offering an alternative explanation. It was becoming increasingly apparent that an experimental test was needed to settle the question of their origin.

A Falsification Test Proposed

Damon's strongest objections to my results centered around two points—the association of polonium halos in granites with primordial polonium and the identification of the Precambrian granites as the primordial Genesis rocks of our planet. It occurred to me there was a laboratory experiment which, if successful, *in theory* would allow scientists to confirm a major prediction of the evolutionary scenario and at the same time falsify my model of creation.

To understand this test readers must remember that in the evolutionary model the proto-earth began some 4.5 billion years ago in a semi-molten condition. A slowly cooling earth supposedly led to the formation of various types of rocks at many different times and places. Geologists think that the Precambrian granites, the crystalline basement rocks of the continents, were among those rocks that formed at different intervals over that long cooling period. According to the *uniformitarian principle* the physical processes which governed the crystallization of the granites in the past are the same as those operable on earth today. The inevitable conclusion is that it should be possible to duplicate the process of granite formation in a modern scientific laboratory. That is, it should be possible—*provided the uniformitarian principle is really valid*.

This was the basis of the laboratory-based test presented to the scientific community in my response to Damon's letter in the May 29, 1979, issue of *EOS*. Two excerpts from my response show how this test was stated:

> . . .Therefore I regard the failure to resolve the long-standing controversy in geology which concerns the origin of the Precambrian granites to be because such rocks are primordial and hence not necessarily explainable on the basis of conventional principles. Even though I think they further qualify for that role in their association as basement rocks of the continents, nevertheless I would consider my thesis essentially falsified if and when geologists synthesize a hand-sized specimen of a typical biotite-bearing granite and/or a similar size crystal of biotite.
>
> I will likewise relinquish any claim for primordial ^{218}Po halos when coercive evidence (not just plausibility arguments) is provided for a conventional origin. . . .and in this respect I will consider my thesis to be doubly falsified by the synthesis of a biotite which contains just one ^{218}Po halo (some of my natural specimens contain more than 10^4 Po halos/cm^3). . . . (Gentry 1979, 474)

Much was and still is at stake in issuing this challenge to synthesize, or produce a duplicate of, a single hand-sized specimen of a piece of granite in the laboratory. The experiment being proposed is quite straightforward. The basic chemical elements of a granite, which are well-known, are to be melted, and then allowed to cool to form a synthetic rock. If my colleagues could do this experiment so that the synthetic rock reproduces the mineral composition and crystal structure of a granite, then they will have duplicated or synthesized a piece of granite. By doing this they would have confirmed a major prediction of the evolutionary scenario—they would have demonstrated that granites can form from a liquid melt in accord with known physical laws. I would accept such results as falsifying my view that the Precambrian granites are the primordial Genesis rocks of our planet. Furthermore, if they were successful in producing just a single ^{218}Po halo in that piece of synthesized granite, I would accept that as falsifying my view that the polonium halos in granites are God's fingerprints.

This test of the creation and evolution models was published in the open scientific literature for all my colleagues to study. In the spirit of free scientific inquiry I hoped they would closely examine my published evidences for creation and be led to respond with contrary evidence, if I was wrong, or else admit there was valid scientific evidence for creation. Neither of these happened.

A Courageous Editorial Decision

Just a few months after Damon's letter and my response were published, another criticism of my work appeared in the August 14, 1979, issue of *EOS* (York 1979). The author was Dr. Derek York of the University of Toronto, a highly respected geochronologist who had also participated in the LSU symposium, *Time: In Full Measure,* mentioned under the heading "Debating the Time Scale" in Chapter 4. His article was not based on any of his own experimental observations about polonium halos. Instead, he promoted Henderson's idea of a secondary origin from uranium and criticized me for not accepting it. He did not mention that he had heard my presentation on halos at the LSU symposium. Initially there was no opportunity for me to rebut York's criticisms, for he never informed me that his article was to be published. My letter of objection (to Spilhaus) concerning this silence is quoted in part below, along with his reply:

(October 23, 1979)

Dear Dr. Spilhaus:

I have spent a great deal of time working on the response to Derek York's direct attack on my research. I could have helped York avoid some embarrassing remarks if he had only shared his article with me prior to publication. . . . But whatever the reason for York's secrecy, I cannot let his misrepresentations of my work go unanswered. Actually, there is much more I could have said—and may yet have to say—about his comments on my work.

The length of this manuscript is about half that of York's article, and, in fact, about the same length as my response to Paul Damon's letter.

Be assured that I have high personal regard for Derek York, even though I have had to take exception to his remarks.

Sincerely,

/s/ Robert V. Gentry

(November 14, 1979)

Dear Dr. Gentry:

I have forwarded your article to one of the *EOS* Associate Editors for review with regard to quality of the substance and for consideration of its suitability for publication in *EOS*. These will be difficult questions. Our decision will rest on whether your present letter makes any substantive addition to the discussion and on the completeness and validity of the work on which it is based. New material may also be rejected by *EOS* as it is not an appropriate medium for original publication of scientific results.

Sincerely yours,

/s/ Fred Spilhaus

Months passed with no further word from Spilhaus about my response to York's article. Finally, after five months had elapsed, I received a letter from Spilhaus, stating that he would be willing to publish a shorter version of my response. However, his suggested version did not include enough detail to properly answer all of York's criticisms; so I wrote Dr. Spilhaus again. Quoted below are both his letter to me and my subsequent letter to him:

(April 14, 1980)

Dear Bob:

I enclose a cut down version of the letter you submitted in response to York's paper on polonium halos. I would be willing to publish this in *EOS* immediately.

I believe that publication of this letter would call attention to the principal exceptions you take to his remarks. In the interests of conducting the scientific process in an orderly way, more extended technical discussion should be directed to journals devoted to the publication of original research and/or reviews.

Sincerely yours,

/s/ Fred

(April 28, 1980)

Dear Fred:

As per your suggestion, I would very much hope that Derek York and others will eventually publish some original research material on radiohalos in specialty journals. And for your sake I am willing to make some significant concessions on the length of my reply and not demand that my original version be published. But I would also hope that you could see why my few brief technical comments need to be incorporated into the revised version.

First, to give Derek the privilege of making technical criticisms of my research while denying me the privilege of specifically responding to those comments constitutes discrimination against a minority view. It would be a case of the establishment attempting to suppress unpopular evidence. You have not struck me as the sort of individual who would agree to this sort of thing.

Second, for me not to specifically respond to Derek's technical comments would leave the impression that I don't have a response, or else it would have been published. After all, a rebuttal is meaningless if it simply says I am right and the other guy is wrong.

Third, it would seem that if this question is ever going to be resolved, those few technical comments need to be put in so that when the next fellow comes along and takes a shot at me, he will at least be firing at the right target. Let me explain. It is conceivable, I think, that Derek read my reports but simply did not catch the significance of the difference in the Po halos in granites and coalified wood. This difference is absolutely crucial to any proposed explanation of Po halos in granites and needs to be briefly spelled out so that other researchers won't go down blind alleys thinking they have solved the problem. Here I want to emphasize that my brief technical response to Derek is not a matter of publishing new data; it is simply that of clarifying data which has already been published but which has been misinterpreted.

So, Fred, I am returning to you a revised version of my reply, which is basically the version you sent to me with the technical comments added. The last sentence or so has been modified to make up for the loss of the background material that has been left out. And one very important citation has been restored to the references along with one or two word changes here and there.

In closing let me again remind you that I did not instigate this discussion and I am not trying to turn it into a cause célèbre. I am of the opinion, however, that there are some individuals who may want to do this if they knew about my difficulties in getting this reply published. In this respect, as volatile as this subject is, there is also a possibility it could turn into a mini-Watergate if some within the news media suspected there was an attempt to suppress or coverup my rebuttal evidence. For your sake I am sincerely hoping this does not happen.

As before, I am requesting that you have the galley proofs sent to me before publication. I have come a long way, and I don't even want a misspelled word to come out under my name, much less an inadvertently omitted word that could change the meaning of a sentence.

I know you have been under great pressure about this situation, and I am trying not to make it any harder on you. Your efforts to be fair are greatly appreciated.

Sincerely,

/s/ Bob

Certainly I still greatly appreciate his efforts. Much was at stake in my work. It was imperative that I be given the right to respond because York had completely ignored the two main features of my letter in the May 29, 1979, issue of *EOS* (Gentry 1979), namely, the challenge to synthesize a piece of granite and the reference to Professor Norman Feather's conclusions relative to the origin of polonium halos in micas.

Polonium Halos: An Independent Evaluation

Professor Feather's interest in polonium halos was apparently traceable to some of my publications. He understood that the ^{210}Po halos discovered in coalified wood were secondarily derived from uranium activity. At the same time, he also saw that the origin of the different types of polonium halos in granites raised some difficult questions. His theoretical investigation, entitled "The Unsolved Problem of the Po-Haloes in Precambrian Biotite and Other Old Minerals," was published in 1978 in the *Communications of the Royal Society of Edinburgh* (Feather 1978). His conclusions are aptly stated in the Synopsis of his article:

> Ever since the discovery of Po-haloes in old mica (Henderson and Sparks 1939) the problem of their origin has remained essentially unsolved. Two suggestions have been made (Henderson 1939; Gentry et al. 1973), but neither carries immediate conviction. These suggestions are examined critically and in detail, and the difficulties attaching to the acceptance of either are identified. Because these two suggestions appear to exhaust the logical possibilities of explanation, it is tempting to admit that one of them must be basically correct, but whoever would make this admission must be fortified by credulity of a high order. (Feather 1978, 147)

Feather's doubts about polonium halos in granitic micas having originated from uranium daughter radioactivity, or from isomers, in essence confirm my earlier investigations. His conclusions were derived from a theoretical investigation of the nuclear properties of the relevant isotopes. My 1968 and 1976 *Science* reports (Gentry 1968; Gentry et al. 1976a; Appendix) and the 1973 *Nature* report (Gentry et al. 1973), to which Feather refers, show respectively that the secondary radioactivity and isomer hypotheses are not valid for polonium halos in granites. Feather did not propose a primordial origin of the Po halos as I have done, yet the results of his investigation greatly strengthened my contention that a conventional explanation of the Po halos in granites is scientifically untenable.

York did not mention this information in his review in *EOS*. I felt it necessary, then, to comment on Feather's work in my rebuttal, finally published on July 1, 1980, almost one year after York's article had appeared. It is quoted in part below:

> York seems to regard even the existence of Po halos as only tentative. But notwithstanding the uncertainties, his article leans heavily toward the proposition that Po halos do exist, at least in micas. York's thesis is that Po halos are most probably explainable within the accepted framework because the interlocking nature of various radiometric dating techniques provides powerful evidence that conventional geochronology is correct. York faults me for ignoring this internal consistency. Contrary to his understanding, I do not ignore these data. But neither do I accept the idea that the presumed agreement between techniques is really coercive

evidence for the correctness of the uniformitarian assumption which undergirds the present model. There was no discussion of the $^{238}U/^{206}Pb$ ratios [Gentry et al., 1976a], which raise significant questions about the accepted geochronological scheme.

While I can appreciate York's desire to emphasize internal consistency, it should be evident that irrespective of how much data has been or yet can be fitted into the present model, the question of its ultimate reliability hinges on whether there exist any observations which falsify the theory. . . .

York's surprise that I would accept Henderson's hypothesis for Po halos in coalified wood [Gentry et al., 1976a] but reject this explanation for mica because of the slowness of solid state diffusion suggests first that the same type of Po halos has been found in both substances and second that my only objection to accepting Henderson's hypothesis in mica was the slowness of solid state diffusion. Here some very important data have been glossed over.

Mica contains three types of Po halos, but coalified wood only one. Much evidence suggests the ^{210}Po halos in coalified wood formed from selective accumulation of ^{210}Po and ^{210}Pb, which have half-lives sufficiently long (138 days and 22 years, respectively) to have migrated to the radiocenters before serious loss occurred from decay. Likewise, the relatively short half-lives of ^{214}Pb and ^{218}Po (27 minutes and 3 minutes, respectively) mean these nuclides generally decayed away before reaching the accumulation sites, which explains the absence of ^{214}Po and ^{218}Po halos. Thus the crucial question is: If Henderson's model results in only ^{210}Po halos being formed under ideal conditions of rapid transport (plus an abundant supply) of U-derived Po atoms, then how can this model account for all three Po halo types in mica, where both the U content and the transport rate are considerably lower? Indeed, the close proximity in clear mica (i.e., without any conduits) of two or more types of Po halos presents what may be incontrovertible evidence against explaining these halos by Henderson's hypothesis [Feather, 1978].

Finally, York failed to mention that my hypothesis that Po halos in Precambrian granites are primordial [Gentry, 1974] could in theory be falsified (and Feather's objections negated) by the experimental synthesis of a biotite crystal that contained at least two dissimilar Po halos in close proximity [Gentry, 1979]. (Gentry 1980, 514)

The publication of this response showed that Dr. Spilhaus was determined to abide by the principles enunciated in *The Affirmation of Freedom of Inquiry and Expression* (see Overview). This was the second time that scientists had been challenged to produce the experimental results that would substantiate the evolutionary view of earth history, and at the same time, in theory, falsify my evidence for creation. I wondered whether there would now be a response, or whether the challenge would continue to be ignored.

Only time would tell.

6

Reaction from the National Science Foundation

The financial support for this research is a story in itself. During my tenure as a guest scientist at ORNL, my salary was provided from grant funds obtained through my affiliation with Columbia Union College. Through the early 1970's these funds came from private sources and the National Science Foundation to cover that expense. The National Science Foundation (NSF) is the government agency entrusted with allocating hundreds of millions annually for research in the scientific disciplines. Like all government agencies, it is publicly funded and legally obligated to disperse those monies impartially. In theory, taxpayers' money should be dispensed without preference for particular views or discrimination against alternative theories.

The earth and planetary sciences receive much support from the NSF through grants to university science departments for research based on the evolutionary model of origins. But has the NSF been equally inclined to support research related to a creation-based model of earth history? This chapter focuses on the reaction of NSF officials after it was more generally recognized that my scientific discoveries supported creation. NSF was supportive of my research *before* they were aware that the implications were damaging to an evolutionary point of view.

In 1974 the NSF awarded me a two-year grant, later extended to mid-1977, of approximately $55,000 for research on both polonium halos and the unexplained dwarf and giant halos. At the time this grant was awarded, the implications of my research had not been revealed to their fullest extent. Quite possibly most NSF officials and reviewers were then unaware that polonium halos provided evidence for an instantaneous creation of the earth.

Several of my scientific reports were published during the 1974-77 grant period: one related to my investigations of the "spectacle halo," a second to my work on halos in coalified wood, and another to the existence of superheavy elements in giant-halo radiocenters. Of these three, the results on giant halos and superheavy elements attracted the greatest attention and the greatest criticism.

The Elusive Superheavy Elements

A background of my research efforts on superheavy elements will be given here to help the reader understand some of the NSF comments about my research proposals. As earlier noted, the primary reason for my affiliation with ORNL in 1969 was to investigate whether the dwarf or giant halos provided evidence of superheavy elements. Consequently, much of my research there concentrated on these unusual halos in collaboration with colleagues at ORNL. In spite of considerable effort, by 1975 none of our investigations of giant and dwarf halos showed any convincing evidence of superheavy elements.

In mid-1975 I learned of a new analytical technique for determining the composition of tiny particles collected in air-pollution studies. In this technique the ion beam from a nuclear accelerator was used to excite the characteristic x rays of the chemical elements composing the particle. Its very high sensitivity seemed ideally suited for searching for superheavy elements in the microscopic-sized radiocenters of the giant halos.

In early 1976 I began a collaboration with physicists at Florida State University at Tallahassee (FSU) and the University of California at Davis (UC-Davis) to search for superheavy elements in giant-halo radiocenters. My main contribution to SHEP (SuperHeavy Element Project) was in supplying the samples to the experimenters at FSU. We conducted our experiments on the Van de Graaff accelerator located in the FSU physics department. A few months after experimentation began, we found what appeared to be indications of superheavy elements in the tiny radiocenters of certain giant halos.

Based on the results of our experiments, we prepared a joint article for *Physical Review Letters*, a rapid-publication physics journal. The article announcing our evidence for superheavy elements was published in the July 5, 1976, issue (Gentry et al. 1976b). This report immediately triggered a greatly intensified worldwide search for superheavy elements. The possible discovery of superheavy elements was featured in all major science news magazines and made the headlines of several newspapers.

Unfortunately, later experiments did not confirm our original interpretation of the evidence. I participated in two elaborate follow-up experiments with colleagues from ORNL, but neither provided any data indicative of superheavy elements. The results of these experiments were subsequently published in two separate reports in *Physical Review Letters* (Sparks et al. 1977 and 1978).

Even though the evidence for superheavy elements was not confirmed in subsequent experiments, our 1976 report sparked enough interest in the topic so that an International Conference on Superheavy Elements was held in Lubbock, Texas, in March 1978. At that Conference my colleagues from FSU and UC-Davis continued to maintain that the giant-halo centers contained superheavy elements. A write-up of that Conference appeared in the April 15, 1978, issue of *Science News.* The following excerpt from that article illustrates the difference between their views and mine at the time of the Conference:

> At the Lubbock symposium, Gentry made clear that while in 1976 he believed the evidence warranted the deduction that the inclusions contained element 126, now he does not. "At present, I do not have evidence for superheavy elements in giant halo inclusions . . . . As the evidence stands today, I will accept the view that the synchrotron radiation experiments did not confirm element 126."
>
> Gentry emphasizes that in making that statement he speaks only for himself: "I don't speak for anyone else and they don't speak for me."
>
> The reason he says that, is that some other co-authors of the original report have not given up the claim. Thomas A. Cahill of the University of California at Davis, for instance, vigorously defends the group's original report and strongly disagrees with Gentry's about-face. "The evidence for 126 in giant haloes has not gone away," he told *Science News.* "It's even stronger" . . . "The lines are there," says Cahill. "Something is there."
>
> Gentry acknowledges that there are some things about the original experiment that even today he does not understand. "But," he told *Science News,* "I have to face it. In my opinion the Stanford work is of a sensitivity that it should see it [any evidence of superheavy elements]." (Frazier 1978, 238)

Ordinarily, a scientist gains some respect from his colleagues when he admits an error. In this instance, however, some opponents of my work later used the above retraction to cast doubt on my published evidences regarding polonium halos and their implications for creation. Generally they ignored my contribution to this Symposium (Gentry 1978a) in which I summarized the technical details of my research on the giant, dwarf, and polonium halos.

Declination of 1977 Proposal

In 1977 I submitted a grant proposal to the NSF which was very similar to the one it had funded in 1974. I requested funds (1) to continue the search for superheavy elements, (2) for additional research on polonium, dwarf, and giant halos in granites, and (3) for further investigations of halos in coalified wood. This time my proposal was declined. My collaborators in the superheavy element experiments at Florida State University and the University of California-Davis were still receiving NSF funding for further work on superheavy elements. But my proposal to continue similar work had been denied. I wrote for an explanation.

Funding decisions within the NSF are based on reviews by six scientists who respond by mail, in addition to panel reviews by six other scientists. Of the six mail reviews of my 1977 proposal, four had actually recommended further funding. The two negative ones cited as their main reason the mistaken identification of superheavy elements.

In contrast to the mail reviews, the panel review evaluation of my proposal was largely negative. Upon my request, Dr. John Hower, then Program Director for Geochemistry at NSF, sent a summary of the panel discussion. It dealt at length with my research on superheavy elements, concluding that "there is little possibility of their detection by proposed techniques." Yet my colleagues at FSU and UC-Davis were using one of the same techniques outlined in my proposal, and the NSF was continuing their funding.

In my case, the panel reviewers decided to reject my entire proposal for what I think were spurious reasons. The following quote from Dr. Hower's letter to me describes the decisive objection found by the panel reviewers in my proposal:

> The most important criticism of the proposal did not, however, have to do with superheavy elements detection. The criticism stemmed from the general nature of *the proposed research on haloes.* The principal investigator has been collecting specimens, examining them petrographically, and reporting their morphology and mineral occurrence for a number of years. The panel considered that these descriptive contributions have been of some value, but felt that more of the same approach had little

potential to contribute something new. The main difficulty with the proposal is that (aside from the superheavy element search) there was no hypothesis concerning the *origin of the haloes* that the principal investigator proposed to test. He has already looked at and described a number of occurrences. The panel felt that it was not justified in recommending funding of a research project that merely proposed to make additional observations of the phenomenon. There seems little possibility that the principal investigator could arrive at a hypothesis by looking *at additional haloes* since he has not been able to propose one at this time. (Hower 1977; Appendix—italics mine)

Initially I understood "haloes" in the first italicized phrase referred to both giant halos and polonium halos as discussed in the proposal. "Haloes" in the second italicized phrase I understood to mean primarily giant halos because of the reference to superheavy elements. And since "haloes" in the last italicized phrase was not qualified, I again assumed it referred to polonium halos in granites. On that basis I felt there were some contradictions in the NSF handling of my case, and I decided to appeal their declination of my proposal.

Appeal to the NSF

The relevant part of my appeal letter, addressed to Dr. Edward Todd, the Assistant Director for Astronomical, Atmospheric, Earth, and Atmospheric Sciences, National Science Foundation, is as follows:

Now with respect to the second criticism of the proposal, the Program Director's letter states that in essence the panel was not able to find that I had any hypothesis to test with respect to the other phases of my research on halos, or that there was any prospect of my finding a hypothesis in the future. I can understand such statements could be made by persons unacquainted with geochemical terminology who might read my published reports. It is, however, very difficult for me to understand how a panel of geochemists could make such statements, especially in view of the fact that I had previously discussed with the Program Director the hypothesis and implications of my research on Po halos *as they have been published in the open scientific literature and referred to in both the previous and the present NSF proposals.* . . .

I specifically refer to the fact that I have proposed that "Po halos" in Earth's basement granitic rocks represent evidence of extinct natural radioactivity and thus imply only a brief period between "nucleosynthesis" and crystallization of the host rocks [Gentry 1975]. ...Furthermore, back in 1973, again in a *Nature* report [Gentry et al. 1973], I pointed out the existence of Po halos "meets with severe geological problems: the half-lives of the polonium isotopes ($t_{1/2}$ = 3 min for ^{218}Po) are too short to permit anything but a rapid mineral crystallization, contrary to accepted theories of magmatic cooling rates." . . .

In fact a person really doesn't have to be a geochemist, or even have training in geochemistry (actually I am a physicist turned aside into nuclear geophysics), to see that in my published reports I am claiming to have found evidence that shakes the foundations of modern cosmology and geochemistry. Thus because I have been very forthright in stating the implications of my research in my published reports, I would like to suggest to you the possibility that, when the Program Director and the review panel indicated they had difficulty finding my hypothesis, what they really meant was that they could not fit the evidence I have reported into any of the popular, currently held geochemical or cosmological theories concerning the origin of the earth.

Much later it occurred to me that the panel reviewers may never have intended any reference to polonium halos in their comments. Perhaps they decided to just ignore this phase of my proposal. In any event, Dr. Todd's response to my appeal letter did not address this issue. It stated:

> . . . It is my conclusion that your proposal received a thorough and fair peer review through the Geochemistry Program Office, a review that included a conscientious and careful consideration of six *ad hoc* mail reviews. As part of the reconsideration process your rebuttal to those reviews has been considered also.
>
> It is my opinion that your proposal was fairly reviewed and that the decision to decline was justified. (Todd 1977; Appendix)

In this response, Dr. Todd ignored the three main points of my appeal letter: (1) NSF support of the other researchers who participated in the original superheavy element experiments, while denying similar support for my research; (2) the panel's refusal to acknowledge that I had proposed a hypothesis for the origin of polonium halos; (3) my claim of finding evidence which challenges the foundations of modern cosmology and geochemistry. Todd's silence on these points led me to believe it would be futile to appeal this decision to a higher level of NSF. Was my proposal rejected because of philosophic bias rather than scientific considerations?

Another Proposal — Another Denial

In 1979 I submitted a new proposal to the National Science Foundation to investigate polonium halos in minerals and other substances. This one was specifically designed to test the NSF pulse on the primordial origin of polonium halos in granites. The implications for creation were clearly stated. In a few words I asked for funds to continue my research. In this brief proposal there were no peripheral issues, such as superheavy elements, for the reviewers to focus on. In this instance they could not escape commenting on my published evidences for creation. They could legitimately criticize the proposal because of its brevity. But if their reactions to the evidence for creation were positive, I could resubmit a longer proposal giving the necessary details. The reviewers' responses would reflect whether they were interested in probing for the truth about creation or in maintaining the status quo of evolution.

Not surprisingly, most of the peer reviews of this new proposal were quite negative. Five reviewers gave it a "poor" rating. I was elated, however, for the one open-minded reviewer who gave a "fair" rating and, in fact, suggested that the proposal should be resubmitted with more details. In general, though, the suggestion of the primordial hypothesis was severely criticized, and the testing of a creation model was referred to as "speculative" and "ridiculous."

One reviewer argued that my primordial polonium hypothesis is "unlikely to be accepted until alternative, conventional interpretations are convincingly shown to be wrong." Another held out the hope that conventional explanations would still be forthcoming: "It is quite likely," argued one reviewer, "that the explanations are to be found in trivial effects involving known phenomena and that explanations already in the literature will suffice." Since my explanation for the data was not conventional, one reviewer commented: "I cannot find any plan . . . to look for alternative explanations of these halos."

The suggestion to look for an explanation within the evolutionary framework was in essence a request to backtrack and head down a dead-end street. Over the previous decade I had already investigated and reported on the two possible explanations for polonium halos in granites that are consistent with evolution, namely, (1) a secondary origin from uranium and (2) the isomer hypothesis. As the earlier parts of this book have shown, the scientific evidences negating these two possibilities had been published in the open scientific literature for many years.

Some reviewers criticized me for not offering "new techniques," "suggestions for new progress," or "a research basis for new progress on the subject." Some of their comments were doubtless inspired by the brevity of my proposal. With others it seems there was an emotional tone. One wrote:

> Gentry merely proposes to do more of the same kind of work he has done before. He does not propose any new technique or approach . . . He does not define any new scientific objectives, except by implication the testing of "a new framework" of cosmology. Therefore, I do not recommend this proposal for support.

These comments express disapproval for my continuing to work with the primordial polonium hypothesis. One reviewer expressed his views as follows: "To me it certainly does not seem worthwhile to further support speculations and ridiculous implications on this subject." Although this reviewer gave no scientific objections to my work, he was not above reacting emotionally to my evidence for creation.

Several of these reviewers had difficulty regarding my hypothesis as genuine and scientific. One felt I was "highlighting personal positions in controversies rather than defining distinct courses of investigation." Another reviewer suggested that the problems I had raised could be solved by other researchers "with greater objectivity."

After first criticizing me for not offering anything new, the most detailed evaluation of my research follows:

> On the plus side, Gentry is . . . probably the world's foremost expert on the observation and measurement of radiohalos. He does his own work, and his financial requirements are quite modest. He is remarkably tenacious in the pursuit of certain observations which are difficult to explain. *His further work will result in publications.* In the past he has seized on several quite new techniques, and arranged to spend several years at ORNL in order to have access to a variety of instruments and scientific associates.
>
> However, his researches seem to have reached a dead end. . . . (Italics mine)

This review exemplifies the contradictory response of the NSF to my work. On one hand, the reviewer downgrades my work, saying that I propose nothing new, yet he acknowledges that I have a record of utilizing new research techniques. My research has reached a dead end, he asserts, yet my future work will be of a quality to warrant publication! If that is so, why did this reviewer oppose funding my work? Publishable research is, after all, exactly what the NSF hopes to obtain from its grant funds.

The answer is not difficult to find. The review continues:

Gentry makes reference in the proposal, and has mentioned in more detail in some of his writings, that the poloniun. halos must be "primordial polonium," which he takes to mean that the polonium was created, along with the host rocks . . . *in a Bible-like instant of creation.* (Italics mine)

Instead of responding to the evidence I had published, the reviewer simply points out that my evidence contradicts the evolutionary framework:

. . . [Gentry] does not discuss the enormous amount of conflicting evidence which ascribes a long process of evolution of the universe, the earth, life on earth, etc. to the present state. If he wishes to propose a new framework for cosmology, he should describe it in detail, with all of its supporting evidence, implications, critical observations which could test it against the "currently accepted cosmological and geological framework, . . ."

This reviewer faults me for not critiquing the entire, comprehensive framework of evolution, as it touches all the scientific disciplines. What he overlooks is that irrefutable evidence for creation invalidates the *uniformitarian principle,* which has been described in this book as the glue binding all the pieces in the evolutionary mosaic together. Where is the logic in evaluating different parts of a theory when all of them are dependent on an erroneous premise? Perhaps the reviewer should have been more concerned that, after many years, evolutionists still failed to explain my widely published evidences for creation.

This reviewer further argues that I needed to detail "critical observations which could test" my hypothesis. This is an interesting but somewhat baffling remark because included with my proposal was a description of such a test—the one discussed at length in the last chapter and published in *EOS* (Gentry 1979). The suggestion that I should propose a new framework of cosmology is something which I had already started and even continued to develop after my proposal was finally rejected.

Both the 1977 and 1979 proposals were thus rejected without any specific, concrete objections to my results on polonium halos. The implications for creation were treated in 1977 with silence and in 1979 with disdain. There was no interest to see whether my observations had pinpointed a critical weakness in the theory of evolution. There was, however, one consolation in all of this. By leaving unchallenged the scientific accuracy of my published experimental work on polonium halos, the reviewers had shown that my evidence for creation must be rather substantial. My scientific colleagues, some of whom were openly antagonistic toward creation, had been exposed to the implications of my research, and their only scientific response to the evidence was silence.

Inquiry by a Member of Congress—1977 Proposal

My interview with the news magazine *Christian Citizen* (Melnick 1981) prompted an individual to contact his U.S. Congressional Representative about the NSF handling of my 1977 proposal. The correspondence between the Congressman and the NSF was forwarded to me.

The NSF gave what appears to be a misleading account of my situation. The first letter, dated June 1982, and addressed to Robert Walker, Representative from Pennsylvania, was written by Francis Johnson, Assistant Director of the Division of Astronomical, Atmospheric, Earth, and Ocean Sciences. It reads in part:

> Mr. Anderson is correct when he states in his letter that Dr. Robert Gentry is the world's leading authority on the observation and measurement of anomalous radioactive haloes. Because of his recognized capabilities, Dr. Gentry's research was funded by the Foundation during the early 1970's. In 1977, however, a proposal presented by Dr. Gentry was declined.
> . . .That action was based upon the recommendations of six of his peer scientists, who found that the proposal did not measure up to either Dr. Gentry's earlier standards, as evidenced by his previously successful proposals, or to the standards of the Foundation. . . .(Johnson 1982; Appendix)

This letter implies that all six reviewers gave negative evaluations of the 1977 proposal, when, in fact, four reviews were actually positive. (The two negative ones focused on my superheavy element research.) It also suggests a decline in the standard of my research. How did the NSF determine that this proposal "did not measure up to" my earlier standards or the standards of the Foundation? Usually the NSF takes the publication record during the preceding grant period as a prime indicator of whether an investigator is making progress in his research. Three reports were published during the 1974-76 grant period, and after the rejection of my 1977 proposal, five additional scientific reports were published in the next five years. Thus as far as my scientific publications were concerned, there certainly had been no decline in my standards. Moreover, the words, "Dr. Gentry *is* the world's leading authority on the observation and measurement of anomalous radioactive haloes," are in the present tense. If, by the NSF's own admission,

I still had that reputation at the time Johnson wrote the letter (June 1982), then my research after 1977 did continue to maintain the standards of my earlier endeavors. And if the NSF classifies someone as an authority in his field, doesn't this imply he has met the Foundation's "standard" of scientific merit?

Representative Walker was not given the full picture. By withholding copies of my correspondence with Dr. Todd, Johnson glossed over the NSF's discriminatory treatment of my proposal. But more to the point, if my appeal letter had been sent to Walker, he could have seen that the NSF avoided responding to my evidences for creation.

To find out if Johnson's failure to send my appeal letters was inadvertent, I called him around July 28, 1982, and pointed out that, in all fairness, his correspondence with Walker left a distinctly erroneous impression. He responded that he was under no obligation to send my appeal letters and drew the conversation to a close. In his letter Johnson assured Walker that the NSF would "be pleased to review and evaluate a proposal from Dr. Gentry at any time. I assure you that any submission will be given a fair, honest and open appraisal by his peers and that if they judge his ideas as worthy of support, he will be funded" (Johnson 1982; Appendix).

The issue is *what standard* will be used to judge whether my ideas are "worthy of support" or not? If scientific credibility hinges upon whether the data support evolutionary ideas, then obviously my research would not measure up to the "standards" of the NSF.

Inquiry by a Member of Congress—1979 Proposal

After hearing me speak at the June 1982 meeting of the American Association for the Advancement of Science (AAAS) in Santa Barbara, another citizen wrote to his Congressional Representative about my funding difficulties with the NSF in 1979. Johnson again responded on behalf of the NSF, writing to Representative Robert J. Lagomarsino of California as follows:

> . . . Only about half of the proposals we receive can be funded. Criteria used are stated in our booklet "Grants for Scientific and Engineering Research" (NSF 81-79, copy of relevant page enclosed). The holding of unorthodox scientific views is not a barrier to the receipt of NSF support, and the best evidence for this is the fact that during the 1970's NSF funded several of Dr. Gentry's proposals including one for $54,900 for the study of "Nuclear Geophysics of Radiohalos."

Please reassure your constituent that NSF funding decisions are based on well-identified criteria and that Dr. Gentry's views have not been a barrier to his receiving NSF support. (Johnson 1983; Appendix)

Johnson cites my previous grants as evidence that the NSF was fair about my earlier proposals. However, he omits relevant information about them: the previous NSF support was given during the early 1970's, a time when the implications of my work for creation were not realized by the scientific community at large.

The creationist implications of my research were published more forthrightly just before and during the periods in which my proposals were refused—in 1977 and 1979. Scientists who had given tentative support to my work in the early 1970's began to give up their hopes that I would discover a conventional explanation of polonium halos in granites, and their attitudes shifted significantly. As one reviewer of the 1979 proposal wrote, "I have previously supported the need for (Gentry's) unorthodox interpretations as a challenge to the rest of the scientific community. Lately, I have concluded that newer, independent efforts are required. . ."

So my experiences show that, contrary to Johnson's words, the publication of "unorthodox scientific views" about creation science did indeed present a "barrier to the receipt of NSF support," once its reviewers understood the issues.

Pro-Evolution at the NSF?

The NSF has a long history of funding proposals which encompass the evolutionary position. Certainly one of the largest grants of this nature was for the Biological Science Curriculum Study (BSCS), which in 1964 published a controversial series of textbooks incorporating evolution as a major theme.

More recently an NSF official has led the way in denouncing the developing creation science movement in America. At the 1981 annual national meeting of the American Association for the Advancement of Science (AAAS), Dr. Rolf Sinclair of the NSF took the opportunity to arouse opposition to creation science. *Science News* reported:

Another topic that provided heat, if not light, was the revived anti-evolution movement. Physicist Rolf M. Sinclair of the National Science Foundation organized a session titled "Views of the Universe: Science versus Tradition." He came out of his ivory tower, he says, and was shocked to find out what the creationists are doing in schools. Their success in getting school districts to teach creationist ideas is restricting and perverting science education, he says.

. . . The session at this year's meeting, explains Sinclair, is just a beginning. The theme of next year's AAAS will be Science Education, and tentatively it will include discussion of ways to combat creationism and the teaching of religion as science. (*Science News* 1981, 19)

Shortly thereafter Sinclair elaborated further on his views in a letter to this news magazine (Sinclair 1981). There he emphasized that the solar system had come into being several billion years ago and that the entire universe began in a "big bang" some tens of billions ago. He also expressed full confidence in the overall record of organic evolution and, in particular, the record of life on earth going back more than a billion years. Both conclusions carry the connotation that evolution is beyond dispute.

Freedom of Inquiry

Since its inception, the NSF has expended vast sums to support research projects based on evolutionary assumptions. It may be argued that the NSF is justified in expending these huge sums because a number of prominent scientists, such as Dr. Sinclair, overwhelmingly endorse evolution as a confirmed theory, or even as fact. If the NSF could prove that evolution is the true description of the origin and development of the cosmos, the earth, and life, then the NSF would be justified in denying funding to scientists whose research proposals question the evolutionary scenario.

But evolution is neither confirmed theory nor fact. If life actually originated by chance, as evolution requires, evolutionary biologists should be able to reproduce that process in laboratory experiments. Still, despite decades of intensive efforts and generous government funding, all attempts to produce life from inert matter have proved fruitless. Likewise, if life evolved by the transformation of one major group into another, where are the numerous transitional forms expected on the basis of evolution? Biologists could long ago have put to rest embarrassing questions about the general absence of transitional forms in the fossil record if they had produced examples of missing links under laboratory conditions. All attempts to create new forms in the laboratory, such as inducing mutations through nuclear irradiation, have produced only variations of existing types. Developing new features in fish, for example, until they begin to develop into amphibians should certainly be simpler than creating life itself and would be the presently observable evidence needed to make evolution a science instead of speculation. There would then be no dispute about its validity.

Since no such demonstration has been accomplished, at best the NSF should consider evolution as a widely held but unproven theory. The NSF is thus morally obligated to treat it as open to challenge, in the spirit of the *Affirmation of Freedom of Inquiry and Expression* (see Overview). Written by evolutionists themselves, it declares that "all discoveries and ideas . . .may be challenged without restriction." I assume that the NSF should also abide by another principle of the *Affirmation:* "Freedom of inquiry and dissemination of ideas require that those so engaged be free to search where their inquiry leads *without fear of retribution in consequence of the unpopularity of their conclusions.* " The reader may decide whether the NSF adhered to this principle in its evaluation of my 1977 and 1979 proposals to continue work on radioactive halos.

The documentation in this chapter shows the reaction of the NSF after they were convinced that my discoveries were contradictory to the "accepted" model of earth history.

7

Creation Science—A Public Issue

Little did I suspect when my work began in 1962 that nineteen years later the results of my research would become a public issue. This all started in the spring of 1981 when the Arkansas state legislature passed Act 590, a bill requiring "balanced treatment of creation-science and evolution in public schools." The American Civil Liberties Union (ACLU) filed suit, challenging the constitutionality of the Act, and a trial date was set for December 7, 1981, at the Federal District Court in Little Rock.

The enactment of Act 590 served as a reminder of the anti-evolution law, the Butler Act, passed in Tennessee in 1925. That law was also challenged by the ACLU, and this led to the famous Scopes trial in Dayton, Tennessee. It is necessary to focus briefly on this earlier trial because its popular legacy as an overwhelming victory for evolution was a key factor in shaping the outcome of the Arkansas trial.

The principals in the 1925 trial were (1) John Scopes, a high school football coach and substitute science teacher, (2) William Jennings Bryan, the great fundamentalist orator and three times presidential candidate, who prosecuted the case against Scopes, and (3) Clarence Darrow, the eminent trial lawyer, who defended Scopes. Scopes' voluntary assent to be arrested for teaching evolution became nationwide news. The reaction to this extraordinary publicity showed that many Americans felt the foundations of their religious beliefs were at stake in this battle. To some the trial was seen as a means of either confirming or denying their understanding of the Scriptures.

The Lessons of Scopes

Generally, popular accounts of the trial picture William Jennings Bryan as a man who feared the truth because of his presumed refusal to permit expert testimony for evolution to be given at the trial. Clarence Darrow, on the other hand, is credited with having outmaneuvered such bigotry when he arranged for his expert witnesses to give their scientific evidences for evolution to the news reporters covering the trial. By this master stroke, Darrow managed to have the theory of evolution disseminated to the remotest bounds of the civilized world. Just as significantly, this was accomplished without any arguments against evolution being mentioned. Thus evolution appeared to be based on incontrovertible evidence.

As a result, evolution is thought to have won the day (even though Scopes lost on a technicality). The perceived outcome of the trial among scientists was both pervasive and self-perpetuating. From henceforth any scientist who openly professed any belief in a literal interpretation of Genesis became suspect among his peers. This disdain for creation was passed on to each new generation of university students by both scientists themselves and educators, many of whom knew no better than to echo their scientific colleagues. Since the Scopes trial, three generations of college-educated Americans have been indoctrinated with the view that evolution represents scientific truth. This widespread indoctrination provided the ACLU with a tremendous psychological advantage as they prepared for the Arkansas trial.

The presiding judge at the Arkansas trial did not live in a vacuum. Even though the Arkansas trial was supposed to be decided solely on the basis of evidence presented in Court, the ACLU well knew the historical impact of the Scopes trial could work to their favor. In addition, the news releases pertaining to the Arkansas trial could be a decisive factor. In general those news reports came from the pens of media representatives who reflected the American cultural scene. It is safe to assume they all had been educated in the mold of contemporary science and its overwhelming preference for evolution. And most likely their image of creation science had been molded by the popular accounts of the Scopes trial. Inevitably their perceptions, and hence their news reports, of the Arkansas trial would reflect the prior conditioning from those accounts.

Unfortunately, it seems much that has been written and filmed in certain of those accounts cannot be substantiated either by historical records or by the transcript of the trial. To illustrate, a recent critique by Dr. David Menton has shown that the account which has received the greatest publicity, namely, the Scopes trial motion picture *Inherit the Wind,* bears little resemblance to the actual events and details of the trial itself. An important part of this critique (Menton 1985) focuses on the circumstances surrounding the arrest of Scopes for presumably breaking the law. According to the historical records, Scopes maintained he never taught evolution during the two weeks he substituted as a science teacher. Thus, in reality, he never broke the law. His arrest was based on a trumped-up charge. It was contrived, with Scopes' assent, by a local mine operator so that the ACLU could challenge the Butler Act.

Did the lawyers who acted in Scopes' defense know of these circumstances? The critique mentioned above provides a clear answer to this question when it refers to L. Sprague de Camp's book, *The Great Monkey Trial.* In this book a remarkable conversation is recorded between Scopes and reporter William K. Hutchinson of the International News Service:

"There is something I must tell you. It's worried me. I didn't violate the law."

"A jury has said you had," replied Hutchinson.

"Yes, but I never taught that evolution lesson. I skipped it. I was doing something else the day I should have taught it, and I missed the whole lesson about Darwin and never did teach it. Those kids they put on the stand couldn't remember what I taught them three months ago. They were coached by the lawyers. And that April twenty-fourth date was just a guess."

"Honest, I've been scared all through the trial that the kids might remember I missed the lesson. I was afraid they'd get on the stand and say I hadn't taught it and then the whole trial would go blooey. If that happened they would run me out of town on a rail."

"Well you are safe now," said Hutchinson.

"Yes, I'm convicted of a crime I never committed," said Scopes. "But my skirts are clear. You know I pleaded 'not guilty.'"

"That will make a great story."

"My god, no!" cried Scopes. "Not a word of it until the Supreme Court passes on my appeal. My lawyers would kill me if it got out now." (de Camp 1968, 432)

Thus, incredible as it seems, those who acted in Scopes' defense apparently not only knew of, but abetted the situation by encouraging some of Scopes' students to commit perjury and testify that Scopes had taught evolution. (Interestingly, deCamp's book (p. 432) singles out Darrow as the lawyer who did the coaching.) In his memoirs Scopes once again disclaimed teaching evolution, which at his trial included a reference to the earth once being "a hot molten mass" (Scopes and Presley 1967, 132-134). At the same time he also attempted to deflect the clear implication of perjury by claiming his students were possibly confused about where they had heard about evolution (Scopes and Presley 1967, 134). These circumstances reveal an aspect of the Scopes trial that is not generally known.

One of the most questionable parts of *Inherit the Wind* relates to its portrayal of William Jennings Bryan as a man who feared the truth because he objected to the introduction of expert testimony for evolution. Quoted below are two paragraphs of Menton's critique which presents a different perspective on this matter:

> MOVIE: The defense is unable to get permission to use their several expert witnesses because Bryan is afraid of their testimony and considers it irrelevant. One by one, Darrow calls his distinguished scientists to the stand but each time, thanks to an ignorant and biased judge, Bryan needs only to say, "objection—irrelevant," and that is the end of it.
>
> FACT: Technically, the only point at issue in the trial was whether or not John Scopes actually taught the evolution of man from lower orders of animals, so naturally the lawyers for the prosecution did question the relevance of the testimony of expert witnesses. The testimony of the evolutionists assembled by the defense was prevented, however, because Darrow adamantly refused to let his scientific witnesses be cross-examined by the prosecution (transcript, pages 206-208). Bryan had asked for, and received, the right to cross-examine the expert witnesses, but Darrow was so opposed to allowing his experts to be questioned that he never called them to the witness stand! Bryan pointed out that under the conditions demanded by Darrow, the evolutionists could take the witness stand and merely express their speculations and opinions on evolution without fear of either perjury or being contradicted. (Menton 1985)

Stacking the Deck Against Creation Science

The negative image of creation science portrayed in the widely viewed *Inherit the Wind* was considerably reinforced by the pretrial publication of some critical reviews of creation science. A good example is the article, "A Response to Creationism Evolves," published in *Science* just a few weeks before the Arkansas trial. This article (Lewin 1981) details the results of two scientific meetings, organized for the purpose of combating the spread of creation science in America. The first was sponsored by the National Academy of Sciences (NAS) and held on October 19, 1981.The second meeting was organized by the National Association of Biology Teachers (NABT) and held on October 20, 1981. At these meetings, both held in Washington, DC, certain influential evolutionary scientists made it appear that creation science was a threat, not just to evolution, but to all of science. They issued a call for opposition to creation science at every opportunity. William Mayer, Director of the Biological Sciences Curriculum Study, Louisville, Colorado, is quoted as declaring:

> The whole structure of science is under attack. And it's not just biology that's in danger, it's all of science: geology, physics, astronomy.The creationists are attempting to mandate what is appropriate for study and what is not. (Lewin 1981, 635)

These alarmist remarks were made before a sympathetic audience. Ironically, these evolutionists failed to see that their own staunch opposition to the teaching of evidence for creation was in itself an attempt to mandate what is and what is not appropriate for study. Note that the emphasis here is not what is truth, but how to maintain the status quo in science. This was further evident when Niles Eldredge, a curator at the American Museum of Natural History in New York, used scare tactics to oppose funding for creation science:

> The creationists have already made moves to secure funding for so-called creation science on an equal footing with evolution science. This should be sufficient to convince my colleagues that the house really is on fire. (Lewin 1981, 635)

Other pretrial articles that provided the ACLU with psychological advantage appeared in the December 1981 issue of the popular monthly *Science 81*. (This issue, devoted primarily to a formidable attack on the "fallacies" of creation science, was deemed so important that copies were given to the National Science Teachers Association for distribution to its members.) An excerpt from the article "Farewell to Newton, Einstein, Darwin. . ." shows how the authors, Allen Hammond and Lynn Margulis, attempt to convey the impression that creation science is in direct conflict with true science:

> All scientific theories, inevitably, are tentative answers to questions about nature. . . . This characteristic of continually revising ideas to reflect the world as it is observed is what makes science science.
>
> In contrast, the creationists start with a "theory" or faith in a particular description of nature drawn not from observation but from the Bible. To argue—as the creationists do—that a theory must be true rather than

that the evidence compels one to it as the best choice is fundamentally antithetical to science. To be unwilling to revise a theory to accommodate observation is to forfeit any claim to be scientific. For it is not facts or theories that are essential to the growth of science but rather the process of critical thinking, the rational examination of evidence, and an intellectual honesty enforced by the skeptical scrutiny of scientific peers. By these standards creationism is not science. Indeed, creationists do not participate in the scientific enterprise—they do not present papers or publish in scientific journals. And it is precisely because creationists present themselves as "scientific" that they do most harm to the educational system. (Hammond and Margulis 1981, 57)

The claim that creationists are unwilling to revise a theory to accommodate observation is nothing more than massive character assassination of all creation scientists. I have already referred to one revision in my own work that occurred in reference to the previously discussed report on superheavy elements. And the claim that creation scientists do not publish in scientific journals is directly contradicted by my own publications.

Another writer, John Skow, also presented uncomplimentary views of creation scientists in his companion article in the same issue of *Science 81*:

The scientific creationists have been on the scene for something more than a decade now, and it is clear that their obduracy is not the result of insufficient education. It is a resolute, structured ignorance, maintained by choice and against odds. . . . They must find "scientific" reasons for the scientifically unreasonable, and by heroic twisting of evidence, they do. . . . Their system of belief resists unwanted information. (Skow 1981, 59)

The question could be raised: Just who is attempting to "twist" the evidence? Skow claims that creationists resist unwanted information. His accusations are quite incongruous, for both his and the previously quoted article fail to mention the persuasive evidence for creation published in my scientific reports. Is it possible that he may have the "system of belief" that "resists unwanted information"?

This widely distributed issue of *Science 81* greatly reinforced the negative view of creation science which had been given such impetus at the Scopes trial. My colleagues at the Oak Ridge National Laboratory who saw this issue were doubtless hoping that I would not be drawn into center stage in this rapidly developing controversy over creation and evolution.

The Arkansas Trial: A Difficult Decision

On the morning of October 27, 1981, a long-time friend called to inform me that he had been asked to testify as an expert witness at the forthcoming Arkansas creation trial. He mistakenly thought an invitation had already been extended for me to testify as well. Upon learning this was not the case, he informed the Arkansas Attorney General's office of my work. Later that same day I received a telephone call from the Attorney General's office requesting me to provide them with reprints and other pertinent data relative to my findings. A few days later the Deputy Attorney General asked me to testify for the State as an expert witness for creation science.

I knew that my involvement in the trial could eliminate any hope of continuing at the Laboratory. If my testimony at the trial were misrepresented in print, it would almost certainly jeopardize my research at ORNL. In September 1981, a few months before the trial, a prominent evolutionist had also called and urged me in a cordial way not to risk incurring the ill will of the scientific community by testifying for the State. He suggested that my reputation as a scientist would undoubtedly suffer if I did, and I was inclined to agree with him. Until I received the October call, it seemed that my presence at the trial would be a needless risk.

Several factors led me to change my mind about testifying for the State. They have all come into much sharper focus since the trial, as will be seen in the remainder of this chapter. First of all, it was and still is difficult for me to follow the logic and consistency of the American Civil Liberties Union (ACLU) in their opposition to Act 590. They claimed that this Act, in requiring balanced treatment of evolution and creation science, constituted the establishment of religion, thus violating the First Amendment. And yet witnesses at court trials swear by God to tell the truth, and the courts themselves are opened with a reference to God. Also, both branches of the U.S. Congress are often begun with reference to the God of the Judeo-Christian ethic. A striking example of this practice was given in the opening prayer for the House of Representatives on April 29, 1982:

The Reverend Ray A. Howe, First Presbyterian Church, Bennetsville, S. C., offered the following prayer:

O Creator God, who created the world and all that inhabit it, we thank You we can so readily see the beauty of Your work. We thank You for all intimations of Your beauty and being in the world of nature and for those pointers of Your being in the lives of Your people. We lift our voices in gratitude for these Halls of free debate, where our leaders can reason together and chart our course. Endow them with a perception of things that endure. May their energies go always to making our land one where freedom and justice find noble expression. May their efforts here for our Nation make noble the idea that their service to people is a high and sacred calling. Amen. (Howe 1982)

This explicit reference to the Creator is recorded in the *Congressional Record.* It is available to all Americans, including public school students studying American government. Thus far, such reference to the Creator has always been deemed consistent with the First Amendment and the academic freedom granted to both student and teacher. And if public schools may refer to the Creator in the context of an American history class, on what basis is it wrong for them to refer to scientific evidences of creation in a science class?

Just as relevant to this issue is the Declaration of Independence itself:

We hold these truths to be self-evident, that all men are *created* equal, that they are endowed by their *Creator* with certain unalienable Rights, that among these are Life, Liberty, and the pursuit of Happiness.

Every state in the union uses these words to inform public school children of the Creator. Why, then, would it be illegal for the states to inform the same students of the scientific evidences of creation?

Repression in the Classroom

I thought about my earlier encounters in the university classroom. There is evidence which suggests that it may be worse now for religiously oriented university students than it was in my day. To illustrate, I refer to a letter published in the October 1982 issue of *Physics Today,* the national news magazine of the American Institute of Physics. This letter was written by an accomplished physicist who is *not* a creationist. It should open the eyes of everyone who advocates basic human rights and academic freedom for those who hold minority views:

After reading a spate of virulently anti-creationist articles and letters in your publication, I decided that something less virulent and more thoughtful should be said.

As we might all easily agree, it isn't very scientific to make assumptions dogmatically and then accept only evidence in favor of these assumptions. It is the practice of this precept that separates the unbelievers from the believers, sheep from goats, and so forth. Most of us, history says, will test as goats. Therefore, a word of caution: How much do we actually know (other than that it has something to do with someone's religion) about this set of ideas we are calling "creationism"? I shall confess that I know next to nothing. Will any of the noisemakers out there also confess?

I do know what we do not know about creation: almost everything. Science, like religion, is not a physical thing itself, but a non-material set of ideas. It is an ideology and is not exempt from the scrutiny to which we subject other ideologies. Science, if it is to progress, must be fed the fuel of inspired thinking—brainstorming, if you will. Religion has generally been the repository of things we felt must be true, in some sense knew were true because we existed ourselves, but which we could not demonstrate rationally or understand. Sometimes the inspiration that sparks great scientific progress has been religious. Other times a dogmatically held religious concept has stifled the development of the very inspiration that it may have been meant to provide. The point is that we have never been very good judges of this and, as scientists living in an age that has history books telling of both atheistic Nazis who purported to worship science, and Spanish Inquisitors, who purported to be doing God's will, that we be a bit more humble and lower our voices.

We have several things to gain by lowering our voices. One is the possibility that paying attention to some radically different ideas, however wacky, may suggest to us an insight into science that we do not expect. For instance, we do not have a thoroughly rational, tested hypothesis about the origin of our species. Indeed, we haven't even been able to agree upon a biological classification system for primates. Somewhere buried in the creationist arguments may be the right question, one that we have been ignoring because it wasn't proper to consider it! *The second thing we have to gain is our decency and humanity. I have myself sat in class after class in the sciences and humanities in which any idea remotely religious was belittled, attacked, and shouted down in the most unscientific and emotionally cruel way. I have seen young students raised according to fundamentalist doctrine treated like loathsome alley cats, emotionally torn apart, and I never thought that this sort of treatment was any better than the treatment that religious prelates, who held authority, gave Galileo. Why scream about the inhumanity of nuclear war if you are also willing to force people of fundamentalist faiths to attend public schools in which their most cherished beliefs will be systematically held up to ridicule and the young children with it? These people are mostly too poor for private schools to be an alternative. The state tries to prevent them from teaching their children at home rather than sending them to school. What choices do they have? Would you call it freedom? Do you call it fair?*

Is it really a terrible thing for a textbook to mention that, aside from the Darwin theory of evolution, there have existed other ideas, many of them religious in nature? Would that not open the mind of students rather than close them to scientific possibilities? Wouldn't it make the fundamentalist student feel a little more welcome and better equip him to take an unbiased view of evolution?

Well now, I've asked a lot of questions and I do not know the answers. I would far prefer to hear physicists discussing such questions than loudly attacking straw men and expressing a Chicken Little attitude that the educational sky is falling because a few creationists want to be heard. (Lane 1982, 15—italics mine)

Repressive treatment of religious students would not be surprising under a totalitarian, atheistic regime. But most readers of this book may be surprised to learn that this kind of religious persecution exists here in America. This letter reveals a side of the story which the ACLU did not tell at the Arkansas trial. The ACLU's opposition to Act 590 was a direct attempt to preserve the exclusive teaching of evolution in public schools. This letter reveals, however, some of the abuses of this arrangement: students who express doubts about the "facts" of evolution are under a potential threat of retribution just for asking questions. This is not academic freedom for all the students. *Real academic freedom should provide opportunity for the whole truth about creation and evolution to be made known to the students.*

This belief, based on my earlier university experiences, became an increasingly strong motivation for me to testify at the trial. A number of well-qualified scientists had already accepted the State's invitation to testify for creation science. Possibly I could assist in their efforts to have the evidences for creation examined more objectively.

Evolution Promoted As Fact

Confirmed evolutionists who have used repressive measures during their teaching careers may have thought they acted in the best interests of science. Perhaps they felt it was their duty to limit inquiries about creation in order to save society from harm. Readers may wonder how many evolutionists really believe that their theory is beyond question: apparently, quite a few.

Reference has already been made in the previous chapter to the strident anti-creation remarks of Dr. Rolf Sinclair at the January 1981 American Association for the Advancement of Science (AAAS) annual meeting. Similar views were also expressed at the same AAAS meeting by another eminent evolutionist, Dr. Porter Kier. The following quote taken from *Science News* shows that Kier's confidence in the certainty of evolution equals that of Dr. Sinclair:

> Discussing the evidence for evolution, Smithsonian Institution scientist Porter M. Kier, former director of the National Museum of Natural History, said there are 100 million facts which support evolution. "In the museums of the world", he says "there are over 100 million fossils that have been identified and age-dated. These fossils have been examined by many thousands of paleontologists and from their investigations we have learned a vast amount about the history of life on the earth." Despite this evidence, Kier admits, "there are many well-educated people still questioning evolution. Part of the problem may be that evolution has been described as the 'theory' of evolution, which gives an erroneous impression—that scientists themselves don't accept evolution as accepted." The word "theory," he says, has done a great deal of damage and should be dropped and the word evolution should stand alone. "Scientists may argue over the details of evolution," he says, "but they agree that evolution is a fact and should be so labeled." (*Science News* 1981, 19)

Nor were Drs. Sinclair and Kier alone in expressing their complete confidence in evolution. As the Arkansas trial drew near, the American Geological Institute, which is comprised of 18 geology-related societies with over 120,000 members, issued the following release:

> Scientific evidence indicates beyond any doubt that life has existed on Earth for billions of years. This life has evolved through time producing vast numbers of species of plants and animals, most of which are extinct. Although scientists debate the mechanism that produced this change, the evidence for the change is undeniable. Therefore, in the teaching of science we oppose any position that ignores this scientific reality, or that gives equal time to interpretations based on religious beliefs only. (American Geological Institute 1981)

Readers may decide for themselves whether the dogmatism expressed in the above statement encourages the kind of intimidation of students referred to in the letter published in *Physics Today*. And they should also reflect on the impact this timely resolution may have had on some of the media representatives assigned to cover the Arkansas trial.

Countdown to the Arkansas Trial

This chapter has recounted a few of the incidents where evolutionists had denied the existence of evidence for creation. It seemed that the only way to settle the issue was to go to the trial. My presence there would guarantee that my work would be scrutinized by the best evolutionist scientists. If errors existed, these would be exposed.

Most importantly, the trial should reveal why there had been no response from the scientific community to the critical falsification test which I had proposed in *EOS* in 1979 and again in 1980. As discussion in Chapter 5 showed, the test was simple. According to the evolutionary scenario, the Precambrian granites had supposedly cooled from a hot magma during a multibillion-year evolution of the earth. If granites had really formed in this fashion, then it should be possible to duplicate the process today; that is, it should be possible to synthesize a hand-sized piece of granite from a hot melt prepared under laboratory conditions. Likewise it should be possible to produce a polonium halo in that piece of synthesized granite. If these experiments were successful, I would withdraw my claims that the Precambrian granites are created rocks and that polonium halos represent primordial radioactivity. The crucial question was whether my colleagues had been able to perform those experiments.

It was time for this issue to be resolved. Scientists had repeatedly claimed no credible evidence for creation existed. At the trial they would have an opportunity to prove that claim by refuting my published evidences for creation. If polonium halos in Precambrian granites were not evidence for creation, then I wanted all my scientific colleagues to know this as soon as possible. Likewise, if my results could not be refuted, I knew this would be of compelling interest to the millions of individuals who are ardently seeking to know the truth about the Genesis record of creation. For these reasons I accepted the invitation from the Attorney General's office to testify. It was one of the most challenging decisions of my life. It is also one I have not regretted.

8

ACLU Strategy Revealed at Little Rock

The Arkansas trial began on a cold Monday morning in December 1981 at the Federal District Court in Little Rock. Judge William Overton presided over the trial where there were more than 200 spectators—including 60 magazine, newspaper, and TV personnel registered as members of the media. They included such metropolitan newspapers as the New York *Times*, Washington *Post*, the *Times* of London, Chicago *Tribune*, Baltimore *Sun*, Kansas City *Times*, Detroit *News*, Milwaukee *Journal*, and Memphis *Press Scimiter*; magazines like *Time, U.S. News and World Report, Harper's, Nature, Science 81, Science News, Discover*, and *Science;* the AP and UPI news services; and of course the national TV networks NBC, CBS, ABC, and PBS, and even the BBC (British Broadcasting Company).

The expert witnesses for evolution gave their testimonies during the first week of the trial. They came out with "guns blazing," a procedure which decidedly reinforced the psychological advantage they already held. The ACLU witnesses had been well coached to make evolution appear invincible. The excellent preparation of these witnesses reflected the efforts of a large, competent ACLU legal contingent. That contingent consisted of two local attorneys, two New York ACLU lawyers, and two more from one of New York's most prestigious law firms, Scadden, Arps, Slate, Meagher, and Flom. In addition, many other lawyers and paralegals from Scadden and Arps backed up the ACLU. Altogether, the ACLU had over 50 lawyers and paralegals working on the case.

In contrast, the Arkansas Attorney General's office could only commit three of its attorneys to the case. This placed the State's case at a disadvantage even before the trial began. To counter the perception that evolution was incontrovertible required that the State strongly confront the expert evolutionist witnesses during their cross-examinations and, just as importantly, be prepared to expose any flaws which might be uncovered. It was impossible for the few State attorneys to completely prepare for the cross-examinations and to also adequately rehearse their own witnesses for direct testimony. They had no choice but to concentrate on the latter; nevertheless, the cross-examinations of the ACLU witnesses were conducted proficiently. On several occasions the State attorneys actually exposed some of the fatal weaknesses in the ACLU case; but this seemed to have little impact on the judge. In fact, in one instance the State was actually thwarted from exploiting a critical weakness in the ACLU case by the judge himself.

The ACLU's Plan for the Treatment of Origins

From my viewpoint as a former evolutionist, it was quite revealing to see how the ACLU treated the subject of origins. In my university courses the theory of evolution encompassed the spontaneous origin of life as well as its diversification. But at the trial the ACLU sought to present the question of origins as something apart from evolution. One of their witnesses, Dr. Francisco J. Ayala, a geneticist from the University of California at Davis, reportedly maintained that even though life had arisen from nonlife by natural laws, this occurrence was *not* a part of evolution. (Ayala's exact words are not quoted here because his testimony had not been transcribed as of this writing in the spring of 1986. However, his testimony is summarized in Norman Geisler's book, *The Creator and the Courtroom* (Geisler 1982, 82-84). There were good reasons why the ACLU wished to avoid directly linking evolution with the spontaneous origin of life.

After decades of research the ACLU knew that evolutionists had been unsuccessful in their attempts to produce life from inert material. But obviously they did not want the State to focus on this glaring failure as evidence that one of the basic tenets of evolution was wrong. On the other hand, the ACLU had to maintain that life could be formed naturalistically; otherwise they would have to consider the possibility of a sudden creation of life, which Act 590 ascribed to creation science. I watched with interest as the ACLU unfolded their strategy to divert attention away from this issue.

Direct Examination of the ACLU Witness for Biophysics

That strategy was unveiled in the direct examination of their expert witness for biophysics, Dr. Harold Morowitz, from Yale University. ACLU Attorney

Jack Novik's examination of Morowitz began with the usual legal formalities and then focused almost immediately on how Act 590 depicted the origin of life from the creation science perspective. According to the official transcript of the trial (Smith 1982a), some of the exchanges between Novik and Morowitz on this topic were as follows (all quotes from the transcripts follow the original wording except where indicated in brackets):

Q   Doctor Morowitz, let me show you a copy of Act 590 marked, I believe, Exhibit 29 in these proceedings. Had you read this Act before?

A   Yes, I have.

Q   Would you look at Section 4 of this statute, particularly Section 4 (a), purporting to define creation science. Do you see any reference in that section to the origin of life?

A   4 (a) (1) refers to sudden creation of life from nothing.

Q   And is "sudden creation" a term that has scientific meaning to you?

A   No. To my knowledge it is not a term in scientific literature or in general use in the scientific community.

Q   Do you know the meaning of the words "sudden creation"?

A   "Sudden creation" assumes a creator, and, as such, implies the supranatural explanation, and, therefore, lies outside the bounds of normal science. [Smith 1982a, p. 495, l. 20, to p. 496, l. 13]

Having presumably established that "sudden creation" is excluded from conventional science because it requires "a creator," Novik subsequently asked:

Q   Does the theory of evolution as used by scientists include the study of the origins of life?

A   Normally that's treated as a separate subject in a technical sense. [Smith 1982a, p. 498, ll. 17-20]

Ordinarily Morowitz's response would have kept the lid on the origin-of-life matter. Yet the ACLU still had to maintain acceptability for the naturalistic origin of life in order to preserve the image that evolutionists have the truth about origins. Thus Novik found it necessary to return to the question of the origin of life on two separate occasions in his later direct examination of Morowitz.

Q   Doctor Morowitz, do you know how life was first formed on this planet?

A   We do not know in any precise way how life was formed. However, it is a very active field of research. There are a number of studies going on, and we are developing and continuing to develop within science a body of knowledge that is beginning to provide some enlightenment on this issue. [Smith 1982a, p. 499, l. 24, to p. 500, l. 6]

Q   Do you know how life was formed precisely?

A   Again, not in precise detail, although as I pointed out, it is an active area of scientific research, and at the moment one, as an enthusiastic scientist always feels, that we are getting close. [Smith 1982a, p. 509, ll. 11-15]

The ACLU and the Origin of Life: A Narrow Escape

Morowitz's bold claim about getting close to knowing how life formed provided State Attorney Callis Childs with a unique opportunity to probe the weaknesses in the ACLU position on the origin of life. As the following exchanges show, Attorney Childs' incisive cross-examination of Morowitz came within inches of exposing the flaw in the ACLU strategy on this matter.

Q   Are you familiar with the work of a fellow named Miller?

A   Stanley Miller?

Q   I believe so, yes, sir.

A   There are a lot of people named Miller.

Q   Are there any Millers other than Stanley Miller that would be working in your particular area of endeavor?

A   Not that I'm aware of.

Q   Did Mr. Miller, or let's say Doctor Miller, did Doctor Miller come up with anything unusual in the 1950's in his research?

A   Yes.

Q   What did he come up with?

A   In Miller's experiments, he took a system of methane, ammonia and water, and in a closed system he provided energy through an electrical, high frequency electrical spark discharge, and he demonstrated the synthesis of amino acids, carbocyclic [sic, carboxylic] acids, and other prebiotic intermediates.

Q   Who was the previous historian, excuse me, the previous scientist in history who dealt with that same subject matter on a significant basis?

A   The origin of life?

Q   Yes.

A   Prior to the Miller experiment, I would say that the leading name in that field was A. I. O'Parin [sic, Oparin].

Q   And prior to that?

A   Prior to that, in a sense, the field didn't really exist.

Q   Why was that?

A   Because people believed through the 1800's that life arose spontaneously all the time; that maggots arose and became meat [sic], and mice old piles of rags [sic], and so forth and so on. And as long as people believed that, there was no need to have a theory of the origin of life.

Q Who put that theory to rest?

A Louis Pasteur.

Q And what were Doctor Pasteur's experiments?

A Basically his final experiments that were most persuasive in this field consisted of flasks of sterile medium to which no organisms were admitted, and these flasks remained sterile for long periods of time.

Q So?

A Meaning no growth of living organisms occurred in them.

Q What work has been done since Stanley Miller's work in the area of generating life in the laboratory?

A Well, there have been some several thousand experiments on the, of tne type done by Miller, follow-up experiments, where various energy sources have been flowed [sic, have flowed]; there has been the flow of various kinds of energy through systems of carbon, hydrogen, nitrogen and oxygen, and there has been a study of the kinds of molecules that are produced in such energy flow systems. These experiments universally show that the flow of energy through a system orders it in a molecular sense.

Q Has anybody created life by the flow of energy?

A Have any of those experiments resulted in the synthesis of a living cell? Is that the question?

Q Yes, sir.

A No. Not to my knowledge, anyway.

Q Would you say that this area has received intensive scientific scrutiny in the scientific community?

A Yes.

Q Do you have any explanation of why you have not been able to synthesize life in the laboratory?

A It's an extremely difficult problem.

Q What is the difficult—

A I would point out to you that we have put far more money into trying to cure cancer, and that is still an unsolved problem, also. We have put far more time, money, effort and human endeavor into that problem, and that is also an unsolved problem because it is a very difficult problem.

Q What is the information you need to accomplish that?

A To accomplish the synthesis of a living cell?

Q Yes, sir.

A Two kinds of information. One is the detailed understanding of the chemical structure of the small molecules, micro molecules, organelles and other structures that make up a living cell. And secondly, one has to know the kinetic processes by which those structures came about in prebiotic systems.

Q   In perusing some of the literature that you've written last night, I came up with an article which would seem to indicate that [you] sincerely believe that given enough time and research, that you or scientists like you can ultimately go back to the ultimate combinations of atoms which led to the formation of molecules.

A   That is not a question.

Q   Do you recall an article to that effect?

A   Well, you said "we can go back to that" and then there should be an 'and' clause, 'and do some things'.

Q   Do you believe that you can go back and ultimately understand how atoms combined to form molecules?

A   That is a branch of chemistry. That is rather well understood.

Q   Well, I'm talking about the first molecules on the surface of the earth. Do you understand my question?

A   No, I don't.

MR. CHILDS: May I approach the witness, your Honor?

THE COURT: Yes.

Q   The article that I have is *Biology as a Cosmeological [sic, Cosmological] Science,* reprinted from *Main Currents and Modern Thought,* volume 28, number 5, May through June, 1972.

Page 50 to, well, the page number I have on this is 615186. The first column is in brackets. I'd like you to read that paragraph, please.

A   "If we are able to obtain the kind of theory of self-order, this kind of theory of self-ordering should challenge us to apply the most profound insights we can muster to link biology to non-equilibrium physical chemistry."

"The job seems very formidable indeed, but the rewards could be very great; the ability to seek out our origins in terms of a law that would promulgate our action. This is truly a new frontier, and one that challenges the maximum intellectual effort of which we are all capable."

Q   Do I understand this paragraph to mean that you believe that you and scientists from the scientific community can explain the origins of man in terms of the laws of atomic interaction?

A   I believe that the origin of life can be explained in terms of the laws of atomic interactions. [Smith 1982a, p. 585, l. 25, to p. 590, l. 25]

Q   Is your theory that—Let me start over. Do you know how life formed on the surface of the earth?

A   I have a theory of how life formed on the surface of the earth.

Q   Have you been able to take that theory and create life in the laboratory?

A   No. [Smith 1982a, p. 600, l. 20, to p. 601, l. 1]

It is most revealing to compare Morowitz's responses in his direct testimony with those given under cross-examination. Note that when ACLU Attorney Novik asked, "Do you know how life was formed, precisely?" Morowitz testified optimistically, "Not in precise detail . . ." but "that we are getting close." However, during Attorney Childs' relentless probing of this matter, a different picture emerges. When Childs asked if Morowitz knew how life had formed on the earth's surface, he responded only that he had a theory. And when asked whether he had been able to use that theory and create life in the laboratory, Morowitz was forced to answer, "No." (Remember that Childs had earlier gotten Morowitz to admit that thousands of experiments designed to produce life had failed.)

The Judge Rescues the ACLU

Morowitz's reponses had established that leading evolutionists have only theories about how life began and, most importantly, that none of them works. Apparently Attorney Childs sensed that he had pinpointed a flaw in the ACLU case, for his next question to Morowitz was designed to refocus and exploit what he had just uncovered. Interestingly, Judge Overton was not about to let Childs probe this defect any further, as will be seen by the trial transcript:

Q   Let me repeat my question. Do you know how life evolved on the surface of the earth?

THE COURT: He just answered that.

MR. CHILDS: I think he said he had a theory.

THE COURT: I think that is the answer. I think he has a theory. He doesn't know for a fact.

MR. CHILDS: I think there has been a blurring in the distinction between a theory and a fact in this lawsuit, and that is the point I am attempting to make, your Honor.

THE COURT: I don't know how it's blurred, but it doesn't seem to me like that answer blurred it.

MR. CHILDS: I will move on, your Honor. [Smith 1982a, p. 601, ll. 2-14].

Thus, whatever climax Childs might have achieved in this phase of his cross-examination was effectively short-circuited by Judge Overton's timely intervention in behalf of this ACLU witness. This was one of the critical points in the trial, generally unnoticed by media representatives.

So the ACLU escaped from having one of the major defects in their case exposed; namely, the numerous failures to synthesize life constitute prima-facie evidence that the *uniformitarian principle* is not now, nor ever has been, a sufficient basis for life to form. If it was, evolutionists long ago should have been able to reconstruct by design that which nature constructed only by chance. Evolutionists continue to fail in synthesizing life from inert matter because they are attempting the impossible—the duplication of a process that lies solely in the hands of the Creator.

The ACLU: No Science but Evolution

Morowitz's cross-examination had established that belief in a naturalistic origin of life, as required by evolution, has no scientific basis. To divert attention away from this truth, the ACLU used the clever strategy of depicting creation science as being unscientific. Generally the ACLU was able to have most of its expert witnesses (Geisler 1982, 92-99) affirm that creation science is not scientific without being challenged in their cross-examinations. This was one of the State's most costly shortcomings. In one instance, though, the State did expose just how far some evolutionists are willing to go in their opposition to creation. This occurred during Attorney General Steve Clark's cross-examination of the ACLU's biology witness, Dr. William V. Mayer, from the University of Colorado. Quoting Geisler (this part of Mayer's transcript was not available), in his direct examination this witness had earlier

> objected to the term "evolution science" in Act 590 on the basis that it implied that there was such a thing as a science which was non-evolutionary, which he said is not true. (Geisler 1982, 99)

This statement effectively mandates that only evolution can be viewed as science. During cross-examination Attorney Clark inquired about this statement. Clark asked Mayer if he had said it "may well be that creationism is correct about origins." To this Mayer agreed, and added that he also had said "even if it were correct, it's not scientific." (Geisler 1982, 102)

This was a revelation. Over twenty-five years earlier I had accepted evolution based on what seemed to be scientific evidence in its favor. At that time I assumed all scientists were searching for the truth, always ready to modify their position if contrary evidence were found. Indeed, my quest for truth was initiated with the hope that evolutionists would fairly evaluate new data even if the outcome conflicted with the status quo, hence my long arduous efforts to inform them of my results in scientific journals. But it is difficult to see how Mayer's view represents an unbiased search for truth. Rather it seems specially geared to preserve the status quo of evolution regardless of how much evidence is discovered for creation.

This was another critical point of the trial—a point where the State could have decimated one of the foundations of the ACLU case. The ACLU had portrayed evolutionists as those dedicated to an open-minded search for truth in science, whereas creation scientists were represented as those who abuse science. But Mayer's responses exploded this myth. If, according to Mayer, there is no science but evolution, then searching for truth in science means that only those evidences in agreement with evolutionary theory will be accepted as scientific. I believe Attorney General Clark should have focused strongly on this issue during his cross-examination of Mayer.

The Age of the Earth: Testimony of the ACLU Geology Witness

The age of the earth was a key issue in the Arkansas trial. The general theory of evolution encompasses both the multibillion-year geological evolution of the earth as well as a multimillion-year evolution of life on the earth. In order to win their case, it was imperative for the ACLU to find a witness who would strongly promote an ancient age of the earth consistent with geological evolution.

To accomplish this task the ACLU called a scientist whom I personally admire very much, one who is considered an eminent authority in the field of radiometric dating, Dr. G. Brent Dalrymple. At the time of the trial Dalrymple held the position of Assistant Chief Geologist of the U.S. Geological Survey in Menlo Park, California. Not surprisingly, Dalrymple testified that the earth is billions of years old, contrary to the generally accepted creation-science position that the age of the earth is less than 10,000 years. He also stated that, on this point, creation science could be falsified and, in fact, had been falsified many times over the last several decades by many different tests.

As indicated in the previous chapter, the popular legacy of the Scopes trial was that true scientists believe in evolution. At every opportunity the ACLU waged a clever psychological war to capitalize on this perception. The following excerpts from the official trial transcript of ACLU Attorney Bruce Ennis' direct examination of Dalrymple illustrate this point quite effectively regarding the age of the earth:

Q   Are you familiar with the creation science literature concerning the age of the earth?

A   Yes, I am. I have read perhaps two dozen books and articles either in whole or in part. They consistently assert that the earth is somewhere between six and about twenty thousand years, with most of the literature saying that the earth is less than ten thousand years old.

Q   Are you aware of any scientific evidence to indicate that the earth is no more than ten thousand years old?

A   None whatsoever. In over twenty years of research and reading of scientific literature, I have never encountered any such evidence.

Q   Are you aware of any scientific evidence to indicate that earth is no more than ten million years old?

A   None whatsoever.

    THE COURT: Wait a second. What is it that the creation scientists say is the age of the earth?

A   They make a variety of estimates. They range between about six and about twenty thousand years, from what I've read. Most of them assert rather persistently that the earth is less than ten thousand years. Beyond that they are not terribly specific.

Q   Are you aware of any scientific evidence to indicate the earth is no more than ten million years old?

A   None whatsoever.

Q   Are you aware of any scientific evidence to indicate a relatively young earth or a relatively recent inception of the earth?

A   None whatsoever.

Q   If you were required to teach the scientific evidences for a young earth, what would you teach?

A   Since there is no evidence for a young earth, I'm afraid the course would be without content. I would have nothing to teach at all.

Q   Is the assertion by creation scientists that the earth is relatively young subject to scientific testing?

A   Yes, it is. It is one of the few assertions by the creationists that is subject to testing and falsification.

Q   Have such tests been conducted?

A   Yes. Many times, by many different methods over the last several decades.

Q   What do those tests show?

A   Those tests consistently show that the concept of a young earth is false; that the earth is billions of years old. In fact, the best figure for the earth is in the nature of four and a half billion years. And I would like to point out that we're not talking about just the factor of two or small differences. The creationists' estimates of the age of the earth are off by a factor of about four hundred fifth [sic, fifty] thousand.

Q   In your professional opinion, are [sic, have] the creation scientists' asser-
     tions of a young earth been falsified?
A   Absolutely. I'd put them in the same category as the flat earth hypothesis
     and the hypothesis that the sun goes around the earth. I think those
     are all absurd, completely disproven hypotheses.
Q   In your professional opinion, in light of all of the scientific evidence,
     is the continued assertion by creation scientists that the earth is relatively
     young consistent with the scientific method?
A   No, it is not consistent with the scientific method to hold onto a
     hypothesis that has been completely disproven to the extent that it is
     now absurd. [Smith 1982b, p. 409, l. 6 to p. 411, l. 19]

I agree that theories which have been shown to be false should be
discarded—that is one of the main themes of this book. But is Dalrymple
correct in claiming those tests disprove or falsify a young age of the earth?
As we shall soon see, Dalrymple's cross-examination showed that the tests
he cited to justify this conclusion all *assume* constant radioactive decay rates.
This assumption is actually just a part of the *uniformitarian principle*—the
glue that holds the evolutionary mosaic together—mentioned many times
earlier in this book.

Dalrymple's claim of certainty about the earth being four and a half billion
years old coupled with his scathing comments about a young age of the
earth were exactly what the ACLU wanted Judge Overton to hear. Certainly Ennis knew beforehand that Dalrymple planned to psychologically attack the young-earth view by linking it with the flat-earth hypothesis. Would
the ACLU ever have allowed Dalrymple to draw this invidious comparison
unless they strongly suspected that Judge Overton had already been primed
in favor of evolution?

Ennis then turned his direct examination to questions concerning various
types of dating techniques. The excerpts from the transcript, shown below,
are some of those linking radiometric dating and the age of the earth.

Q   How do geochronologists test for the age of the earth?
A   We use what are called the radiometric dating techniques. [Smith 1982b,
     p. 411, ll. 20-23]
Q   Why did geochronologists rely upon radiometric dating techniques rather
     than other techniques?
A   Because radioactivity is the only process that we know of that's been
     constant through time for billions of years.
Q   Is radioactive decay affected by external factors?

A   No, radioactive decay is not affected by external factors. That's one reason we think it's been constant for a long time. [Smith 1982b, p. 413, l. 24, to p. 414, l. 6]

Q   Have any tests ever shown any change in the decay rates of any of the particular isotopes geochronologists use in radiometric dating?

A   None. They've always been found to be constant.

Q   Are changes in decay rates of various isotopes at least theoretically possible? [Smith 1982b, p. 416, ll. 7-12]

A   . . .There have never been any changes affecting any of the decays being used for radioactive dating. [Smith 1982b, p. 417, ll. 13-14]

Note here that my respected colleague asserts that radioactivity is the only process *known* to be constant for billions of years and then affirms this assertion by saying there have *never* been any changes in the decay rates of the isotopes used in radioactive dating. There is no question that his absolutist remarks were crucially needed by the ACLU to bolster their case for an eons-long evolutionary development of the earth. But the truth is that Dalrymple was not around during the period when he claims to have certain knowledge of radioactive decay rates. As we shall see in the next chapter, his great assurance about this matter is, in fact, nothing more than what evolution assumes to be true. This was brought out clearly in Deputy Attorney General David Williams' cross-examination of Dalrymple. And it is in this cross-examination that the topic of radioactive halos comes to the fore.

9

Confrontation in the Courtroom

Regardless of how fervently the ACLU expert biology witnesses might propound the merits of biological evolution, the ACLU well knew that neither the geological evolution of the earth nor the biological evolution of life was even remotely credible without the support of vast amounts of time. So in the last analysis, the cornerstone of the ACLU case rested on establishing the plausibility of an ancient age of the earth. This is why Dalrymple's testimony about the reliability of radioactive dating was crucial to their intent of overturning Act 590.

The State Challenges Radiometric Dating Techniques

Under cross-examination Dalrymple was asked to affirm his confidence in uniform radioactive decay for increasingly greater times in the past. Although he expressed belief in a uniform decay process for the past 4.5 billion years, continued cross-examination brought him to admit his uncertainty about this assumption for earlier periods in the history of the universe, saying that his area of expertise did not extend to that time. Quoted below from the official trial transcript are segments of the cross-examination by Deputy Attorney General David Williams:

Q   Is constancy of the rate of radioactive decay a requirement for radiometric dating?

A   Yes. It is required that radiometric dating be based on constant decay rates, at least within limits of significant areas, and what I mean by that is that if the decay rates were to change a percent or two, that would probably not significantly alter any of our major conclusions in geology.

Q   To the best of your knowledge, has the rate of radioactive decay always been constant?

A   As far as we know from all the evidence we have, it has always been constant. We have no, either empirical or theoretical, reason to believe it is not.

Q   So as far as you know, it would have been constant one billion years ago, the same as it is today.

A   As far as we know.

Q   Five billion years ago?

A   As far as we know.

Q   Ten billion years ago?

A   As far as we know.

Q   Fifteen billion?

A   I don't know how far back you want to take this, but I think for the purposes of geology and the age of the solar system, we are only interested in using radiometric dating on objects we can possess in our hand, so we only need to take that back about four and a half or five billion years.

I think whether it's been constant fifteen billion years is irrelevant. We have no way of getting samples that old. We can only sample things that have been in the solar system. [Smith 1982b, p. 449, l. 8 to p. 450, l. 13]

Notice the change in Dalrymple's position on the constancy of the decay rate. In his direct testimony (see Chapter 8) he claimed to know that radioactive decay rates had been constant without any time qualification whatsoever. Likewise during the intial part of his cross-examination, he affirmed they had been constant at least ten billion years. However, at the fifteen-billion-year mark he apparently senses that Williams is approaching the presumed time of the Big Bang. He then begins to backtrack and suddenly reveals that "we only need" them to be constant for the last four and a half billion years. In other words, at this point in the cross-examination, it appears that the requirement, or "need," to establish credibility of the evolutionary model determines how far back in time evolutionists are willing to affirm constant decay rates.

Williams must have realized this was a startling revelation, for he continued pressing Dalrymple to find out more about the decay rates and, in addition, to probe just what evidence he had for their constancy over the last several billion years.

Q   How old is the solar system, to the best of your knowledge?

A   As far as we know, it is four and a half billion years old.

Q   The solar system itself?

A   The solar system itself. Now, when we talk about the age of something like the solar system, you have to understand that there was a finite period of time over which that system formed, and we may be talking about a period of a few hundred years, so it is not a precise point in time, but some interval. But compared with the age of the solar system, it is thought that that interval was probably rather short—a few percent.

Q   Are you aware of when those scientists hypothesized or when the so-called big bang occurred, how many years ago?

A   No, I am not sure exactly when that was supposed—

Q   Would the rate of radioactive decay have been constant at the time of the big bang?

A   I am not an astrophysicist. I don't know the conditions that existed in the so-called primordial bowl of soup, and so I am afraid I can't answer your question.

Q   So you don't have any opinion as to whether it was constant then?

A   That's out of my field of expertise. I can't even tell you whether there were atoms in the same sense that we use that term now.

Q   But you did state that it had always been constant as far as you knew, but now you state you don't know about the big bang, whether it was constant then; is that correct?

A   Well, what I said, it's been constant within the limits in which we are interested. For the purposes of radiometric dating it hardly matters whether it was constant at the moment of the big bang. Let me say this—

Q   I don't want to interrupt you.

A   That's all right.

Q   You say as far as you are concerned, for the purposes of your concern it has been constant as far as you know, and your purposes go back to the age of the earth for four point five billion years; is that correct?

A   Yes, that's correct.

Q   But you base that age of the earth on the assumption or on this require-ment that it has always been constant; is that correct?

A   That is not entirely—That's correct, but it is not an assumption. It is not fair to calculate it that way. In a certain sense it is an assumption, but that assumption has also been tested.

For example, if you look at the ages of the oldest, least disturbed meteorites, these objects give ages of one point five to four point six billion years. A variety of different radioactive decay schemes, schemes it at [sic, that have?] different half lives. They are based on different elements. They would not give those identical ages if the rate of decay had been [sic, had not been?] constant.

Q   But do those schemes that you mentioned there rely upon the require-
     ment that the rate of radioactive decay has always been constant as well?
A   Yes, they do.
Q   So all methods you know would rely upon this, what you termed a re-
     quirement and what I termed an assumption; is that correct?
A   That is correct.
Q   The rate of decay is a statistical process, is it not? I think you testified
     yesterday to that.
A   Basically, it is.
Q   Would you agree that any deviation in the rate of decay would have
     to be accompanied by a change in physical laws?
A   As far as we know, any change in decay would have to be accompanied
     by a change in physical laws, with the exceptions that I mentioned yester-
     day. There are small changes known in certain kinds of decay, specifically
     in electron capture, a tenth of a percent.
Q   What do you consider the strongest evidence for the constant rate of
     radioactive decay?
A   Well, I don't think I could give you a single piece of strongest evidence,
     but I think the sum total of the evidence, if I can simplify it, is that
     rates of decay have been tested in the laboratory and found to be essen-
     tially invarient [sic, invariant].

     Theory tells us those rates of decay should be invarient [sic, invariant].
     And when we are able to test those rates of decay on undisturbed systems;
     that is, systems that we have good reason to presume have been closed
     since their formation clear back to the oldest objects known in the solar
     system, we find we get consistent results using different decay schemes
     on isotopes that decay at different rates.

     So that is essentially a synopsis of the evidence for constancy of decay.
     [Smith 1982b, p. 450, l. 14, to p. 454, l. 6]

It is most informative to compare the responses that Dalrymple gave in
his direct testimony and cross-examination. The last chapter revealed that
in his direct testimony Dalrymple claimed that "tests consistently show
. . . that the earth is billions of years old." When questioned what those
tests were, the response was "radiometric dating techniques." And when
asked why "geochronologists rely upon" these techniques, the reply was
that "radioactivity is the only process that we know of that has been constant through time for billions of years."

The above cross-examination reveals, however, that Dalrymple's confidence
in constant decay rates in the distant past rests on his belief that the assumption of constant decay has been tested. For one test he cites the fact that

decay rates are observed to be constant at present. Of course this is not a test about events in the past, but simply an observation about the present. His only other test for constant decay rates was that certain samples give consistent results when they are analyzed for different radioactive elements having different decay schemes. Attorney Williams, apparently perceiving there was a flaw in this presumed test, continued to press Dalrymple on this point.

Q   Did you say— But is it not true that as long— Well, if the rate of decay has varied and as long as the variation would have been uniform, would you still get these consistent results?
A   It is possible to propose a set of conditions under which you could get those consistent results.

THE COURT: Excuse me. I didn't understand that.

THE WITNESS: I think what he is saying is, is it possible to vary the decay rate in such a way that you could still get a consistent set of results by using different decay schemes, and I think it is always possible to propose such a set of circumstances, yes.

So that question is in the nature of a "what if," and one can always come to the conclusion that you can restructure science in such a way to make that "what if" happen. But that is not the sort of thing we usually do unless we have good reason to presume the physical laws have changed, and we presume they have not.

The same is true with things like the speed of light, gravitational constant and so forth. May I elaborate just a little bit more? We are not talking about small changes in decay. If the creation scientists are correct and the earth is only ten thousand years old, we are talking about many orders of magnitude, thousands of times difference. The difference between the age of the earth that scientists calculate and the age that the creationists calculate are different by a factor of four hundred and fifty thousand.

So you don't have to perturb the constancy of decay laws a little bit; you have to perturb them a lot. [Smith 1982b, p. 454, l. 7 to p. 455, l. 11]

In the above responses my respected colleague now admits that consistent results obtained by different decay schemes do not actually prove constant decay in the past after all. He then attempts to reduce the impact of this admission by noting that varying decay rates would involve changes in physical laws. His only argument against this possibility is plainly stated: scientists "presume they [physical laws] have not" changed. But the presumption that physical laws have not changed over the course of time is just the *uniformitarian principle*. Thus, his entire testimony concerning constant decay rates and an ancient age of the earth was hinged on his faith in this unproven *principle*. No proof was given for constant decay rates and an ancient age of the earth because no proof exists.

Indeed, when Dalrymple said, "If creation scientists are right and the earth is only ten thousand years old . . . ," his main argument against an earth this young was that this meant decay laws had to be perturbed "a lot." But Dalrymple provided no evidence to show this had not occurred sometime in the past. In short, he was unable to scientifically counter the possibility of a young earth. Unfortunately for the State, this was not generally understood at any time during his cross-examination. And there was something else of equal importance for the State's case that went undetected.

We have already noted the change in my colleague's stance on the decay rate: from his confident, opening assertion that the decay rate has always been constant to his somewhat defensive position that its constancy beyond a certain point is irrelevant. We now refocus on that part of the cross-examination when State Attorney David Williams asked whether the radioactive decay rate had been constant at the time of the Big Bang. The reply was:

> I am not an astrophysicist. I don't know the conditions that existed in the so-called primordial bowl of soup, and so I am afraid I can't answer your question. [Smith 1982b, p. 451, ll. 9-11]

Dalrymple's reticence to commit himself fully on the one assumption that supports the entire evolutionary framework—the *uniformitarian principle*—together with his earlier assertion that it was irrelevant whether the decay rate was constant beyond a certain time, placed the ACLU case in a very vulnerable position. If the State had drawn attention to the implications of these statements, this would have greatly reduced the credibility of the ACLU position for the remainder of the trial. His response also allowed him to avoid questions about the supernatural nature of the Big Bang. Quite possibly the ACLU realized such questions could open up the proverbial Pandora's box. Any witness who testified about the evolutionary beginning of the universe would give the State the opportunity to focus on the Big Bang as an event not subject to known scientific laws. It would then have been evident that evolution as well as creation requires a supernatural beginning. Thus the cornerstone of the ACLU strategy would have crumbled!

The Granite Synthesis Experiment: An Evolutionary Perspective

As the cross-examination continued, Dalrymple first confirmed what has been stated earlier: dating of fossils as well as rocks depends on geochronology. Having established that geochronology is of preeminent importance to evolution, Williams then delves into this subject preparatory to asking about my work and the falsification test.

I had looked forward to this part of the trial with great anticipation. Everyone there would soon learn whether Dalrymple or some other geologist had succeeded in synthesizing a piece of granite. For this information I had decided to risk everything and come to the trial. If someone had been successful in the synthesis experiment, then I was going to withdraw my claim that the Precambrian granites were the Genesis rocks of our planet. If this had not been accomplished, then it was certain that the polonium halo evidences for creation were not going to be refuted at the trial. I listened carefully as Attorney Williams proceeded with the cross-examination.

Q   Mr. Dalrymple, is it correct that you think that geochronology establishes an age of the earth, not only that the earth is several million years old, but also establishes the age of the fossils which are enclosed in the rocks?

A   Yes. That's correct. [Smith 1982b, p. 458, ll. 4-9]

Q   Now, do you understand that biologists consider these fossils enclosed in these rocks to be the relics or the remnants of some evolutionary development?

A   Well, I think the fossils are the relics of an animal.

Q   Would that be the evidence of the evolutionary development?

A   Well, as far as I know, yes.

Q   Then would it be fair to say in your mind that the ages for the various types of fossils have been most precisely determined or measured by radioactive dating or by geochronology?

A   That sounds like a fair statement.

Q   Since geochronology does play such an important role on the ages of the rocks and the fossils, would you agree that it would be important to know whether there is any evidence which exists which would bear on the fundamental premises of geochronology?

A   Of course. Let me add that that's a subject that's been discussed considerably in scientific literature. We're always searching for that sort of thing. That's a much debated question.

Q   I think you said yesterday that anyone who believes in a young age of the earth, in your opinion, to be not too bright scientifically, and are in the same category as people who believe that the earth is flat?

A   Yes.I think if we are talking about people who profess to be scientists and insist on ignoring what the actual evidence is for the age of the earth, then I find it difficult to think that their thought processes are straight.

Q   Is it true that you do not know of any scientists who would not agree with you, with your viewpoint on this radioactive dating and of the age of the earth and fossils?

A   Will you rephrase that? I'm not sure I understand it.

Q   Is it true that you stated, I think in your deposition, that you do not know of any scientists—

MR. ENNIS: Excuse me. If you're referring to the deposition, please identify it, what page.

MR. WILLIAMS: I'm not referring to a page at this point, I'm asking a question.

MR. WILLIAMS: (Continuing)

Q   Is it true that you do not know of any scientist who does not agree with you and your viewpoint and opinion as to the age of the earth and the fossils?

A   It depends on who you include in the word "scientist." I think if you want to include people who categorize themselves as creation scientists, then that would not be a true statement. I know that some of those do not agree.

As far as my colleagues, geologists, geochemists, geophysicists and paleontologists, the ones that I know of, I don't know of any who disagree that the earth is very old or that radiometric dating is not a good way to date the earth.

Q   Are you aware of any creation scientist, then, who has published evidence in the open scientific literature who has questioned the fundamental premises of geochronology by radioactive dating?

A   I know of one.

Q   Who is that?

A   That's Robert Gentry. I should say that Robert Gentry characterizes himself as a creation scientist, if I understand what he's written.

Q   Are you familiar with Paul Damon?

A   Yes. I know him personally.

Q   Who is Mr. Damon?

A   Mr. Damon is a professor at the University of Arizona at Tuscon [sic, Tucson]. He specializes in geochronology.

Q   Are you aware that Mr. Damon has stated in a letter that if Mr. Gentry's work is correct, that it casts in doubt that entire science of geochronology?

A   Which letter are you referring to?

Q   Do you recall the letter which you gave to me from *EOS* by Mr. Damon?

A   Yes. I recall the general nature of that letter.

Q   And do you recall that Mr. Damon said that if history [sic, Gentry] is correct, in his deductions it would call up to question the entire science of geochronology?

A   Well, I think that's the general sense of what Paul Damon said, but I think it's an overstatement. I'm not sure I would agree with him on that. [Smith 1982b, p. 459, l. 19, to p. 463, l. 1]

Here we see that Dalrymple was so anxious to minimize the implications of primordial polonium halos in granites that he was willing to take issue with Damon's published statement. This prompted Attorney Williams to focus on Damon's qualifications as he continued the cross-examination.

Q   Mr. Damon is not a creation scientist, is he?

A   No. Doctor Damon is not a creation scientist, by any means.

Q   Would you consider him to be a competent scientist and an authority in this field?

A   Yes. He's extremely competent.

Q   Are you aware as to whether Mr. Gentry has ever offered or provided a way for his evidence to be falsified?

A   I am aware that he has proposed one, but I do not think his proposal would falsify it either one way or the other.

Q   Have you ever made any attempts, experiments that would attempt to falsify his work?

A   Well, there are a great many— I guess you're going to have to tell me specifically what you mean by "his work." If you could tell me the specific scientific evidence you're talking about, then let's discuss that.

Q   Well, first of all, do you like to think you keep current on the scientific literature as it may affect geochronology?

A   Well, I keep as current as I can. There's a mass amount of literature. In the building next to my office, there are over two hundred fifty thousand volumes, mostly on geology. It's extremely difficult to keep current. But I am currently relatively up on the mainstream, anyway.

Q   Certainly the most important points?

A   I do my best.

Q   And if someone had issued a study which would, if true, call up to question the entire science of geochronology, would you not want to be made aware of that and look at that closely yourself, as an expert in the field?

A   Oh, yes, I would.

Q   And as a matter of fact, your familiarity with Mr. Gentry's work is limited, is it not, to an article that he wrote in 1972 and a letter that he wrote in response to Mr. Damon's letter, in terms of what you have read, is that correct?

A   Those are the things I can recall having read, and the reports that I have some recollection of. I have never been terribly interested in radioactive haloes, and I have not followed that work very closely. And that is the subject upon which Mr. Gentry has done most of his research.

   As I think I told you in the deposition, I'm not an expert on that particular endeavor. I'm aware that Mr. Gentry has issued a challenge, but I think that challenge is meaningless.

Q   Well, let me ask you this. You stated in the deposition, did you not— Let me ask you the question, can, to your knowledge, granite be synthesized in a laboratory?

A   I don't know of anyone who has synthesized a piece of granite in a laboratory. What relevance does that have to anything?

Q   I'm asking you the question, can it be done?

A   Well, in the future I suspect that it will be done.

Q   I understand. But you said it has not been done yet?

A   I'm not aware that it has been done. It's an extremely difficult technical problem, and that's basically what's behind it. [Smith 1982b, p. 463, l. 2 to p. 465, l. 13]

A long awaited moment of truth had come. Dalrymple did not have a piece of synthesized granite to present at the trial. The ACLU had failed to respond to the challenge of creation, and they badly needed to minimize the impact of this failure. The best Dalrymple could do for them was to say he suspected that the granite synthesis would be done in the future and that I had proposed a meaningless test. We shall later discuss both comments in more detail. For the present we continue with the cross-examination as Williams begins to ask more specifically about Dalrymple's knowledge of my work.

A Very Tiny Mystery

Q   To the extent that you are familiar with Mr. Gentry's work and that as you have reviewed it, would you consider him to be a competent scientist?

A   I think Mr. Gentry is regarded as a competent scientist within his field of expertise, yes.

Q   And you would agree with that?

A   From what I've seen, that's a fair assessment of his work, yes. He's a very, did some very careful measurements, and by and large he comes to reasonable conclusions, I think, with the possible exception of what we're hedging around the fringes here, and that is his experiment to falsify his relatively recent inception of the earth hypothesis. We have not really discussed what his hypothesis is and what his challenge is, we've sort of beat around the edges.

Q   Well, you haven't read his articles that he wrote since 1972, have you?

A   No. That's true.

Q   So if his hypothesis were in those articles, you really wouldn't be able to talk about it, at any rate, would you?

A   His hypothesis, I believe, is pretty fairly covered in those letters between, exchange of letters between Damon and Gentry, and I can certainly discuss that part.

    That's a very current exchange of letters. It is just a few years old. And it is in that letter that he throws down to [sic, a] challenge to geology to prove him wrong. What I'm saying is, that challenge is meaningless.

Q   Are you familiar with his [Gentry's] studies of radio haloes?

A   No, I'm not familiar with that work at all.

Q   But to the extent that work shows that evidence that these formations are only several thousand years old, you're not familiar with that?

A   I'm not familiar with that, and I'm not sure I would accept your conclusion unless I did look into it.

Q   If you're not familiar with it, I don't want to question you about something you're not familiar with.

A   Fair enough. [Smith 1982b, p. 465, l. 14, to p. 467, l. 1]

Q   I think you stated earlier that you reviewed quite a bit of creation-science literature in preparation for your testimony in this case and also a case in California, is that correct?

A   Yes. I think I've read either in whole or in part about two dozen books and articles.

Q   But on the list of books that you made or articles that you have reviewed, you did not include any of Robert Gentry's work as having been reviewed, did you?

A   That's right. I did not.

Q   Although you consider Gentry to be a creation scientist?

A   Well, yes. But, you know, the scientific literature and even the creation science literature, which I do not consider scientific literature— It's outside the traditional literature—there is an enormously complex business. There is a lot of it. And we can't review it all.

    Every time I review even a short paper, it takes me several hours to read it, I have to think about the logic involved in the data, I have to reread it several times to be sure I understand what the author has said; I have to go back through the author's references and sometimes read as many as twenty or thirty papers that the author has referenced to find out whether what has been referenced is true or makes any sense; I have to check the calculations to find out if they are correct. It's an enormous job. And given the limited amount of time that I have to put in on this, reviewing the creation science literature is not a terribly productive thing for a scientist to do.

Q   How many articles or books have you reviewed, approximately?

A   You mean in creation science literature?

Q   Creation science literature.

A   I think it was approximately twenty-four or twenty-five, something like that, as best I can remember. I gave you a complete list, which is as accurate as I can recall.

Q   And if there were articles in the open scientific literature—Excuse me—in refereed journals which supported the creation science model, would that not be something you would want to look at in trying to review the creation science literature?

A   Yes, and I did look at a number of those. And I still found no evidence.

Q   But you didn't look at any from Mr. Gentry?

A   No, I did not. That's one I didn't get around to. There's quite a few others I haven't gotten around to. I probably never will look into all the creationists' literature. I can't even look into all the legitimate scientific literature. But I can go so far as to say that every case that I have looked into in detail has had very, very serious flaws. And I think I've looked at a representative sample.

And also in Gentry's work, he's proposed *a very tiny mystery* which is balanced on the other side by an enormous amount of evidence. And I think it's important to know what the answer to that little mystery is. But I don't think you can take one little fact for which we now have no answer, and try to balance, say that equals a preponderance of evidence on the other side. That's just not quite the way the scales tip. [italics mine]

Q   If that tiny mystery, at least by one authority who you acknowledge his [sic, is an] authority, has been said [sic, has said], if correct, [it would] call [in] to question the entire science of geochronology.

A   Well, that's what Damon said. And I also said that I did not agree with Paul Damon in that statement. I think that's an overstatement of the case by a long way. I think that Paul in that case was engaging in rhetoric. [Smith 1982b, p. 467, l. 20, to p. 470, l. 14]

The above responses vividly illustrate the ACLU's attempts to demean the significance of my reports. Certainly my colleague could have studied them before the trial if the ACLU had wanted this to be done. Apparently the ACLU reasoned that it was safer to ignore them than to risk admitting that they had been studied without successfully refuting the evidences contained therein.

On the surface it would seem that having polonium halos in granites labeled a very tiny mystery—something scientifically insignificant—was one of the cleverest achievements of the ACLU at the trial. But it also involved a serious contradiction which, unfortunately for the State, slipped by unnoticed during Dalrymple's cross-examination. My colleague generally claimed ignorance of the details of my work, saying he hadn't read any of my scientific reports published since 1972. But if he hadn't read them, he couldn't possibly know much about the scientific evidences for primordial polonium halos. *How then could he testify that polonium halos in granites were irrelevant to the issue of creation?*

Even though the State didn't capitalize on this opportunity to pinpoint a contradiction in the ACLU's case, the State's incisive cross-examination did expose the inability of the ACLU to refute the evidence for primordial polonium halos and the falsification test. This had damaged the ACLU case and made it imperative for Attorney Ennis to conduct a redirect-examination of Dalrymple. As we shall see in the next chapter, my colleague gave some remarkable testimony during this redirect-examination and subsequent recross-examination by the State.

Readers should understand that it was imperative for me to respond to the various phases of Dalrymple's testimony if this book was to have any meaning. These responses have not lessened my personal respect for him.

10

Creation's Test on Trial

We pick up Dalrymple's redirect-examination at the point where ACLU Attorney Ennis begins to question him about the falsification test. Dalrymple's second answer spurs Judge Overton to interrupt the proceedings with some of his own questions about my work. Dalrymple has no choice but to respond, and he does so in an amazing way. He provides such a superb explanation of the implications of polonium halos in granites that, for the moment at least, it seems he is about to convince Judge Overton of the evidence for creation.

ACLU Witness Explains Evidence for Creation

Quoting from the transcript, Ennis continues his redirect-examination of Dalrymple as follows:

Q   During cross examination Mr. Williams asked you if Mr. Gentry's argument or hypothesis could be falsified. Has Mr. Gentry proposed a method for falsifying his hypothesis?

A   Yes, he has proposed a test and that is the one I characterized as meaningless.

Q   Why would it be meaningless?

A   Let me first see if I can find a statement of the test, and I will explain that. I have it now. [Note: Here Dalrymple refers to the statement of the falsification test that I published in 1979 in *EOS*. The publication of this test (Gentry 1979, 474) and (Gentry 1980) was earlier discussed in Chapter 5.]

     THE COURT: May I read what you quoted from the newsletter before you go to that?

A   (Continuing) Okay, sir.

The experiment that Doctor Gentry proposed—

THE COURT: Let me ask you a question. As I understand it, that's his conclusion. I still don't understand what his theory is.

THE WITNESS: [Dalrymple]: He [Gentry] has proposed that it is either a theory or a hypothesis that he says can be falsified.

THE COURT: What's the basis for the proposal? How does he come up with that?

THE WITNESS: Well, basically what he has found is there is a series of radioactive haloes within minerals in the rocks. Many minerals like mica include very tiny particles of other minerals that are radioactive, little crystals of zircon and things like that, that have a lot of uranium in them.

And as the uranium decays, the alpha particles will not decay, but travel outward through the mica. And they cause radiation damage in the mica around the radioactive particle. And the distance that those particles travel is indicated by these radioactive haloes. And that distance is related directly to the energy of the decay. And from the energy of the decay, it is thought that we can identify the isotopes.

That's the kind of work that Gentry has been doing.

And what he has found is that he has identified certain haloes which he claims are from Pollonium-212 [sic, polonium-218; correct form of the chemical elements used hereafter]. Now, polonium-218 is one of the isotopes intermediate in the decay chain between uranium and lead.

Uranium doesn't decay directly from [sic, to] lead. It goes through a whole series of intermediate products, each of which is radioactive and in turn decays.

Polonium-218 is derived in this occasion from radon-222. And what he has found is that the polonium haloes, and this is what he claims to have found, are the polonium-218 haloes, but not radon-222 haloes. And therefore, he says that the polonium could not have come from the decay of radium, therefore it could not have come from the normal decay change [sic, chains].

And he says, how did it get there? And then he says that the only way it could have gotten there unsupported by radon-222 decay is to have been primordial polonium, that is polonium that was created at the time the solar system was created, or the universe.

Well, the problem with that is polonium-218 has a half-life of only about three minutes, I believe it is. So that if you have a granitic body, a rock that comes from the melt, that contains this mica, and it cools down, it takes millions of years for a body like that to cool.

So that by the time the body cooled, all the polonium would have decayed, since it has an extremely short half-life. Therefore, there would be no polonium in the body to cause the polonium haloes.

So what he is saying, this is primordial polonium; therefore, the granite mass in which it occurs could not have cooled slowly; therefore, it must have been created by fiat, instantly.

And the experiment he has proposed to falsify this is that he says he will accept this hypothesis as false when somebody can synthesize a piece of granite in the laboratory.

And I'm claiming that that would be a meaningless experiment. Does that—I know this is a rather complicated subject.

THE COURT: I am not sure I understand all of this process. Obviously I don't understand all of this process, but why don't you go ahead, Mr. Ennis?

MR. ENNIS: Yes, your Honor. Obviously, your Honor, these subjects are somewhat complex, and if the Court has additional questions, I'd hope that the Court would feel free to ask the witness directly. [Smith 1982b, p. 476, l. 21 to p. 480, l. 2]

At this point I suspect Attorney Ennis was more than just a little nervous about Judge Overton's comments. Ennis had just heard my arguments for creation summarized extremely well by his own star witness. In the light of Dalrymple's lucid commentary, it seems that Judge Overton was somewhat perplexed—perhaps he didn't quite understand why my conclusions were wrong and why the falsification test was meaningless.

Remember that in his earlier cross-examination Dalrymple deftly sidestepped the challenge of creation by saying that polonium halos are a tiny mystery, which some day would be solved; and he did likewise with the falsification test, saying he suspected that a granite would be synthesized in the future. The ACLU claimed that evolution represented the true picture of the origin of the earth, but they had signally failed to defend their position in two major encounters.

Confronting the Falsification Test

This repeated postponement of confronting these issues, I believe, had come close to placing the ACLU's case in jeopardy. The ACLU was on the verge of becoming victims of their own strategy—namely, someday, somewhere, someone was going to find a solution to the evidence for creation. To re-establish credibility in their case, Ennis may have thought that Dalrymple needed to present something tangible to back up his assertion about the falsification test, and so in his next question he again asks about it:

Q   Why, in your opinion, would the test proposed by Mr. Gentry not falsify his hypothesis?

A   Let me read specifically first what his proposal is. He said, "I would consider my thesis essentially falsified if and when geologists synthesize a hand-sized specimen of a typical biotite barium [sic, bearing] granite and/or a similar sized crystal of biotite."

And if I understand what he's saying there, he's saying that since his proposal requires that granite form rapidly, instantly, by instantaneous creation, that he does not see any evidence that these granites, in fact, cool slowly; his evidence said they cool rapidly. And he would accept as evidence if somebody could synthesize a piece of granite in the laboratory.

There are a couple of problems with that. In the first place, we know that these granites did form slowly from a liquid from the following evidence: These rocks contain certain kinds of textures which are only found in rocks that cool from a liquid. And we can observe that in two ways, these textures. They are called ligneous [sic, igneous] and crystalline textures.

We can observe these textures by crystallizing compounds in the laboratory that we are able to crystalize [sic, crystallize]. And they always form these crystalline textures. We can also observe things like lava flows and watch them cool today and see what kind of textures they produce.

There has been an experiment since 1959 going on in the Kilauea-Iki lava lake. Now, Kilauea-Iki is a small volcano event on the top of the Kilauea volcano, which is one of the five volcanoes which make up the island of Hawaii.

And in 1959, Kilauea-Iki erupted, it not only threw up fountains of lava, lava flows, but it formed a large pool of lava that was captured in a crater. And that lava is hundreds of feet thick. Since 1959, scientists have been drilling down through that lava, watching it crystallize. Every few years they go back and drill another hole and watch the degree to which that lava lake is cooled. It takes a long time for this to cool. This is a fairly thick one.

And we see that in the case of lava lakes and lava flows and these things, when they cool from their melt, from their liquid, they form these textures that are unique to all rocks that pool [sic, cool] from a liquid. When we go to a granite and we see these same textures, then I think we are entitled to presume that these rocks also formed from a liquid. There is no other way that they could have formed.

The other problem with Gentry's proposal is that the crystallization of granite is an enormously difficult technical problem, and that's all it is. We can't crystallize granite in the laboratory, and he's proposing a hand-sized specimen. That's something like this, I presume.

In the first place, the business of crystallizing rocks at temperatures, most of them crystallize at temperatures between seven hundred and twelve hundred degrees centigrade. The temperatures are high. And in the case of granites and metamorphic rocks, sometimes the pressures are high, many kilobars. So it takes a rather elaborate, sometimes dangerous apparatus to do this.

And the apparatus is of such a size that usually what we have to crystallize is very tiny pieces. I don't know of anyone who has developed an apparatus to crystallize anything that's hand-sized.

So he's thrown down a challenge that's impossible at the moment, within the limits of the present technical knowledge.

The second thing is that the crystallization of granite, the reason we have not been able to crystallize even a tiny piece in the laboratoray [sic, laboratory] that I know if [sic, of], unless there has been a recent breakthrough, is essentially an experimental one. It's a kinetic problem.

Anyone who has tried to grow crystals in a laboratory knows that it's very difficult to do if you don't seed the melt. That is, you have to start with some kind of a little tiny crystal to begin with. And when the semiconductor industry, for example, grows crystals to use in watches like this, they always have to start with a little tiny seed crystal. And once you have that tiny seed crystal, then you can get it to crystallize.

So it's basically a problem of getting the reaction to go, it's a problem of nucleation, getting it started, and it's a problem of kinetics, getting the reaction to go on these viscous melts that are very hot under high pressure.

And what I'm saying is that even if we could crystallize a piece of hand-sized granite in the laboratory, it would prove nothing. All it would represent would be a technical breakthrough. All of a sudden scientists would be able to perform experiments that we cannot now perform.

But in terms of throwing down a challenge to the age of the earth, that's a meaningless experiment. So he's thrown down a challenge that has no meaning, hand-sized crystallized granite. And he's saying, "If you don't meet it, then I won't accept you [sic, your] evidence." Well, it's a meaningless challenge. It's not an experiment. [Smith 1982b, p. 480, l. 4 to p. 483, l. 25]

This is incredible! Evolutionists claim they have the truth about the origin and age of the earth, and yet when they have an opportunity to provide experimental evidence to substantiate their views, they call it a "meaningless" challenge. This forces me to ask a penetrating question: If evolutionists really believe that the granites formed by slow cooling instead of instantaneous creation, why are they reticent to put their theory of granite formation to the test? It is inescapable that the granite synthesis test is at the center of the creation/evolution controversy. For that reason we need to carefully examine Dalrymple's lengthy commentary about it. Doing this also provides an opportunity to explain a facet of my creation model that has not been previously discussed.

Primordial Rocks Derived from a Primordial Liquid

Dalrymple begins his response by referring to my statement of the falsification test. From this he concludes that my *"proposal requires that granite form rapidly, instantly, by instantaneous creation, that he [Gentry] does not see any evidence that these granites, in fact, cool slowly; his evidence said they cool rapidly."* This statement, which contrasts slow cooling of the granites with their rapid cooling and instantaneous creation, suggests that Dalrymple perceives that my creation model may involve a liquid precursor for these rocks. This is correct. Just because Precambrian granites are considered primordial or created rocks does not preclude the possibility that they were formed from a liquid. The Creator, after calling the chemical elements into existence, might well, in the next instant of time, have formed those elements into a liquid, and then immediately cooled that liquid so that it crystallized into the granites containing the polonium halos. These granites would have been created instantly and yet still show the characteristics of rocks that crystallize from a liquid or melt.

Dalrymple presents no direct evidence to refute the possibility of instantaneous cooling but instead begins to build a case for the granites having formed by slow cooling in accord with the evolutionary scenario. In support of this view, he testifies that the texture of rocks, known to have cooled slowly from a liquid, is the same as granite. Here the term texture refers to the size, shape, and arrangement of the particles of which a rock is composed. In particular, he compares the textural similarity of granites to specimens taken from the Kilauea-Iki lava lake.

Imitation Granite

Since the trial I have obtained some Kilauea-Iki lava lake specimens from
the U.S. Geological Survey in Reston, Virginia. In bulk composition and
mineralogy the lava specimens are olivine-rich basalt, grossly different from
any granite. Dalrymple did not testify about these major differences—he
only said that the texture was the same. But in examining the lava specimens,
I found that there is an essential difference in the texture which Dalrymple
did not mention. In the Kilauea-Iki samples the minerals have grown together
in the interlocking, intergranular manner characteristic of rocks which have
crystallized from a melt. The minerals in Precambrian granites also exhibit
an intergranular, interlocking arrangement, and thus are texturally similar
to the Kilauea-Iki specimens *in this one respect.* However, another aspect
of texture is the *size* of the minerals composing the rock. The Kilauea-Iki
samples are fine-grained, meaning that the different mineral grains in them
are very small, often microscopic in size. The Precambrian granites, on the
other hand, are generally characterized as being coarse-grained, having
mineral grains large enough to be identified visually without magnification. This means the *only similarity between the granites and the lava
specimens is the interlocking, intergranular arrangement of the crystals making up the rocks.* This characteristic can be accounted for naturally by slow
cooling of the lava in the case of the Kilauea-Iki specimens—or by rapid
or instantaneous cooling from a primordial liquid in the case of the granites.
Thus Dalrymple is incorrect in claiming that the Kilauea-Iki lava specimens
show that the Precambrian granites formed by slow cooling. And his reference
to slow cooling brings up a most important point concerning a basic assumption of evolutionary geology.

It is a fact that hot fluid rock, such as that produced at Kilauea-Iki,
can cool over a period of a few years to form fine-grained volcanic rocks
composed of microscopic-sized crystals. The same is true of rocks that form
when granites deep in the earth are melted. The granite melt may extrude
onto the surface and cool rapidly to form a glassy rock; or it may cool more
slowly beneath the surface to become rhyolite, a fine-grained rock (which
in certain instances contains unmelted fragments of sidewall rocks broken
off in the upward passage of the magma). Both the glassy rock and the
rhyolites are intrinsically different from the coarse-grained granites. The last
section of the Radiohalo Catalogue illustrates the considerable difference
between a biotite-rich, coarse-grained granite and a slowly cooled rhyolite
specimen, extracted from a depth of 1683.3 feet at Inyo Domes, California

(Eichelberger et al. 1985). *This difference pinpoints another reason why granite synthesis remains a crucial challenge to evolutionary geology: even though the laboratory of nature has repeatedly provided a suitable environment for granites to crystallize from a granite melt, still there is no evidence of this taking place.* Geologists say this is because temperature, pressure, and length of cooling must be different. It appears, however, that evidence exists, independent of polonium halos, which long ago should have led geologists to doubt their theory of granite formation.

For example, the tiny crystals of which rhyolite is composed bear no comparison in size to the very large crystals found in certain regions within granites known as pegmatites. Some pegmatites contain crystals of biotite, the mineral in which polonium halos are most easily found, that are several feet in length. Evolutionary geology assumes that these extremely large biotite crystals are evidence of a very long period of crystallization—the larger the size, the longer it took to form. The problem is that no one has yet synthesized even a penny-sized crystal of biotite in the laboratory; so the assumption that large crystals of biotite have grown from small ones is actually a leap of faith without a point of departure. In other words, there is no evidence from the laboratory of nature or of science to show that pegmatitic biotite crystals, as shown in the Radiohalo Catalogue, attained their large size by evolutionary processes. Moreover, the existence of polonium halos in these biotites provides clear evidence that these large crystals were the product of instantaneous creation. (Most of the polonium halos in mica shown in the Radiohalo Catalogue were found in specimens of biotite taken from pegmatites.)

The above analysis shows, I believe, that Dalrymple's comparison of granites with the Kilauea-Iki lava specimens did not provide a scientifically valid basis for rejecting the falsification test. I do not know whether Dalrymple realized the weaknesses in making this comparison, but I do know that about midway in his response he began to address the granite synthesis challenge directly.

He claims that granite synthesis is impossible—*but only because of technical reasons.* At first he emphasizes the monumental difficulties in trying to synthesize a *hand-sized* piece of granite. Then he says—unless there had been a recent breakthrough—no one had yet succeeded in synthesizing a *tiny piece.* After protesting at length that I had proposed an unreasonably large-sized piece of granite to synthesize, the truth emerges: experimenters have difficulties in even getting the the granite synthesis reaction started.

Polonium Halos Revisited

Attorney Ennis continued his re-examination by returning to the topic of polonium halos.

Q   Doctor Dalrymple, if I understand correctly, polonium-218 is the product of the radioactive decay of radon-222, is that correct?
A   Yes, that's correct.
Q   And does polonium-218 occur through any other process?
A   Not as far as I know. I suspect you could make it in a nuclear reactor, but I don't know that. I'm not sure, but I don't think polonium-218 is a product of any other decay chain.
Q   So if there were polonium-218 in a rock which did not have any previous radon-222 in that rock, then that existence of polonium-218 would mean that the laws of physics as you understand them would have had to have been suspended for that polonium to be there; is that correct?
A   Well, if that were the case, it might or it might not. But there are a couple of other possibilities. One is that perhaps Gentry is mistaken about the halo. It may not have been polonium-218. The second one is that it's possible that he's not been able to identify the radon-222 halo. Maybe it's been erased, and maybe for reasons we don't understand, it was never created.

This is why I say it's just a tiny mystery. We have lots of these in science, little things that we can't quite explain. But we don't throw those on the scale and claim that they outweigh everything else. That's simply not a rational way to operate.

I would be very interested to know what the ultimate solution to this problem is, and I suspect eventually there will be a natural explanation found for it.

Q   Does Mr. Gentry's data provide scientific evidence from which you conclude that the earth is relatively young?
A   Well, I certainly wouldn't reach that conclusion, because that evidence has to be balanced by everything else we know, and everything else we know tells us that it's extremely old.

The other thing that I should mention, and I forgot to make this in my previous point, if I could, and that is that Mr. Gentry seems to be saying that the crystalline rocks, the basic rocks, the old rocks of the continents were forms [sic, formed] instantaneously. And he uses granite.

But the thing that he seems to overlook is that not all these old rocks are granites. In fact, there are lava flows included in those old rocks, there are sediments included in those old rocks. These sediments were deposited in oceans, they were deposited in lakes. They [sic, There are even Pre-Cambrian glacial deposits that tells [sic] that the glaciers were on the earth a long, long time ago.

So it's impossible to characterize all of the old crystalline rocks as being just granite. Granite is a very special rock type, and it makes up a rather small percentage of the Pre-Cambrian or the old crystalline rocks that formed before the continents. [Smith 1982b, p. 484, l. 1 to p. 486, l. 3]

In the above testimony Dalrymple suggests I might be mistaken about the identification of the polonium-218 halo. As we shall shortly see, however, the recross-examination by Attorney Williams showed these comments were only speculation. Dalrymple also misunderstands how various rock types fit into my creation model and thus arrives at incorrect conclusions about my views on the origin of the granites. A brief discussion of my creation model is necessary to clarify this misunderstanding.

Primordial and Secondary Rocks in a Creation Perspective

I agree with Dalrymple that granite "is a very special rock type," but I have not said that "all of the old crystalline rocks" are granites. Neither do I necessarily consider all rocks that geologists classify as Precambrian to be primordial. What I have said is that the polonium halos in Precambrian granites identify these rocks as some of the Genesis rocks of our planet— created in such a way that they cannot be duplicated without the intervention of the Creator. The creation episode described in Genesis outlines a lot of geological activity on this planet during creation week. The earth, after having been created on Day 1, was left covered with water. On Day 3, the "dry land" emerging from this watery environment may well have included, in addition to the primordial crystalline rocks of Day 1, certain sedimentary strata, presently considered Precambrian by geologists. The sudden appearance of "dry land" also suggests tremendous upheavals on or beneath the earth's surface and might even have included vulcanism and the formation of some rocks which geologists classify as intrusive rocks. Possibly there was some mixing of the primordial rocks of Day 1 with other rocks created on Day 3. Many possibilities for mixing are viable since Day 1 and Day 3 may also have included the creation of some non-Precambrian granites and metamorphic rocks. This discussion shows that my creation model is not governed or restricted by the conventional geological classification of various rock formations.

Here I should emphasize that creation week and the duration of the flood were special periods, both characterized by events beyond the explanation of known physical laws—periods when the *uniformitarian principle* was not valid. Each or both of these periods may have been accompanied by an increased, nonuniform radioactive decay rate.

Recross-Examination

We now turn our attention to the last phase of Dalrymple's testimony: his recross-examination by State Attorney David Williams.

Q   You state that the challenge which Mr. Gentry has issued, if I understand you, is essentially impossible?

A   It is presently impossible within our present technical capability. There have been people working on this, and I suspect someday we'll be able to do it.

Q   Is it not true that you can take a pile of sedimentary rocks and by applying heat and pressure just simply convert that to something like a granite?

A   Something like a granite, yes, that's true. But it's something like a granite, but they have quite different textures. When you do that, you now have a metamorphic rock, and it has a different fabric, and it has a different texture, which is quite distinct from an igneous texture. They are very easily identified from both a hand specimen and a microscope. Any third year geology student could tell you if you handle a piece of rock whether it's igneous or metamorphic. It's a very simple problem.

Q   But it is quite similar to a granite, but you just can't quite get it to be a granite, can you?

A   Well, granite sort of has two connotations. In the first place, in the strict sense, granite is a composition only. It's a composition of an igneous rock. Granite is a word that we use for rock classification.

It is also used in a looser sense, and that looser sense includes all igneous rocks that cool deep within the earth. And they would include things like quartz, diorite— I won't bother to tell you what those are, but they are a range of composition.

Sometimes granite is used in that loose sense. People say that the Sierra Nevada is composed primarily of granite. Well, technically there is no granite in the Sierra Nevada. They are slightly different compositions.

It is also used to describe the compositions of certain types of metamorphic rocks. So you have to be a little careful when you use the term 'granite' and be sure that we know exactly in what sense we are using that word.

Q   Now, you stated that you think, in trying to explain why Gentry's theory might not be correct or not that important, you said that perhaps he misidentified some of the haloes, and I think you also said that perhaps he had mismeasured something, is that correct?

A   Well, I think those were the same statement. I'm just offering that as an alternative hypothesis.

Q   Do you know that's what happened?

A   Oh, no, no.

Q   You have not made any of these studies and determined that yourself, have you?

A   No, no. [Smith 1982b, p. 486, l. 26 to p. 488, l. 24]

In my view these answers constitute a marvelous testimony for creation. Here we have the noted ACLU witness for geology again testifying that granite synthesis is essentially impossible for what he claims are only technical reasons. But if nature had gotten the reaction started endless numbers of times throughout the presumed vast expanse of evolutionary time, why would it be so difficult to get it started now? *Moreover, since granite synthesis has never been done in the laboratory, how could my colleague possibly know that the obstacles are only technological?* To be sure, the above responses also exposed the fact that he had no scientific data whatsoever to support his criticisms of my identification of polonium halos.

Dalrymple's references to the different connotations of the word granite necessitate that I provide additional details of my creation model, for it encompasses many more possibilities than he perceives to be the case. These details are given in "Vistas in Creation" at the close of Chapter 14.

Reflections on the First Week of the Trial

The State's cross-examinations had revealed a number of serious flaws in the ACLU's case, but it seems these were usually overlooked by the media personnel. For example, the cornerstone of the ACLU case rested on establishing the scientific credibility of a multibillion-year age of the earth. The State's cross-examination showed, however, that the evidence for an ancient earth was based on nothing more than an unproven assumption. The numerous reporters covering the trial seemed oblivious to this revelation.

Their reaction to the labeling of the polonium halos as a "tiny mystery" also seemed curious. One of the world's foremost authorities in evolutionary geology did this while admitting that he was unable to explain my published evidences for creation by conventional scientific principles. This hardly caused a stir among the reporters. Ordinarily nothing attracts the attention of scientists and reporters more than a scientific mystery, especially a tiny one. A "tiny mystery" should be solvable, and every scientist likes to work on problems he feels can be solved. At the trial the ACLU was given the opportunity to resolve the question of the "tiny mystery" and its implications for creation by responding to the granite synthesis test. Their only response to this challenge was to call it a "meaningless experiment."

I have reflected on this evaluation many times since the trial. Certainly the ACLU wanted Judge Overton to believe it. But is it really a meaningless experiment from the standpoint of the American taxpayer? Each year the Federal Government, through the National Science Foundation, grants millions of dollars for research based on evolutionary ideas, and over the years of its operation possibly hundreds of millions have gone for the same purpose. With this much money at stake, it is not easy to understand why the media did not seek to find out more about this "tiny mystery" which the ACLU had failed to explain on the basis of evolutionary principles.

In any event, the first week's media coverage left the impression that the evolutionary witnesses were infallible. Over sixty years ago in the Scopes trial, evidence for evolution was promoted nationally and internationally without mention of the weaknesses and flaws in the theory. It happened again in Arkansas. Why was there so little about these counter arguments? Was it because the issues were not made plain or because the reporters were unfamiliar with them?

Taking the Stand

The issues were clarified during the second week of the trial. In my four-hour-long testimony given during the last two days, I reviewed most of the evidence for creation and a several-thousand-year age of the earth that I had published during the sixteen years of my research. I utilized an overhead projector and showed over a hundred transparencies as well as fifty 35mm color slides of radioactive halos (see Radiohalo Catalogue). In several of the transparencies I outlined a creation model, showing how creation and the flood provide a credible framework for incorporating the data of earth history. More details about this creation model are given in Chapter 14. In particular, I testified at length about polonium halos in granites as evidence of creation and emphasized the falsifiability aspect of my creation model. During my testimony Judge William Overton was given pieces of Precambrian granite and biotite to inspect, to help him comprehend what was involved in the proposed granite and/or biotite synthesis.

What was the reaction to my testimony? What was the judge's decision about Act 590? Is there evidence that some people at the trial resisted "unwanted" information?

11

The Trial Decision

On December 18, 1981, the day after the close of the creation trial in Little Rock, ACLU Attorney Bruce Ennis was quoted on page 18A of the Arkansas *Democrat* as saying:

> "The state tried to prove there is scientific evidence for creation. They failed not because of a lack of effort but because that evidence does not exist."

In the same article ACLU Attorney Jack Novik said:

> "We've made that point several times. If creation science had any credibility, they wouldn't need a law to get it taught in classrooms."

These statements reflect the ACLU objectives throughout the trial. They tried to portray evolutionary scientists as objective, honest seekers for truth and to blacken the State's witnesses as religionists who only masqueraded as scientists. Novik's comment implied that my testimony for creation on the last two days of the trial had been completely discredited. Actually, during my cross-examination ACLU Attorney Bruce Ennis didn't even attempt to challenge my scientific evidence for creation. Instead, he first asked whether I accepted the Genesis account of creation—an attempt to brand me as a religionist rather than a scientist. Interestingly, he never referred to the falsification test that I had proposed.

Ennis' other line of questions focused on two scientific mistakes in an attempt to undermine my qualifications as a scientist. One involved the previously discussed work on superheavy elements; the other was a misidentification of a certain halo in the mineral fluorite. Both of them had been corrected in print years before the trial. Those mistakes were rectified as a natural part of my ongoing scientific research—an endeavor that involves testing new ideas over and over again, modifying and/or recanting as demanded by further experimentation and peer evaluation. In summary, the cross-examination seemed to be directed toward diverting the attention of the judge and the media away from my discoveries.

Evolutionists Win the Game

On January 5, 1982, Judge Overton ruled against Act 590. In his *Memorandum Opinion,* Judge Overton evaluates creation science as follows:

> The proof in support of creation science consisted almost entirely of efforts to discredit the theory of evolution through a rehash of data and theories which have been before the scientific community for decades. The arguments asserted by creationists are not based upon new scientific evidence or laboratory data which has been ignored by the scientific community. (Overton 1982, Section IV.(D))

Such statements are not consistent with the evidence I presented to the court. Virtually none of my testimony consisted of a rehashing of previous data or theories. On the contrary, it visually portrayed how my recently discovered evidences for creation were based on laboratory experimentation and how, for the most part, they had been ignored by the scientific community.

In another part of his *Opinion,* Judge Overton states:

> Creation science, as defined in Section 4(a) [of Act 590], not only fails to follow the canons defining scientific theory, it also fails to fit the more general descriptions of "what scientists think"; and "what scientists do." The scientific community consists of individuals and groups, nationally and internationally, who work independently in such varied fields as biology, paleontology, geology and astronomy. Their work is published and subject to review and testing by their peers. The journals for publication are both numerous and varied. There is, however, not one recognized scientific journal which has published an article espousing the creation science theory described in Section 4(a). (Overton 1982, Section IV.(C))

It is difficult to understand these remarks. In my own case Judge Overton was given references to twenty or more scientific publications. Wasn't this ample evidence that, for sixteen years, my work had been tested and subjected to review by my peers in the scientific community? Is it possible that the judge's designation of who is a scientist was based on one's position on origins rather than one's actual scientific associations and work?

Court Judgment Reveals Evolutionary Bias

Perhaps the most revealing of the above comments was the statement, "There is, however, not one recognized scientific journal which has published an article espousing the creation science theory described in Section 4(a)." This should not be surprising since Section 4(a) of Act 590 covers a broad spectrum of creation science. Just as no one evolutionist is expected to expound on all the different aspects of evolution, neither does any one creation scientist have the expertise to write about all the diverse aspects of creation science. The real issue centers on the *different* scientific articles for creation science which had been presented before the Court. Why were these discounted when the judge wrote his *Opinion*?

Judge Overton gives his evaluation of my work as follows:

> . . . Mr. Gentry's findings were published almost ten years ago and have been the subject of some discussion in the scientific community. The discoveries have not, however, led to the formulation of any scientific hypothesis or theory which would explain a relatively recent inception of the earth or a worldwide flood. Gentry's discovery has been treated as a minor mystery which will eventually be explained. It may deserve further investigation, but the National Science Foundation has not deemed it to be of sufficient import to support further funding. (Overton 1982, Section IV.(D))

Here Judge Overton greatly minimizes my difficulties with the National Science Foundation, which were presented in great detail before the Court. Readers may decide for themselves whether Judge Overton's comments about those experiences, as described in Chapter 6, represent an objective evaluation of the facts.

Radiohalos: Tiny Mystery or Block to Evolution?

Judge Overton's conclusion that my work was "ten years old" and that my discoveries were only a "minor mystery," which eventually would be explained, leaves the impression that the scientific community had found nothing significant in my work. In essence, he interprets silence about my results as showing they are insignificant. Was this conclusion justified?

Through Professor Ray Kazmann's letter (Chapter 4), the judge was shown that this silence about my results was because they seriously conflicted with the evolutionary time scale. He also was shown the article entitled "Mystery of the Radiohalos,'' which featured letters of evaluation about my research from a number of eminent scientists here in America, Europe, and the Soviet Union. One of these letters, from an internationally known American geochemist, reads in part:

> His [Gentry's] conclusions are startling and shake the very foundations of radiochemistry and geochemistry. Yet he has been so meticulous in his experimental work, and so restrained in his interpretations, that most people take his work seriously . . . I think most people believe, as I do, that some unspectacular explanation will eventually be found for the anomalous halos and that orthodoxy will turn out to be right after all. Meanwhile, Gentry should be encouraged to keep rattling this skeleton in our closet for all it is worth. (Talbott 1977, 5; Appendix)

This is a very significant letter. At the time of the Arkansas trial, about five years had passed since it was written. During that span I had endeavored to ''keep rattling this skeleton...for all it is worth.'' In this five-year period I had challenged my evolutionist colleagues to duplicate a hand-sized piece of granite or biotite as a means of confirming the basic premise of their theory. The evidence for creation that had been rattling in the evolutionary closet for many years was now knocking more loudly than ever, but for some reason Judge Overton and the ACLU contingent had a difficult time hearing it. Was this a case of resisting ''unwanted information''?

Judge Overton's dismissal of my scientific discoveries as a ''minor mystery'' echoed the ''tiny mystery'' designation given by the ACLU's expert geology witness. By doing this the judge effectively denied the existence of valid evidence for creation science. To have done otherwise would have destroyed the logical basis of his entire *Opinion*.

Evolutionary Article of Faith

In Section IV.(C) Judge Overton gives what he considers to be the five qualifying characteristics of science:
(1) It is guided by natural law;
(2) It has to be explanatory by reference to natural law;
(3) It is testable against the empirical world;
(4) Its conclusions are tentative, i.e., are not necessarily the final word; and
(5) It is falsifiable.

Judge Overton states that creation science fails to meet these essential characteristics, noting that the Arkansas creation law ''asserts a sudden creation 'from nothing.'" He maintains that "such a concept is not science because it depends upon a supernatural intervention which is not guided by natural law, is not testable and is not falsifiable." By applying this line of reasoning only to creation science, Judge Overton ignores part of the evidence presented to him. In my testimony I showed that evolution also requires a supernatural beginning.

I testified that the most widely accepted evolutionary scenario of the beginning of the universe, the Big Bang model, begins with an article of faith. Evolutionary scientists postulate that all matter in the universe emanated some 17 billion years ago from a gigantic primeval explosion. The ultimate cause for such a beginning is not a matter capable of scientific investigation. That event is not presumed to be guided or explainable by natural law, nor is it testable against the empirical world. If the court had consistently applied its own description of science, it would have been as critical of "the beginning" postulated for evolution as of the supernatural beginning for creation science.

While writing this book, I found a comment which summarizes my testimony before the court about this mythical event. It was made by the well-known British astronomer, Professor Paul Davies. In one of his books the comment is made that the creation of the universe by the Big Bang

> . . . represents the instantaneous suspension of physical laws, the sudden abrupt flash of lawlessness that allowed something to come out of nothing. It represents a true miracle—transcending physical principles...(Davies 1981, 161)

This forthright statement by an eminent evolutionist admits that evolutionary science requires as much of a "miracle" in the beginning as does creation science—"something to come out of nothing." Such was the essence of my testimony about the Big Bang. If Judge Overton had recognized this fact in his *Opinion*, it would have invalidated his contrast between creation and evolution.

True Science Defined by the Court

Perhaps the most blatant contradiction in Judge Overton's *Opinion* occurs when he criticizes the methodology of creation scientists:

> The methodology employed by creationists is another factor which is indicative that their work is not science. A scientific theory must be tentative and always subject to revision or abandonment in light of facts that

are inconsistent with, or falsify, the theory. A theory that is by its own terms dogmatic, absolutist and never subject to revision is not a scientific theory. (Overton 1982, Section IV.(C))

This is truly an incredible statement. Apparently the judge decided to ignore a large part of my testimony relating to the proposed falsification experiment. My testimony about this experiment encompassed an in-depth review of the material discussed in the earlier chapters of this book. It included (1) my presentation at the 1978 Louisiana State University symposium on the measurement of geological time (when I presented evidence that polonium halos in Precambrian granites suggested a very rapid formation of those rocks), (2) Professor Damon's letter about my contribution to that symposium, along with my responses to Damon and York as published in *EOS*, and (3) a discussion of the comments of Professor Norman Feather (see Chapter 5) concerning the exceeding difficulty of explaining polonium halos in granites by conventional scientific principles. I specifically stated that the synthesis of a hand-sized piece of granite or biotite would suffice to render my creation model invalid. Thus, Judge Overton's comments that creation science is not testable or falsifiable were contradictory to the testimony presented at length in his own court.

Throughout his *Opinion* Judge Overton seems to have accepted the ACLU's position on most of the issues that were argued at the trial. I must pay my respects to the ACLU contingent for that achievement. Admittedly, they won a tremendous psychological victory when the judge ruled in their favor. This was no small accomplishment. They were sitting on Pandora's box, and throughout the trial they ran the risk of having it come open. Legally, it was an impressive victory. But what an empty victory it was! Scientifically, they were confronted with evidence for creation, and they didn't even try to refute it. Make no mistake—if the ACLU had found a flaw in that evidence they certainly would have brought it out during my cross-examination. Their only recourse was to treat the evidence for creation presented at the trial as a "tiny mystery." On this occasion their strategy worked very well. Whether it would ever work again remains to be seen.

By now the reader may realize that the events described in this chapter draw attention to the question arising in my mind when I first encountered the Big Bang concept in a graduate physics course. That question centered on how matter and energy could be formed in the Big Bang when a fundamental law of physics prohibited it. In reality, as Professor Davies' statement so cogently reveals, the laws of physics have never been sufficient to account for the Big Bang. Thus, ironically, even the most resolute evolutionists are, in the end, forced to admit to an incredible contradiction—a miracle of *creation* must be invoked to start this mythical scenario.

12

Media Reaction to the Arkansas Trial

Along with a few other cherished concepts in science, evolution enjoys superstatus. It is tacitly understood that any scientist who wants to maintain a reputable standing within the scientific community must never publicly challenge such a theory. Up until the Arkansas trial my research was not generally considered in that category because I had stayed within the norm of publishing my results in the scientific literature. At the Arkansas creation trial I stepped outside of that accepted norm and issued a public challenge to one of the superstatus theories of science. For that reason alone my participation at the Arkansas trial was certain to evoke discussion among my scientific colleagues. To a large extent, their reaction would depend on whether my testimony and evidences for creation received favorable or hostile reviews in reputable scientific magazines.

Effects of Journalism on Research Funding

Government laboratories are sensitive to any evaluation of their staff activities published in respected scientific journals. A positive evaluation of a project or scientist at a national laboratory provides an incentive for the parent agency, such as the Department of Energy, to recommend a high level of support when budgets are prepared for Congress. At the same time, a government laboratory must be wary of supporting a scientist who is criticized in one of those journals. Support of controversial research could produce a negative reaction from the Congress and in turn affect funding for that laboratory. This chapter focuses on two accounts of the trial published in the January 1 and January 8, 1982, issues of *Science* and how they adversely affected my status as a professional scientist.

This journal had always given my technical reports fair treatment. Their chosen reporter, Roger Lewin, was expected to provide an evenly balanced account of the trial proceedings. But as I read his reports (Lewin 1982a and 1982b), it seemed that the creation position at the trial, my testimony in particular, was considerably minimized for the benefit of evolution. A few months later I learned firsthand of his strong preference for evolution when he was featured as an invited speaker at the American Physical Society meeting in Washington in April 1982 (Lewin 1982c). In that presentation he upheld the standard evolutionary scenario. That same year he authored a book on evolution (Lewin 1982d). Conceivably, some other staff reporter might have given a different perspective of the trial, and this chapter might not have been written.

Reporting from an Evolutionist Perspective

On the surface Lewin's two reports of the trial appear to be a simple reviewing of the important events. But a close examination reveals a different picture. By omitting and minimizing crucial parts of the trial testimony, while emphasizing other phases, he favors the evolutionary position and leaves the impression that the creation science position was in shambles. Lewin accomplishes this feat by building up the ACLU contention that evolution is truly scientific whereas creation science is religion in disguise.

In his first account Lewin refers to the testimony of one of the ACLU witnesses:

> Science has to be testable, explanatory, and tentative, said Michael Ruse, a philosopher of science at the University of Guelph, Canada, and he made it plain that in his mind creation science was none of these. (Lewin 1982a, 34)

A few paragraphs later the build-up continues as these witnesses are allowed the privilege of defining the scientific status of their own theory:

> Each [evolutionist] testified that yes, evolutionary theory was thoroughly scientific even though there were problems with it; and that no, creation science (Ayala could hardly bring himself to mouth the phrase) most definitely was not. (Lewin 1982a, 34)

Note that Lewin is not content to report the evolutionists' evaluation of their own theory. Here he uses a parenthetical comment to inject his own appraisal of Ayala's reaction to creation science. From this one could easily conclude that the ACLU witnesses were intellectual heroes, the brave defenders of scientific truth.

In contrast, Lewin pictures the creation science position as being confused and fearful:

> The attorney general presented six science witnesses, two more than had testified for the ACLU, presumably on the grounds that quantity made up for evident lack of quality. There would have been more had not a serious case of disappearing witnesses set in as the second week wore on. Dean Kenyon, a biologist from San Francisco State University, fled town after watching the demolition of four of the State's witnesses on day 1 of the second week. (Lewin 1982a, 34)

True enough, one of the planned witnesses for the State did leave town very hurriedly after observing how the ACLU tried to intimidate the State's witnesses during their cross-examinations. Lewin cannot be faulted for reporting this occurrence. But to imply this was because four of the State's witnesses were demolished is an opinionated statement. It leaves the impression that creation science was not up to the challenge of the day. Near the end of Lewin's first commentary my work is described as follows:

> Defense witness Robert Gentry, a physicist associated with the Oak Ridge National Laboratory, brought the trial to a close with 4 hours of excruciating detail about an anomalous result in the radiometric dating of the age of the earth that Dalrymple had described as a "tiny mystery."
>
> Judge Overton left the bench at 10:46 on Thursday, still holding his head from Gentry's massive presentation. . . . (Lewin 1982a, 34)

Readers should note that after Lewin heard those four hours of evidence, which encompassed years of research and many publications in respected scientific journals, the most perceptive comments he can offer in this first write-up are that my testimony was "massive" and involved "excruciating detail" of an "anomalous result." No mention is made of my scientific publications or of the granite synthesis experiment which I had proposed. Lewin's greatest assist for the evolutionary position, the one most needed by the ACLU to maintain a posttrial image of scientific invincibility for evolution, is his repeated silence about this critical falsifiability test.

Where is the Science in Creation Science?

The second account of the trial, entitled "Where Is the Science in Creation Science?," is another deft attempt to establish that creation science is not science. The author sets the stage for his drama in this article by noting that only seven out of sixteen potential creation science witnesses actually testified at the trial. Next he states:

"... in their pretrial depositions *many* creation scientists admitted that what they practiced was not scientific." ... (Lewin 1982b, 142—italics mine)

This appears to be a very damaging admission. But there is more than one way to interpret such a statement. In fairness the context should have been included so that the reader could evaluate just what was meant by these remarks. True, the work of *many* creation scientists involves the observation and interpretation of existing geological data. They also utilize a flood model of earth history in interpreting that data. Since geologists generally exclude a worldwide flood from their scientific perspective, possibly these creation scientists are only admitting their interpretive framework of science differs in some respects from the orthodox view of science. To illustrate, I quote a recent statement from Dr. Ariel Roth, one of the creation scientists who was the target of Lewin's thrust:

> ... the question of whether creation is science is trivial. It revolves around varied definitions of science and conflicting scientific practices. By promoting the proposition that creation is not scientific, evolutionists are directing their energies to a non sequitur that distracts from the more basic question of origins. C'est magnifique, mais ce n'est pas la guerre! (This is magnificent but this is not the war.) The real question is whether evolution or creation is true. (Roth 1984, 64)

I suggest this statement throws a different light on the issue. This kind of information Lewin could easily have obtained to give a balanced perspective, but he chose not to do so.

Neither does Lewin mention that my views on this topic were necessarily different. As a scientist whose work has dealt mainly with experimental data obtained in the laboratory, I have consistently maintained my work is scientific and have invited my colleagues to test my results. This was made clear in my pretrial deposition and in my court testimony, but Lewin is silent about it. Anyone reading his second account of the trial may erroneously think that all creation science witnesses, including me, had admitted that creation science, even my experimental work, was not scientific. This one misunderstanding alone would have been sufficient to raise serious questions among my scientific colleagues. And the damage does not stop there.

Lewin then refers to an assessment of creation science held by Duane Gish, a well-known creation scientist who was not a witness at the trial:

> In admitting that creation science is not a science, Gish and his colleagues are quick to point out that, in their opinion, neither is evolutionary theory scientific. ... (Lewin 1982b, 142)

The phrase "Gish and his colleagues" suggests that all creation scientists think alike on this point, which again invites a misunderstanding about my experimental results.

In the next paragraph Lewin says:

> . . . creationist literature, Act 590, and defendants' counsel, avoid the term "theory" in reference to creation and evolution explanations, because of its implied property of testability, tentativeness, and explanation. (Lewin 1982b, 142)

This statement implies that creation scientists cannot stand to have their ideas put in the marketplace of science for critical scrutiny. This is just the opposite of what I had done for a decade and a half of research. And it was contrary to the testimony which Lewin heard me give before the court. This is the third instance that Lewin remained silent about my position. And there is more to come. A few paragraphs later we find:

> In addition to the pretrial proclamation that creation science is not science, the defense opened its scientific case with a second distinct disadvantage. (Lewin 1982b, 142)

Where is the "pretrial proclamation" Lewin mentions? To my knowledge no such proclamation was made. Is this a reference to the pretrial depositions of the other creation scientists? If so, this was no proclamation. How could it be when my pretrial deposition emphasized the opposite view? This is the fourth instance where the author, by his silence, left a cloud over my experimental results and cast doubt upon my reputation as a professional scientist. As earlier noted, this need not have been the case at all for the other creation science witnesses, who were possibly utilizing a different definition of what is scientific.

The "second distinct disadvantage" in the last quote refers to several creation science witnesses (including me, at that time) who held membership in the Creation Research Society (CRS). Lewin correctly notes that CRS members affirm faith in the Genesis account of creation and the flood as well as in the widely held Christian belief that Jesus Christ is the Savior of mankind. I ask: Is the "disadvantage" Lewin mentions here a reflection of his own attitude toward these beliefs? At the trial I asked my evolutionist colleagues to show where my evidence for creation is wrong and theirs for the *uniformitarian principle* is correct. This they failed to do either at the trial or since then. Instead of faulting the evolutionists for this failure, Lewin casts aspersion on the CRS members who testified for the State:

One after another these five witnesses agreed that the work they did and the conclusions they felt able to draw were inspired by these beliefs. (Lewin 1982b, 143)

This statement contains factual information, but the whole truth is not evident. As one of those five witnesses, I must take exception to this characterization of my work. As a scientist I have worked to uncover the truth about the origin and history of the earth. At the trial my conclusions unequivocally supported creation, but those conclusions were based on scientific evidence. What Lewin does in the above statement is to confuse the motivation for my research—wanting to know the truth about Genesis—with the scientific results achieved in that search.

Discounting the Evidence

After recounting my testimony (which he garbles), Lewin refers to my cross-examination by ACLU Attorney Bruce Ennis and focuses on my motivation as if it were a detriment to my work:

In 10 minutes of cross-examination Ennis showed that the principal motive for Gentry's work was his literal reading of the Bible—in particular, Genesis. . . . (Lewin 1982b, 146)

More precisely, Ennis' question about my motivation was in reference to one of the Ten Commandments:

Q   Isn't one of the primary reasons that you began to rethink the entire issue of evolution and creation is because of the moral perspective of the Fourth Commandment? (Merkel 1981; Appendix—1.28 to 1.30)

Early in this book I discussed how my complacent acceptance of theistic evolution was shattered when I realized this Commandment refers to the six days of creation in the context of six literal days; thus my response to this question was, ''Absolutely.''

The reader may wonder why Ennis chose to ask this question. What did it have to do with the issues before the court? The problem was that the ACLU had no way of directly countering the published scientific evidences for creation which I had discovered. So during my cross-examination Attorney Ennis steered clear of challenging my claim that polonium halos in Precambrian granites represent evidence for creation. To obscure his inability to confront this evidence required that he somehow try to discredit me, or some facet of my work. As a matter of tactics, he utilized two separate strategies. First, as we have just noted, he focused on my motivation — it was almost as if the ACLU would like to have blamed the existence of polonium halos in granites on my motivation.

Ennis' second strategy was to raise doubts about my credibility as a scientist. To accomplish this he referred to the superheavy element report mentioned earlier in Chapter 6. This was no surprise as I had fully expected the ACLU would do this in an attempt to undermine the credibility of my results pertaining to creation.

A considerable surprise, however, was Lewin's recounting of this phase of my cross-examination. There appeared to be serious variances between what I remembered and what was reported. Yet for over four years after the trial I was unable to challenge Lewin's version of this phase of the trial because my testimony had never been transcribed. Fortunately, the required information was obtained just in time to be included in this book. For the sake of chronological order my comments about this important material are deferred until near the end of this chapter.

Lewin's second write-up closes with the following comment:

> The combined testimony of the creationists' scientific witnesses was, it has been acknowledged, not impressive. Anyone who was hoping for a body of science to stand in equal force against conventional evolutionary biology, and the background of geology, chemistry and physics, would have been disappointed. (Lewin 1982b, 146)

Who acknowledged that the combined testimony of the creationists' scientific witnesses was not impressive? Roger Lewin? The ACLU witnesses? This was the theme of the ACLU case. Thus Lewin permits the ACLU itself to pass judgment on evidence for creation presented at the trial. Then Lewin assumes the role of final arbiter of the trial; he pictures supporters of creation science as a disheartened lot because their position could not withstand the force of evolutionary evidence. I grant there was much disappointment about the trial. But was it because of lack of evidence for creation or because it wasn't accurately and fully reported?

Lewin's brief discussion of my testimony would have been the opportune time to describe the pivotal granite synthesis experiment. By synthesizing a piece of granite in the laboratory, evolutionist scientists could in theory falsify my creation model and show their *uniformitarian principle* to have some basis in fact. An explanation of the falsification test by Lewin would have enabled other scientists to see that my testimony had a credible scientific foundation. But in his write-up, my deduction about granites being primordial, created rocks appears to be left hanging, as if it could not be tested. The creation model I proposed as a scientific framework to incorporate the evidences for creation and the flood is not mentioned. [The creation model presented later to the American Association for the Advancement of Science symposium (see Chapter 14) is similar to the one presented at the trial.] As a result, my testimony at the Arkansas trial is placed in the framework of a religious ad hoc hypothesis without scientific merit.

Lewin's silence about my credentials portrays me as a scientist outside or, at best, on the fringes of the scientific community, rather than one who had carried on recognized scientific research for sixteen years. If he had forthrightly admitted that I had published evidence for creation and the flood which had not been refuted (even though a challenge to refute it had been in the scientific literature for several years), this would have shed a different light on my participation at the trial. But this did not happen, and the readers of *Science* were left with the impression that creation science was indefensible.

Correction Attempt Fails

The preceding accounts of the trial had a pronounced, negative impact on my position as guest scientist at ORNL. The attitude of certain colleagues toward my work changed. The following response to Lewin's remarks about my testimony was submitted to *Science* on March 2, 1982, in an effort to allow my colleagues at the Laboratory and elsewhere an opportunity to see in print where Lewin had failed to represent my position correctly:

MY RESPONSE TO ROGER LEWIN'S TRIAL ACCOUNT

In Roger Lewin's summation "Where is the Science in Creation Science?" (8 Jan. p. 142), it was clearly his prerogative to report that some creation scientists testified that they did not believe that creation science is testable or scientific. But it hardly does me justice before my scientific colleagues for him not to also mention that I represented a different position at the trial. The fact that I explained how the one-singularity Big Bang Model and the two-singularity Creation Model (ref.1) both involve prediction and are in theory capable of falsification makes it doubly curious why Lewin chose not to give the readers of *Science* an opportunity to evaluate my thesis for themselves. (I define a singularity as a set of events requiring more than known physical laws to explain.)

In support of the Creation Model I referred to my results (ref.2) on halos in coalified wood as evidence for the Flood singularity. Such data also imply that certain coals should have formed within a few months to a few years (but not instantaneously as Lewin reported). I suggest these predictions about the relative rapidity of coal formation can be tested in the laboratory by subjecting water-saturated samples of wood to elevated temperatures (150-300 °C) and then analyzing the residue for coal-like properties. And speaking of predictions, on the basis of this Creation Model I have also suggested that newly developed accelerator techniques should be used to search for small amounts of ^{14}C in coal and amber (ref.1). Conventional geological theory predicts that the amount of ^{14}C in such materials should be infinitesimally small, and hence undetectable.

As evidence for the initial creation singularity (ref.1) I referred to my results (refs.3,4) suggesting that polonium halos in Precambrian granites are primordial, hence implying that the granites must themselves be primordial rocks, or rocks that were created. This hypothesis would be scientifically meaningless had I not also proposed the following experiment which in theory I will accept as falsifying that hypothesis if it is successful.

Briefly, I testified that since the standard Big Bang Model predicts the Precambrian granites formed slowly over geological time with nothing more than conventional physical laws to govern their crystallization, then it should be possible to synthesize in the laboratory a small (hand-sized) piece of such granite to confirm that hypothesis. My testimony was that I would accept the synthesis of a piece of granite as a falsification of my thesis that the Precambrian rocks are primordial rocks, and further that the subsequent synthesis of a single ^{218}Po halo in such a piece of granite would also be sufficient to falsify my view that Po halos in granites are primordial.

I anxiously await the critical response of my scientific colleagues to these proposals. The issues are clearly too important for them to be ignored any longer.

Robert V. Gentry

References
1. Robert V. Gentry, EOS, Trans. Am. Geophy. Union 60, 474 (1979);_____, 61, 514 (1980).
2. Robert V. Gentry et al., *Science* 194, 315 (1976).
3. Robert V. Gentry, *Science* 184, 62 (1974).
4. Robert V. Gentry et al., *Nature* 252, 564 (1974)

As noted by the following reply from the Letters Editor, my attempt to provide a rebuttal was refused. Such arbitrary rejection was difficult to understand.

(March 9, 1982)

Dear Dr. Gentry:

Thank you for your letter of 2 March, which has been studied by the editorial staff. I regret that we do not plan to publish it.

While it is understandable that you might have preferred a different emphasis or different details in Lewin's account of your testimony, we do not find that, in this case, his presentation needs clarification or amplification. *Science*'s staff writers must present material in very limited space and can not usually include all of the details that individuals featured in articles would like.

We note that much of what you have written has appeared in other publications and has therefore been made available to your colleagues.

Sincerely,

/s/ Christine Gilbert

Christine Gilbert
Letters Editor
Science   (Gilbert 1982)

My situation at the Laboratory might have been rectified had I been afforded the customary professional right to defend myself in *Science*. My credibility as a scientist had been called into question, but obviously this had no effect on the decision not to publish my rebuttal. This letter of rejection seems contrary to the lofty aims of *Science* as displayed on the editorial page of every issue:

> *Science* serves its readers as a forum for the presentation and discussion of important issues related to the advancement of science, *including the presentation of minority or conflicting points of view,* rather than by publishing only material on which a consensus has been reached. Accordingly, all articles published in *Science*—including editorials, news and comment, and book reviews—are signed and reflect the individual views of the authors and not the official points of view adopted by the AAAS or the institutions with which the authors are affiliated. (italics mine)

Lewin's considerable coverage of the Arkansas trial proves that *Science* considered the outcome of the Arkansas trial as an important issue "related to the advancement of science." Why then was not my response accepted for publication? Certainly it qualified as a "presentation of minority or conflicting points of view." First, it is certain that my rebuttal letter, if published, would have alerted the worldwide readership of *Science* to the credibility of the evidence for creation. This might have led to some penetrating questions about why such important information was missing from Lewin's published accounts of the trial. We must also ask whether the official position of the AAAS toward creation science could have been partially responsible for suppressing my response.

AAAS and Evolutionary Presuppositions

At the 1982 AAAS annual meeting, held soon after the Arkansas trial, the Council of the AAAS and its Board of Directors issued a joint resolution condemning creation science. That resolution reads as follows:

Whereas it is the responsibility of the American Association for the Advancement of Science to preserve the integrity of science, and

Whereas science is a systematic method of investigation based on continuous experimentation, observation, and measurement leading to evolving explanations of natural phenomena, explanations which are continuously open to further testing, and

Whereas evolution fully satisfies these criteria, irrespective of remaining debates concerning its detailed mechanisms, and

Whereas the Association respects the right of people to hold diverse beliefs about creation that do not come within the definitions of science, and

Whereas Creationist groups are imposing beliefs disguised as science upon teachers and students to the detriment and distortion of public education in the United States,

Therefore be it resolved that because "Creationist Science" has no scientific validity it should not be taught as science, and further, that the AAAS views legislation requiring "Creationist Science" to be taught in public schools as a real and present threat to the integrity of education and the teaching of science, and

Be it further resolved that the AAAS urges citizens, educational authorities, and legislators to oppose the compulsory inclusion in science education curricula of beliefs that are not amenable to the process of scrutiny, testing, and revision that is indispensable to science. (American Association for the Advancement of Science 1982, 1072)

This resolution shows the AAAS hierarchy picture themselves as guardians of the integrity of science. In this self-appointed role they assert that creation science has no scientific validity. But was it scientific integrity for *Science,* the publishing arm of the AAAS, to suppress a letter that directly contradicted that assertion? (Later I learned more about why my response was rejected, and this is discussed in Chapter 15.)

Audio Tapes Reveal Factual Account

Earlier in this chapter I reviewed how Roger Lewin had raised doubts about my credibility as a scientist in his accounts of the Arkansas trial. Also mentioned was a more serious matter: in certain places his version of my cross-examination seemed to differ from my own recollection. For over four years my suspicions about this material could not be confirmed because my testimony had not been transcribed. Since there were no plans for this to be done, it seemed that the matter would rest as Lewin had pictured it. Then, as this book neared completion, I remembered that the court reporter had made an audio recording of my testimony in addition to her own record. Contact was made and duplicates of the audio tapes were sent to me in time for this new material to be incorporated into this chapter.

Quoted below from the audio tapes are the NSF and superheavy element-related questions and my responses to ACLU Attorney Ennis:

Q   You testified at some length about a letter from the National Science Foundation, July 11, 1977, which denied your application for a particular grant.

A   Yes.

Q   Is it not fair to say that that letter concluded that one of the reasons they denied your grant application at that time was that the panel felt that you and your colleagues were to be faulted for the techniques you used in coming to your initial conclusion that there were superheavy elements?

A   Yes, I believe it did say that.

Q   Did not that rejection letter go on to say that the panel felt that the principal investigator and his colleagues should have checked out all such possible reactions before publication because we know that that technique might produce the results you found? Is it not true?

A   I think what you are saying is generally true. [Merkel, 1981; Appendix—1.167 to 1.182]

The reader should understand that I agreed to Ennis' second and third questions only because he asked *whether the NSF letter (see Chapter 6) had made those criticisms.* Yes, "the [Geochemistry] panel did fault the principal investigator [Gentry] and his colleagues for the techniques they used to try to detect superheavy elements" (Hower 1976; Appendix). However, with due respect to the NSF panel—all of whom were evolutionists—nothing was at fault with the technique which we used to generate our published report on superheavy elements. And the NSF's objection to another method (see Hower's letter) was just a red herring because our report contained nothing based on it.

The technique used in the superheavy element experiments, proton-induced x-ray fluorescence, is routinely used to determine the elemental composition of an almost unlimited variety of specimens. It is based on one of the most reliable elemental identification methods in experimental physics. In our experiments we did misinterpret some of the x-ray lines, but contrary to the implication of Ennis' third question, we did check other reactions before publication. I agreed to the third question only because Ennis asked whether the NSF letter made that criticism, not because I believed that criticism was valid. Indeed, the reader may remember from Chapter 6 that some of my associates remained adamant that our original results did show evidence of superheavy elements long after other experiments indicated otherwise. They reasoned that the other nuclear reactions had been so thoroughly investigated that the evidence for superheavy elements still remained. In this respect, it is always possible to misinterpret the results of a single set of experiments, regardless of the technique used. This is why continued experimentation is necessary until a proposed interpretation is confirmed or denied.

We now quote Lewin's version of the superheavy-element part of my cross-examination:

> . . . Ennis also established that Gentry had shown poor judgment in using a certain technique in looking for primordial superheavy elements.
>
> Q You referred to the grant rejection letter of 11 July 1977. Isn't it fair to say that one reason the request was turned down was because the panel felt you were to be faulted for using a technique that was known to give false results?
>
> A Yes. (Lewin 1982b, 146)

A scientist who uses techniques that are "known to give false results" is incompetent or untrustworthy, and this is the inference that can be drawn about me from the above information. The audio-tape quotes given above show that Lewin's highly incriminating phrase, "known to give false results," is nowhere to be found in Ennis' questions. Neither is it found in the National Science Foundation letter (Hower 1976; Appendix) to which Ennis referred. This means (1) Lewin had no factual basis to claim that Ennis "established" that I had shown "poor judgment in using a certain technique" in the superheavy-element experiments, and (2) Lewin's version of the superheavy-element part of my cross-examination deviates, much to my detriment, from the actual courtroom proceedings. Lewin has my agreeing under oath to something essentially different from what Attorney Ennis actually asked during my cross-examination; moreover, I would not have agreed to the question if it had been worded as Lewin claimed.

Lewin's last comment about my work occurs near the end of his second report:

> Ennis closed his cross-examination by asking Gentry if other people working in the field thought that conventional explanations would be found for the anomalous results he had. Gentry said "yes." . . . (Lewin 1982b, 146)

This statement lends great credence to the idea that a conventional explanation will be found for my "anomalous results" because of its appeal to the authority of "people working in the field," which in this case must refer to scientists doing research on halos. I didn't recall that Attorney Ennis had made such a reference, for this would have provided me with the opportunity to take exception to their proposed explanations; and I knew this had not occurred. So I highly suspected that Lewin's version of Ennis' question was incorrect, and that again he had pictured me as agreeing to something different from what actually transpired in the courtroom.

The audio tapes confirmed my suspicions. They reveal Ennis' question and my response as follows:

Q   And Anders. Is it not true that Wheeler and Anders and other scientists who have read your material think that a conventional natural law explanation will be found for the existence of other polonium halos in granites?
A   Yes, they do.
Q   I have no further questions. [Merkel 1981; Appendix—1.223 to 1.228]

The above quotes show that the scientists whom Ennis cited are those "who have read" my material. This is quite distinct and different from Lewin's characterization of them as "other people working in the field," because obviously this phrase denotes scientists actually doing research. Thus, the "other people" to whom Lewin referred had no tangible scientific evidence which would support a conventional explanation of polonium halos in granites. In fact, it would have been much to my advantage if Lewin had reported exactly what Dr. John Wheeler and Dr. Edward Anders had said about my work (Talbott 1977; Appendix). Indeed, Anders' evaluation was quoted in the last chapter as evidence to show that Judge Overton had ignored some important information in arriving at his decision. The last part of that evaluation reads:

. . . I think most people believe, as I do, that some unspectacular explanation will eventually be found for the anomalous halos and that orthodoxy will turn out to be right after all. Meanwhile Gentry should be encouraged to keep rattling this skeleton in our closet for all it is worth. (Talbott 1977, 5; Appendix)

Perhaps the publication of this material, showing how Lewin's accounts deviate from the actual court proceedings, may yet rattle another skeleton buried within the scientific establishment.

Figure 12.1 Gentry Meets the Press
This photograph was taken shortly after his testimony for the State of Arkansas at the creation/evolution trial in Little Rock.

Another Viewpoint

A positive account of my participation in the Arkansas trial was published in the January 16, 1982, issue of *Science News*. It was entitled "They Call It Creation Science," with the subtitle, "Why would any reputable scientist agree to testify on behalf of the state of Arkansas in last month's creationist trial? Two who did tell *Science News.* '' These two were Professor N. C. Wickramas-inghe, Chairman of the Department of Mathematics and Astronomy at the University College at Cardiff, Wales, and I. The first paragraph of this interview, quoted below, shows that the writer, Janet Raloff, provides a much different perspective of my contribution at the Arkansas trial:

> Not everyone in science shares the view that ''creation science'' has no scientific validity. Among them are two who testified on behalf of the defending Attorney General's office as its key witnesses during the creation science trial last month in Little Rock, Ark. (*Science News:* 1/2/82, p. 12). About the only things these scientists have in common are the respect of the scientific community for the meticulous quality of their primary pursuits and their shared belief that life's grand scheme may be the product of ''a creator.''(Raloff 1982a, 44)

I was gratified that my research, when fairly evaluated, was recognized for adhering to the scientific method and that this was published in a national news magazine. But in practical terms, this subsequent account of my research was insufficient to override the negative impact of the articles in *Science*.

The history of science reveals that certain cherished theories have always been considered immune to criticism. Scientists who refused to acknowledge this immunity, openly challenging those theories, were on occasion ''excommunicated'' from the scientific establishment. Irrespective of how much evidence I had accumulated, I had openly challenged a superstatus theory which certain scientists felt should be immune from attack. Repercussions were bound to follow.

13

The Aftermath of the Arkansas Trial

At the time of the Arkansas creation trial in December 1981 I had been at the Laboratory as a guest scientist for twelve and a half years. It was a most cordial and productive arrangement. During this period I had published research papers in collaboration with my colleagues at ORNL, and had undertaken cooperative research projects with scientists at other laboratories and universities, some of them overseas. Yet by the summer of 1981, my main purpose of coming to the Laboratory—finding superheavy elements—still had not been accomplished. One final year was given me to do this. My superheavy search involved several different experimental approaches; all of them were quite time-consuming.

Along with these investigations I turned some of my attention to a new research project: the long-term storage of nuclear wastes in granite. Two months prior to the Arkansas trial some colleagues and I had already obtained definitive experimental results concerning waste storage in those rocks. I hoped that the discovery of these new data might provide a basis for the Laboratory to extend my stay beyond June of 1982, irrespective of my results on superheavy elements.

Conventional Nuclear Waste Containment

It is well known that many individuals within and without government circles perceive the long-term storage of nuclear wastes to be one of the more important technological problems of our time. The goal of nuclear waste research is to determine (1) what type of storage container will best withstand nuclear radiation effects so as to prevent leakage during a several-thousand-year storage period, and (2) the geological site best suited to minimize nuclear waste leakage into the environment in case of accidental rupture of the primary containers. This involves a prediction of the long-term geological stability of the site based on both present-day geological assessment and an estimate of the geological age of the formation.

The standard approach to the problem of site selection assumes that the geological formations best suited for storage are those thought to have remained stable over long geological periods. The U.S. Department of Energy estimates geological age by reference to the *presumed* geological development of the earth. Site selection procedures thus depend partly on the assumption of uniformitarian geology. If uniformitarian geology does not provide a correct timetable for the earth's geological history, then one of the basic criteria for nuclear waste site selection is called into question. We have already discussed (in Chapter 4) how the results on the coalified wood from the Colorado Plateau provide evidence that those formations are only several thousands of years old instead of several hundred million. Professor Kazmann's article (Kazmann 1978) focused attention on the nuclear waste implications of these results.

Although it is possible to fill metal containers with radioactive wastes and bury them in some underground cavity, common sense tells us we must take additional precautions. There is always the possibility that container rupture might occur, due either to corrosion or to some disaster such as an earthquake. Thus it would be unwise to select burial sites near the earth's surface, with its higher risk of waste leakage into the environment.

The leakage hazard can be reduced by burial in granite. Granite formations, extending far below the earth's surface, would obviously permit waste storage at much greater depths. However, at greater depths the temperature rises sharply, again raising the possibility of waste-container rupture. One additional precaution would be to first encapsulate nuclear wastes within some type of impervious matrix, which would resist leakage even at higher temperatures. A most important goal of nuclear waste research is to identify what type of matrix would safely retain radioactive elements under high-temperature conditions.

In recent years nuclear waste specialists have investigated a variety of substances which could serve as the primary encapsulation medium. Certain types of glass have been investigated, and initially some of them seemed to hold great promise. The radioactive material was incorporated into the molten glass mixture and then allowed to cool in the form of a cylinder. Subsequent studies have shown, though, that after a few years the radioactive emissions had damaged the glass structure, making it more susceptible to corrosion. This raises questions about the long-term stability of nuclear wastes in this matrix.

An alternative approach is to investigate various types of synthetic minerals whose natural forms contain significant amounts of the radioactive elements uranium and thorium. By ascertaining which natural radioactive minerals have retained these elements over the course of the earth's history, we can identify the most suitable synthetic counterparts for long-term nuclear waste encapsulation.

There was also the question of where the waste containers themselves would be placed. One plan was to bury the waste containers in deep granite holes. The rationale was: even if the primary container did rupture, the radioactive hazard to the environment would be reduced. Prior to our studies, scientists had only investigated the retention of radioactive minerals taken from granite-rock formations near the earth's surface. But if nuclear waste containers were to be encased in granite, they would need to be buried in 15,000-foot-deep granite holes, where temperatures would be quite high. How much these higher temperatures would affect leakage of radioactivity from the minerals was a crucial question. The only solution was to analyze natural radioactive minerals from deep granite cores. But where were such specimens to be found? Holes deeper than 15,000 feet had been drilled in search of oil but always through sedimentary rocks such as limestones and sandstones.

An Innovative Approach to the Nuclear Waste Problem

In mid-1981 I learned of a 15,000-foot-deep hole in a granite formation drilled by the Department of Energy in New Mexico in the late 1970's. The purpose was to explore the possibility of using high-temperature rock at the bottom of the hole as a heat exchanger to generate steam energy. In this hot-dry-rock experiment (as it was called), water injected into one drill hole at the top would cascade to the bottom and be heated to steam. The steam would then return through a separate hole to a power-generating station on the surface.

Core sections were taken at five different depths from about 3000 feet down to about 15,000 feet during the drilling operation. Fortuitously, each of these granite cores contained many small crystals of the radioactive mineral zircon. These cores were exactly the samples needed to determine how well the radioactive zircons had resisted leakage under the increasing temperatures (up to 313 °C at the bottom of the hole). My affiliation with the Laboratory proved invaluable in obtaining pieces of each one of these priceless cores.

The advantage of analyzing the zircons from these cores was clear: they had already experienced the exact environmental conditions anticipated for nuclear waste storage in granite. By determining the amount of diffusion or leakage of radioactivity out of these zircons, we could accurately determine whether it would be safe to encapsulate nuclear wastes in synthetic zircons of the same type. These experiments also had the potential of providing critical information about the age of the granites.

Remember that the element lead is the end product of uranium and thorium decay chains (and hence is known as radiogenic lead). Since zircon crystals contain small amounts of both uranium and thorium, there will be a constant accumulation of this element in zircons located on the earth's surface. That is, lead diffuses out of zircons very slowly at surface temperatures. With increasing depth, however, the temperature rises considerably, and the lead diffuses out of the zircons far more rapidly.

Now the age question enters the picture. If the granites in New Mexico are over a billion and a half years old, as uniformitarian geology supposes, this would be time for considerable amounts of lead to be lost from the zircons taken from the deepest (highest temperature) sections of the drill hole. In fact, in this scenario the lead should steadily diminish with increasing depth (due to steadily increasing temperatures). However, if the earth is only several thousand years old, only negligible lead loss is expected. In this case the amount of radiogenic lead in the zircons should be about the same regardless of depth. Here was a clear-cut test.

Experimental Results Reach the U.S. Congress

The results of our investigations were definitive. We found that the radioactive zircon crystals extracted from the granite cores had lost essentially none of their radiogenic lead, even at the bottom of the hole where the temperatures were highest. This is exceptionally strong evidence that the presumed 1.5-billion-year age of these granites is drastically in error. Specifically, the data are consistent with a several-thousand-year age of the earth. I realized, however, that these startling implications for a young earth would never pass peer review if they were clearly stated in any report submitted for publication. They would have to be de-emphasized and take second place to the implications for nuclear waste in order for them to ever be published.

Thus, when the results were written up, I emphasized that new evidence had been found, showing that nuclear wastes encapsulated in synthetic zircons would constitute a very safe mode of containment. Our report was submitted for publication to *Science* a month or so before the Arkansas trial and, coincidentally, was being reviewed for publication around the time I was testifying in Little Rock. The report did pass peer review and was subsequently published in *Science* under the title, "Differential Lead Retention in Zircons: Implications for Nuclear Waste Containment" (Gentry et al. 1982a; Appendix). Later some geologists criticized certain aspects of this report. Fortunately, I was given the opportunity to respond (Gentry 1984b).

Janet Raloff, the writer who had published an interview about my testimony at the Arkansas creation trial (Raloff 1982a; Appendix), now publicized the implications of this report for nuclear waste storage in the May 1, 1982, issue of *Science News* (Raloff 1982b). Just before this date I learned that the U. S. Senate was considering an amendment to a nuclear waste bill. It would require the Department of Energy to investigate nuclear waste storage sites other than the tentatively selected salt-dome repositories in Louisiana and Mississippi. Senator Thad Cochran of Mississippi was informed of our recently published report and expressed immediate interest, the extent of which can be judged by his actions when the nuclear waste amendment came before the Senate on April 30, 1982.

On that date he introduced an amendment to the National Nuclear Waste Policy Act of 1982. In doing this Senator Cochran brought our results to the attention of the Senate and had our entire report reproduced along with his comments in the *Congressional Record.* (I was later informed that his office had also written to the Secretary of the Department of Energy about our report and its implications for alternative storage sites.) Some of Senator Cochran's remarks before the Senate are quoted below:

> . . .There is a great deal of controversy and concern, as has already been expressed, about the [nuclear waste storage] sites the Department of Energy is now considering for possible site characterization. There is no hard evidence that any of them will prove suitable for a permanent repository.
>
> Past problems with hasty site selection have caused delays and undermined public confidence. As an example, Mr. President, in 1972, the Atomic Energy Commission had to abandon a salt site in Lyons, Kans., that they were planning to use for a waste repository because water was discovered leaking into the mine, and scientists decided the mine had too many holes in it.

Salt, despite serious problems associated with it, has been a favorite geologic medium with the Department of Energy up to this point because it has been the most extensively studied medium. Even though many experts believe that granite and other forms of crystalline rock may be very promising media, they are not being aggressively investigated. . . .

The fact is that the time that would be required for characterization of granite falls behind the timetables set by DOE and the schedule that this bill contains as it is now drafted, and it arbitrarily, therefore, eliminates granite from consideration in the selection process.

This decision flies in the face of scientific evidence that granite may be the best possible medium for a site for nuclear waste disposal.

As evidence, Mr. President, I cite an article contained in a recent edition (April 16, 1982) of *Science* magazine. The article is authored by scientists affiliated with the chemistry division of the Oak Ridge National Laboratory addressing the question of using natural rock granite as a site to insure the maximum possible degree that radioactive material can be stored in a way that would not permit escape or create any hazard.

The authors used an innovative ultrasensitive technique for a lead isotope analysis in a natural site of granite at Los Alamos National Laboratory in New Mexico.

The results showed, Mr. President, that lead, which is a relatively mobile element compared with nuclear waste, has been highly retained at elevated temperatures under conditions that are similar to those that would apply to the storage of high-level nuclear wastes in deep granite holes.

This study is crucial and it is important because it was based not just on laboratory work but on an analysis in a natural site under adverse environmental conditions.

The Department of Energy should be able to incorporate this kind of finding and this research immediately in its review process. But to follow the dictates of this legislation and the predisposition of the Department to continue studying other kinds of formations would result in their not being able to take advantage of this kind of research.

Mr. President, I ask unanimous consent that a copy of this article I have just referred to be printed in the *Record*. (Cochran 1982, S4307)

Senator Cochran was not the only senator to show interest in this report. On the day prior to the Senate vote on the amendment, I was contacted by Mr. Peter Rossbach, legislative aide to Senator Jim Sasser of Tennessee, about the implications contained therein. Some Tennesseans had expressed concern about the possibility of hauling nuclear wastes across the state down to the salt repositories in Louisiana and Mississippi. According to Mr. Rossbach, Senator Sasser wanted a better understanding of our results so that he could vote more intelligently on the amendment. Even though Senator Cochran's amendment did not pass, Mr. Rossbach wrote me a letter of appreciation and ended by saying, "If there is anything we can do for you from here, please let me know."

Appeal to Continue Research

Mr. Rossbach's offer of assistance was appreciated, and I decided to ask if Senator Sasser would consider appealing to the Department of Energy for a continuation of my guest position. In the next few weeks I received copies of the following letters:

Mr. William S. Heffelfinger   (May 18, 1982)
Assistant Secretary for Management
   and Administration
Department of Energy
Washington, D. C. 20585

Dear Mr. Heffelfinger:

This letter is written on behalf of Robert V. Gentry, Associate Professor of Physics at Columbia Union College and currently Guest Scientist at Oak Ridge National Laboratory.

Mr. Gentry has been a Guest Scientist at ORNL for the past 13 years. During this time, he has published nearly 20 scientific reports, some of which have received national recognition. I have enclosed two published commentaries concerning Mr. Gentry's work which testify to the depth and importance of the research he has been able to conduct while at ORNL.

In addition, Robert Gentry has been particularly helpful to me and my staff on energy-related matters, particularly nuclear waste site selection issues. He has provided valuable evaluations and technical expertise, which has assisted us in ascertaining the full implications of various energy policies.

It is my understanding that Mr. Gentry has been notified that his current dollar-a-year consultant contract will be terminated on June 30, 1982. I also understand that he has recently discovered new evidence relating to nuclear waste containment about which he would like to conduct experiments and further research. However, he will be unable to do this if his contract is terminated on schedule.

I wanted to take this opportunity to bring my interest in Mr. Gentry to your attention and to request that he be allowed to continue his work at Oak Ridge National Laboratory, if at all possible. I am sure that an extension of his contract would allow him to finish his research and prepare conclusions based on those experiments.

I would greatly appreciate any assistance you can offer Mr. Gentry in this regard, and I look forward to hearing from you at your convenience.

Sincerely,

/s/ Jim Sasser

Jim Sasser
United States Senator   (Sasser 1982a; Appendix)

I was grateful for this cordial response, but as the following letters show, it was ineffective in securing a renewal of my research contract.

(June 16, 1982)

Dear Robert:

I wanted to bring you up to date on the latest information I have received concerning your contract with the Department of Energy as a Guest Scientist at Oak Ridge National Laboratory.

You will recall that I contacted Mr. William S. Heffelfinger, Assistant Secretary of Energy for Management and Administration, Washington, D. C., on your behalf. As a result, I have received the enclosed letter from Mr. Heffelfinger, which is for your information.

Robert, it was a pleasure for me to make this inquiry, and I regret that a more favorable response was not received. However, I want to encourage you to contact me again in the future whenever I may be of service to you on matters of mutual concern.

Sincerely,

/s/ Jim

Jim Sasser
United States Senator   (Sasser 1982b)

* * * * * * * *

Honorable Jim Sasser   (June 14, 1982)
United States Senate
Washington, DC 20510

Dear Senator Sasser:

This is in reference to your letter dated May 18, 1982, on behalf of Robert V. Gentry, a guest scientist at the Oak Ridge National Laboratory (ORNL) operated by Union Carbide Corporation for the Department of Energy.

At the time of his assignment at ORNL 13 years ago, Mr. Gentry's supporting sponsor was Columbia Union College. The original purpose of his research was to study pleochroic halos, an area of interest to ORNL at that time, but a field of less significance to the Laboratory's mission in recent years.

Mr. Gentry's more recent efforts in nuclear waste containment referenced in your letter are quite peripheral to the primary thrust of ORNL's ongoing waste isolation programs.

When ORNL entered into its current subcontract with Mr. Gentry, effective July 1, 1981, it was for him to continue his own research on halos, using Laboratory facilities. It was anticipated that he could finish his work during the year; no other work was authorized under the subcontract. He was advised in June 1981 that he should seek other arrangements under which to pursue his research interests beyond June 30, 1982.

Diminishing ORNL budgets require marked cutbacks in activities not directly related to its priority program areas. Unfortunately, Mr. Gentry's work does not fall in that category. Accordingly, we cannot be encouraging about an extension of his agreement at ORNL.

Thank you for your continuing interest in Department of Energy programs.

Sincerely,

/s/ William S. Heffelfinger

William S. Heffelfinger
Assistant Secretary
Management and Administration
Department of Energy   (Heffelfinger 1982; Appendix)

The message in Heffelfinger's letter was quite clear. The recent attention given my work in the U. S. Senate was not a sufficient basis for the Laboratory to renew my guest-scientist status.

Final Results Support Young Age of Earth

Another report on the safety of long-term nuclear wastes in granite was completed just prior to my contract expiration date. It was based on collaborative research with two colleagues and was published after I left ORNL under the title "Differential Helium Retention in Zircons: Implications for Nuclear Waste Containment" (Gentry et al. 1982b; Appendix). As the title of the report suggests, we again analyzed microscopic-sized zircons from the same five depths as were used in the lead-retention studies. However, in these experiments the zircons were analyzed for their content of the rare gas helium.

These experiments provided even stronger evidence for a several-thousand-year age of the earth than did the lead-retention experiments.

To understand this we must remember that alpha particles emitted in the radioactive decay of uranium and thorium, are in reality nothing more than helium atoms stripped of their electrons. So it follows that helium is produced wherever uranium and thorium occur. This is the source of the helium in the zircons. However, being a gas means that helium can diffuse or migrate much more rapidly than the solid element lead. Indeed, studies have shown that helium migrates out of various minerals, such as zircon, even at room temperatures. Because of this continual loss, scientists have generally given up using the helium content to estimate the radiometric age of zircons found at or near the earth's surface. Thus, according to the evolutionary model, it would be senseless to attempt to measure the helium content of the zircons taken from the deep granite cores. Presumably almost all the helium should have migrated out of the tiny zircons during the billion or so years they were exposed to the higher temperatures at greater depths.

However, on the basis of my creation model I expected something different. That model is based on the occurrence of primordial polonium halos in Precambrian granites as evidence that all such rocks were created on Day 1 of creation week about 6000 years ago. On this basis I thought helium might still be retained in the zircons taken from some of the deep granite cores. Here was one of the clearest and most stringent tests of the creation and evolution models in regard to the age of the earth.

The experiments showed amazingly high retention of helium even at 197 °C, directly contradicting the expectation based on the evolutionary model of earth history. These startling results (Gentry et al. 1982b; Appendix) are in complete agreement with my creation model; moreover, they constitute what seems to be the strongest scientific evidence yet discovered for a several-thousand-year age as opposed to a several-billion-year age of the earth. And they complement perfectly the results of my earlier studies on the Colorado Plateau coalified wood specimens. Those studies (Chapter 4) provided evidence for a young age of sedimentary formations previously thought to be several-hundred-million years old.

Paradoxically, just when my research opportunities were about to be withdrawn at ORNL, my long-term goals were being realized with more certainty than ever before. To outward appearances I was losing everything I had worked so diligently to gain—friendship and respect of scientific colleagues and access to the finest of research facilities. In reality, I was succeeding in discovering striking evidence for a young age of the earth, evidence which accords perfectly with the view that the Precambrian granites were all created about the same time. My first and latest scientific discoveries were complementing each other, and my two-decade quest for truth about the origin and age of the earth was being fulfilled. The cost was high in loss of friends, and my financial support remained erratic until it completely disappeared soon after my departure from ORNL. My long association with Columbia Union College came to an end as well. Providentially, I believe, concerned persons made it possible for this book to be written.

End of an Era—A Summary

My original 1969 appointment as guest scientist at the Laboratory was prompted by my research on unusual types of radiohalos. At that time, several laboratories around the world were gearing up their research facilities to search for chemical elements heavier than any previously known. Theoretical studies suggested the existence of superheavy elements, and the search for them was to intensify over the next decade.

The invitation to join ORNL had provided an exceptional opportunity not only to search for superheavy elements but also to utilize their unparalleled research facilities in the investigation of polonium halos. My research endeavors continued to warrant publication in respected scientific journals; thus I was invited year after year to continue as a guest scientist until the time of the trial, twelve and one-half years after beginning my affiliation with the Laboratory. If my research endeavors had been inferior, if my work had not been published in the open literature, or if I had shown prejudicial bias in my publications, the Laboratory management rightly would have terminated my research contract long before they did on June 30, 1982. I had not found superheavy elements in my research efforts, and the Laboratory was justified in terminating my research contract. However, had it not been for the negative reporting of my testimony at the Arkansas trial, I think my most recent research activities regarding nuclear waste storage might have been deemed of sufficient value to warrant continuation of my research at ORNL.

And so my work at ORNL came to an end. My hopes of continuing the search for the elusive superheavy elements apparently had evaporated. I had invested many years looking for them, and despite the ill-fated results of the giant halo experiments at Florida State in 1976, I am still convinced that superheavy elements do exist.

The Case of the Unmailed Letter

A few weeks before my departure from ORNL I learned that Steve Clark, the Arkansas Attorney General, was considering writing a letter to several Congressmen about my situation. A year and a half later, in the spring of 1984, I asked the former Deputy Attorney General (who handled the State's case for the Arkansas creation trial) to investigate whether such a letter was ever sent from the Attorney General's office. In his investigation this former Deputy Attorney General found a letter to Senator Bumpers in the state archives in Little Rock. According to him, the plan was for identical copies of this letter to be individually addressed to each member of the entire Arkansas Congressional delegation after the copy to Senator Bumpers was signed and the date affixed. Curiously this letter, which was apparently signed about the time of my departure from the Laboratory, was never dated or sent. No one seems to know exactly how this happened. The letter is copied below so that readers may ponder for themselves what events may have transpired had it been sent.

The Honorable Dale Bumpers
United States Senator
New Senate Office Building
Washington, D.C. 20515

Dear Senator Bumpers:

In my recent defense of Act 590 of 1981 (better known as the Creation-Science Law), I had the opportunity to become acquainted with several of the world's leading scientists who testified on behalf of both the State and the American Civil Liberties Union. Of all the scientists involved on both sides of the lawsuit, no one impressed me anymore than Robert Gentry, who for the past several years has been a guest scientist at the Oak Ridge National Laboratory in Oak Ridge, Tennessee. This letter is written to bring to your attention Mr. Gentry's work and to enlist your aid on his behalf.

Mr. Gentry's testimony at trial concerned the presence of radioactive polonium halos in granite. The significance of these halos is that their presence in the granites is fundamentally inconsistent with the conventional wisdom that the granites underlying the earth's structure cooled over thousands of years. Mr. Gentry is acknowledged as the world's foremost authority on this particular subspecialty.

From every indication available to me, Gentry's work at the National Laboratory has been of a uniformly high quality and has added significantly to the progress made at the facility. Furthermore, as a guest scientist, Gentry has been paid only $1.00 per year by the government. (A college of which he is a faculty member has paid his salary.) Thus, the government has been able to avail itself of his services essentially free of charge.

However, Mr. Gentry has recently learned that his contract as a guest scientist will not be renewed for next year. As one admittedly viewing these events from afar, it appears to me that Gentry is being penalized for his generous offer of assistance to help the State of Arkansas and his own religious beliefs. Bob Gentry is very frank and forthright in stating his religious beliefs, of that there can be no doubt. His religious beliefs are, however, irrelevant to the work which he performs at Oak Ridge. His work in studying granites was recently quoted in the Congressional Record in connection with a discussion of possible sites for storage of low level radioactive wastes. Obviously, this is an important issue and one on which Gentry has been on the cutting edge.

I want to ask for your assistance to assure that Robert Gentry will not be a victim of religious discrimination at the hands of his supervisors. The Oak Ridge National Laboratory, although operated by a private corporation under a contract, is, as I understand it, under the jurisdiction of the U.S. Department of Energy. I solicit your help in contacting the Energy Department through appropriate channels and requesting that the decision to not renew Gentry's contract be reviewed personally by the Secretary of Energy to assure that this decision was based solely upon the merits of his work, and not upon the subjective prejudices of his supervisors. It will be a sad day, indeed, if the First Amendment's guarantee of freedom of religion and the supposed freedom of scientific inquiry have both become hollow promises for men like Bob Gentry.

If I can supply you with any additional information regarding this matter, please call upon me at your convenience.

Yours truly,

/s/ Steve Clark

Steve Clark   (Clark 1982; Appendix)

Final Inquiry by a Member of Congress

Attorney General Clark's letter never reached Congress. However, in 1984, Don Strother, a Baptist minister in Johnson City, Tennessee, whom I did not know, wrote to U.S. Representative James H. Quillen, First District of Tennessee, and asked for an inquiry into the circumstances surrounding my departure from the Laboratory. The following letter relates the outcome of the investigation by the U. S. Department of Energy in Oak Ridge, Tennessee:

Honorable James H. Quillen   (September 4, 1984)
United States House of Representatives
Cannon House Office Building
Washington, D. C. 10515

Dear Mr. Quillen:

This is in reference to your letter dated August 6, 1984, to Secretary Hodel concerning Dr. Robert V. Gentry, a former guest scientist at the Oak Ridge National Laboratory (ORNL).

Our records reflect that Dr. Gentry's association with ORNL began in July 1969 with Columbia Union College as his supporting sponsor. The original purpose was to conduct his own research on radioactive halos, which was an area of interest to ORNL at the time, but during the late 1970's became less significant at ORNL.

Since his work in the Department of Energy's Waste Isolation Program involved moderately low priority supporting research, Dr. Gentry was advised in June 1981 that he should seek other arrangements under which to pursue his research interests beyond June 30, 1982. This decision was the result of diminishing ORNL budgets that required a cutback in activities not directly related to high priority program areas. We have found no evidence to suggest that Dr. Gentry's religious beliefs influenced this decision in any way.

We appreciate your interest in this matter.

Sincerely,

/s/ Joe La Grone

Joe La Grone
Manager, Oak Ridge Operations
Department of Energy   (La Grone 1984)

This is a carefully worded letter. I never said that my religious beliefs per se were responsible for my termination, but I do believe that the negative publicity from the Arkansas trial was a factor.

After this letter was sent, I had a cordial visit with two officials at the Oak Ridge National Laboratory whom I hold in the highest esteem. I expressed gratitude for the thirteen years I was allowed to remain at ORNL and asked about the possibility of resuming my search for superheavy elements. While the response was negative at that time, nevertheless a change in circumstances may yet result in a favorable decision. In the meantime my research continues using other facilities.

14

Creation Confronts Evolution

A climactic event in the twenty-year history of my research was the invitation to speak before the Pacific Division of the American Association for the Advancement of Science (AAAS) in June 1982. *"Evolutionists Confront Creationists"* was the title of a symposium held at the Santa Barbara campus of the University of California. Two biologists from San Diego State University, Drs. Frank Awbrey and William Thwaites, organized the symposium and invited eight scientists to present the evolutionary view. Two scientists from the Institute for Creation Research in San Diego were originally scheduled to present the creation perspective. Subsequently, one of them withdrew, and I was invited to take his place. This was a new day in the annals of the AAAS, for creation scientists had been excluded from a similar symposium held at the AAAS annual meeting in 1981. Dr. Rolf Sinclair later justified their exclusion from this meeting, saying that the organizers were at a loss to know whom they should choose to represent the creation position (Sinclair 1981).

Creation/evolution symposia were similarly held at the American Physical Society (APS) meeting in Washington, DC, (April 1982) and the Geological Society of America (GSA) meeting in New Orleans (November 1982). Again only scientists representing the evolutionary position were allowed to speak. My request to contribute a paper was turned down by the organizers of both meetings.

But the forthcoming Santa Barbara meeting was different, and the prospects were exciting. The very title of the symposium suggested that all evidences for creation would be confronted by opposing scientific evidence. If my work was to be refuted, the most likely speaker to do so would have been Dr. G. Brent Dalrymple of the U.S. Geological Survey in Menlo Park,

California. Seven months earlier he had been the main ACLU witness at
the Arkansas creation trial in support of a 4.5-billion-year age of the earth.
There he had labeled the evidence, which I had presented in behalf of creation, a "tiny mystery." What would be his position at this symposium?

A Geologist Evaluates Creation Science

The title of Dalrymple's presentation, "Radiometric Dating and the Age
of the Earth—A Reply to 'Scientific' Creationism," suggested his views about
creation science had not changed. In his presentation Dalrymple essentially
repeated what he said at the Arkansas trial—that radioactive decay rates
have been "effectively constant through time" and hence that radiometric
dating methods are "the most reliable sources of geological information
available today" (Dalrymple 1982, 4).

As noted earlier, evidence for creation invalidates the *uniformitarian
principle*—the basis of the constant decay rate assumption used in radiometric
dating. However, despite the title of the symposium, none of the speakers
chose to "confront" the evidence for creation. There was no explanation
given for polonium halo formation in granites, nor was there a response
to the challenge of duplicating a piece of granite. Neither was there discussion of the evidences for a young earth obtained in the coalified wood and
the zircon investigations. Instead, Dalrymple chose to (1) focus on what
are perceived to be weak arguments for creation, (2) again label polonium
halos in granites a "tiny mystery," and (3) define creation and science as
being mutually exclusive. His feelings about creation science were vividly
expressed at the end of his talk:

> I think that it will be a sad day for civilized humanity if science, that
> magnificent field of objective inquiry whose only purpose is to decipher
> the history and laws of the physical universe, is allowed to fall victim to
> the intellectual fraud of the creation-science movement. (Dalrymple 1982,
> 27)

I do not defend everything that is called creation science. Nevertheless,
those who condemn all of creation science on the basis of weak or irrelevant
arguments advanced in its favor should consider that their perceptions may
not be entirely without bias. They should also remember that most creation scientists have been shut away from obtaining the research funds and
equipment which would have allowed them to do better work. Often they
have had to rely on the data that evolutionists have collected and placed

in the evolutionary framework. True, the process of fitting those same data into a creation science framework may at times be in error. But there is no field of science without some errors and misconceptions in its formative stages, and efforts to develop a practicable creation model are no exception. The progress of science depends on proposing and testing ideas and hypotheses in support of various theories. Scientists do not discard a theory just because weak or erroneous arguments were once used to support it. On the contrary, if they are genuinely interested in knowing the truth about a theory, they seek to test the strongest arguments in its favor.

Were those in attendance at the AAAS symposium seeking to do this? Or were there attempts to dismiss creation science on other grounds? The answer is found in Dalrymple's introductory remarks of his published contribution to the symposium:

. . .Even a cursory reading of the literature of "scientific" creationism, however, reveals that the creation model is not scientifically based but is, instead, a religious apologetic derived from a literal interpretation of parts of the book of Genesis. Indeed, this literature abounds with direct and indirect references to a Deity or Creator, and citations of the Bible are not uncommon. . . .(Dalrymple 1984, 67)

Here my colleague advocates a great loophole for evolution. *To disqualify the creation model because it refers to Genesis means that no amount of data supporting that model would ever be accepted, regardless of its empirical foundations.* On this basis evolutionists would never have to respond to any scientific discoveries for creation—they may choose to relegate all such evidence to the confines of a religious apologetic. Followed to its logical conclusion, this reasoning would permit scientists to label these evidences as mysteries which will someday be found to fit into the evolutionary framework. This is precisely how Dalrymple referred to my work near the end of his presentation:

The exact way in which the enigmatic Po halos were formed is not yet known.The Po halos are, I'm afraid, one of science's abundant tiny mysteries. As a scientist, I am confident that the halos will eventually be explained as the result of natural processes. Certainly, I see no logical reason whatsoever to seek explanations outside of physical processes, or to entertain for even a moment Gentry's creationist model, which requires us to suspend the laws of physics and chemistry, to call upon intervention by an unknown and unknowable supernatural agent, and to ignore overwhelming and conclusive evidence that the Earth, as we see it now, formed and evolved by natural processes over billions of years. (Dalrymple 1982, 26)

In Chapter 11 I quoted Professor Davies' description of the mythical Big Bang to show that even evolutionists recognize it to be beyond explanation by the laws of physics and chemistry. That my respected colleague would mention the suspension of those laws as a criticism of the creation model is therefore inconsistent with his own acceptance of Big Bang cosmology. True enough, creation cannot be explained by known laws of physics and chemistry, and it does require the intervention of God. In this respect a faith factor enters the picture. But the same is true with evolution. In fact, in the evolutionary scenario all the important events—the Big Bang, and therefore the origin of galaxies, stars, the sun, the earth, and life on it—have always been a matter of faith. In a number of instances faith in evolutionary origins is held even when evolutionists themselves have been unable to find the crucial evidence to support their beliefs. To illustrate, in a book review a noted astrophysicist has recently commented on the origin of stars:

> The universe we see when we look out to its furthest horizons contains a hundred billion galaxies. Each of these galaxies contains another hundred billion stars. That's 10^{22} stars all told. The silent embarrassment of modern astrophysics is that we do not know how even a single one of these stars managed to form. There's no lack of ideas, of course; we just can't substantiate them. (Harwit 1986)

Not being able to substantiate those ideas is an understatement. As Harwit's review explains, the fundamental premise of all modern theories of star formation involves the contraction of interstellar dust clouds into dense, massive objects. This violent process should be marked by three distinct astrophysical processes. Harwit notes that astronomical evidence for those processes has not been found.

I suggest that astronomers have failed to find the critical evidences predicted by their model because stars did not originate with evolutionary processes—but instead were called into existence by the same God who created the earth.

My Presentation at the AAAS Symposium

The creation-based organization, Students for Origins Research (SOR), previewed the AAAS creation/evolution symposium and featured a discussion of my research in the Winter-Spring 1982 issue of their publication, *Origins Research*. About this time Drs. Awbrey and Thwaites, symposium organizers, sent a letter (dated March 1, 1982) to SOR containing the statement, "It would certainly make the Santa Barbara meetings the most important meetings of the century, if even one piece of *bona fide* evidence *for* creation could be presented" (emphasis theirs). They further explained they wanted to see hard data from properly controlled experiments or observations, not meaningless extrapolations, out-of-context quotations, or vague generalities.

With this challenge in mind, I set out for Santa Barbara to present my published scientific results in the context of a creation model of origins. A lively crowd of about 200 scientists was present in the amphitheater where I spoke on the first afternoon of the day and a half of lectures; my presentation was video-taped the following day for wider distribution (Battson, 1982). The symposium was billed as a collision encounter—"Evolutionists Confront Creationists." Believing that my published evidences for creation might satisfy the demand "to see hard data," I decided to reverse the emphasis of the symposium and make the theme of my talk "Creation Confronts Evolution." The Abstract of my talk, published in the *Proceedings* of the symposium, shows how this theme was developed:

ABSTRACT

If the earth was created, it is axiomatic that created (primordial) rocks must now exist on the earth, and if there was a Flood, there must now exist sedimentary rocks and other evidences of that event. But, if the general uniformitarian principle is correct, the universe evolved to its present state only by the unvarying action of known physical laws, and all natural phenomena must fit into the evolutionary mosaic. If this fundamental principle is wrong, all the pieces in the evolutionary mosaic become unglued. Evidence that something is drastically wrong comes from the fact that this basic evolutionary premise has failed to provide a verifiable explanation for the widespread occurrence of Po halos in Precambrian granites, a phenomena which I suggest are in situ evidences that those rocks were created almost instantaneously in accord with Psalm 33:6,9: "By the word of the Lord were the heavens made; and all the host of them by the breath of his mouth. For he spake, and it was done; he commanded, and it stood fast." I have challenged my colleagues to synthesize a piece of granite with ^{218}Po halos, as a means of falsifying this interpretation, but have not received a response. It is logical that this synthesis should be possible if the uniformitarian principle is true. Underdeveloped U halos in coalified wood having high U/Pb ratios are cited evidences for a Flood-related recent (within the past few thousand years) emplacement of geological formations thought to be more than 100,000,000 years old. Results of differential He analyses of zircons taken from deep granite cores are evidence for a recently created, several-thousand-year age of the earth. A creation model with three singularities, involving events beyond explanation by known physical laws, is proposed to account for these evidences. The first singularity is the *ex nihilo* creation of our galaxy nearly 6000 years ago. Finally, a new model for the structure of the universe is proposed, based on the idea that all galaxies, including the Milky Way, are revolving about the Center of the universe, which from Psalm 103:19 I equate with the fixed location of God's throne. This model requires an absolute reference frame in the universe whereas modern Big Bang cosmology mandates there is no Center (the Cosmological Principle) and no absolute reference frame (the theory of relativity). The motion of the solar system through the cosmic microwave radiation is cited as unequivocal evidence for the existence of an absolute reference frame. (Gentry 1984a, 38; Appendix)

As the Abstract reveals, I suggested how the evidences for creation discussed in this book can be embodied within a viable model of origins based on the Genesis account of earth history. This tentative creation model postulates three special periods, or singularities, which cannot be explained on the basis of known laws. These singularities are the creation, the fall of man, and the flood—events marked in a major way by the intervention of the Creator.

The last part of the Abstract refers to my most recent investigations involving astronomy. Technical comments on the interpretation of galactic red shifts, the cosmic microwave radiation, and its surprising implications about the theory of relativity are given in the full article (Gentry 1984a; Appendix). This report elaborates on my discovery that the mathematical basis for the Big Bang model of an expanding universe is based on erroneous assumptions. My alternative model postulates that the galaxies in the universe are revolving in different orbital planes around a fixed Center, the Creator's throne. This Center is calculated to be several million light-years away from our galaxy, the Milky Way. (These results formed only a small part of my talk and thus were not included in subsequent discussions at the symposium.)

During the question and answer session, doubts were expressed that my proposed creation model could account for all the data adapted into the evolutionary framework. I reminded all those present that their own model involves at least *one* singularity, the Big Bang, and then complete uniformity to the present. In contrast, my proposed creation model involves *three* singularities, with uniformity between those events. I suggested that whatever data can be fitted into a one-singularity model must also fit into a model with three singularities, for in this case there is much greater latitude.

Still, many of those in attendance seemed to think that evolution must be true because of the abundance of data already fitted into this framework. I improvised a parable to show that these numerous points of agreement in no way confirm evolution. The quest for truth was analogized to the "Parable of the Grand Design," which is featured in the Epilogue of this book.

A National Forum

In the same month that the AAAS symposium was held, the nationally circulated physics journal, *Physics Today,* opened the pages of their Letters section to the creation/evolution topic. From those Letters it was quite apparent that many physicists were still unaware of the implications of my work for creation. Taking advantage of this new forum, I published a letter describing the results of my research in the October 1982 issue (Gentry 1982). This first letter precipitated objections from a geologist. His comments and my response (Gentry 1983a) were both published in the April 1983 issue of this journal. Other objections and my responses (Gentry 1983b, 1984c, 1984d) were published in the November 1983, April 1984, and December 1984 issues.

Most of these objections reasoned from the assumption of the *uniformitarian principle;* hence it was argued that my interpretation of polonium halos must be incorrect. Significantly, none of those letters attempted to directly refute the evidence for creation. And most significantly, there was no mention of the crucial granite synthesis experiment.

Creation/Evolution Newsletter Attacks Polonium Halo Evidence

Publications of much less significance than *Physics Today* are also involved in the creation/evolution controversy. A notable example of this is the *Creation/Evolution Newsletter,* edited by Karl Fezer of Concord College, Athens, West Virginia. This newsletter is reputed to be "dedicated to defending and enhancing the integrity of science education." Its contents include newspaper clippings supportive of evolution, news of the activities of certain creation scientists, and comments putting down scientists or theologians who support biblical creation. One issue of this newsletter printed a letter about my work preceded by these editorial remarks:

GENTRY'S PLEOCHROIC HALOS

Robert V. Gentry is widely regarded as one of the more conscientious and scholarly creationists. His research on radioactive halos is in a field outside the expertise of most scientists. Gentry's arguments are criticized by G. BRENT DALRYMPLE, U.S. Geological Survey, Menlo Park, CA, in the following letter to Kevin H. Wirth, Director of Research, Students for Origins Research, Santa Barbara, CA: (Fezer 1985, 12)

Dalrymple charges in his letter (Dalrymple 1985; Appendix) that my creation model is "unscientific" and "ridiculous," that my interpretation of the polonium halo evidence for creation is "absurd" and "naive," and that my challenge to the scientific community to falsify my conclusions by the synthesis of a hand-sized piece of granite is "silly," "inconclusive," and "nonsense." Another evolutionist (Osmon 1986) used Dalrymple's comments when he published a follow-up letter in the same newsletter. My response (Gentry 1986; Appendix) to Dalrymple's criticisms, given at the end of this book, also serves as a rebuttal to Osmon's technical comments.

Elsewhere in his letter Osmon urges that my "creation hypothesis" should be given a "thorough review," to see whether it fits the canons of science as defined by an evolutionist philosopher (Kitcher 1982). Kitcher's book serves two functions for all those who are adamantly opposed to creation: (1) it attempts to establish that creation science is not true science; and (2) it constructs a philosophy of science in which evolutionists will never have to be placed in a position where they would be forced to substantiate the basic premise of their theory with experimental evidence.

Applying one of Kitcher's criteria to my work Osmon concludes that:

> . . . neither [Gentry's] hypothesis or the [his] theory provides any problem-strategy at all. If a geologist asks how does rock with the properties of granite form, Gentry's answer is "Kazam." . . . (Osmon 1986)

This is somewhat ironic—I thought "Kazam" was the onomatopoetic description of the Big Bang!

In another place Osmon surmises that I might have proposed the falsification experiment because I knew "it would be very expensive to perform. . ." Here Osmon unwittingly reveals a basic contradiction in his argument. Over the last several decades countless millions of government funds have been spent on incredibly "far-out" ventures specifically designed to test a number of evolutionary predictions—one prime example being the costly unmanned space mission to Mars to look for evidence of the evolutionary beginning of life. This mission failed to find any trace of even the most primitive forms of life. Despite this failure, evolutionists continue to obtain funds for almost any experiment which they feel is important. We must conclude that until now evolutionists have not been inclined to launch a full-scale effort to perform the falsification test.

But why would confirmed evolutionists want to continually postpone a confrontation based on experimental evidence produced in the laboratory? After all, success in this experiment would be the desperately needed evidence to show that evolution has some basis in fact, for it would substantiate the evolutionary origin of the granites based on the *uniformitarian principle.* With everything at stake, why are there not scores of dedicated evolutionists seeking to vindicate the fundamental premise that holds all of the evolutionary scenario together? As a first step, why do they not show how polonium halos can be experimentally produced in granite that already exists, instead of just hypothesizing about how these halos might have formed in accord with conventional laws?

By minimizing the crucial importance of the granite synthesis experiment, Osmon has in effect deflected attention from some important truths: all models of origins—whether based on a biblical framework, an atheistic framework, or any combination of religious/atheistic beliefs—involve a faith factor. I have already discussed how the Big Bang cosmological model is dependent on this faith factor. The theory of punctuated equilibrium (quantum jumps from one species to another) also involves an immense faith factor for biologists mainly because its basic premises are little more than idealized speculation.

The important point is that all scientific models of origins rest on certain basic premises. Thus the ultimate scientific test of any model of origins hinges on whether its basic premises are true or false. If data are discovered which contradict either a model's basic premises, or an undeniable consequence of those premises, then the model is false regardless of how many pieces of data can be fitted into it. Polonium halos in Precambrian granites falsify the entire theory of evolution because they contradict its basic premise, the *uniformitarian principle.* The only way this statement can be refuted is by providing laboratory evidence showing that granites with polonium halos can form naturally.

I do not believe that a report will ever be published describing the synthesis of a granite containing even a single ^{218}Po halo, much less one containing all three types. (By comparison, some natural specimens of biotite contain thousands of ^{218}Po halos in just one cubic centimeter.) My confidence is based on experimental data obtained from the laboratory of nature, the ultimate proving ground for all models of origins.

As detailed in Chapter 4, the secondary polonium halos in coalified wood provide demonstrative evidence that, even under ideal conditions of high uranium concentrations and rapid transport, only the ^{210}Po halo type will develop secondarily from the accumulation of uranium daughter activity. In contrast, three types of polonium halos occur in granites where both the uranium concentration and the transport conditions necessary to produce secondary polonium halos are missing. Consequently, I maintain that all attempts to duplicate a granite containing the three types of polonium halos will meet with failure.

In brief, the laboratory of nature has provided both positive, unambiguous evidence for a primordial origin of polonium halos in granites as well as decisive, independent evidence against their secondary origin.

Vistas in Creation

This book has shown a number of instances where evolutionists have misunderstood my creation model. That model, based on the Genesis record of creation and the flood, is not restricted or at all governed by the uniformitarian concept of a worldwide geologic column, which is based on radiometric dating and index fossil classification. Rather it begins by connecting "In the beginning . . ." with the primordial Earth being called into existence on Day 1 of creation week about 6000 years ago. More specifically, I envision a continual series of geologically oriented creative events occurring throughout the 24-hour period of Day 1, with each of those events beginning with the appropriate matter being called into existence from nothing. As mentioned in Chapter 10, the initial state of that matter may have been a primordial liquid, which was instantly cooled to form primordial rocks.

The Precambrian granites show evidence of an instantaneous creation and hence are identified as part of the primordial rocks of the earth; further investigations are needed to determine which additional rocks should be classified as primordial. Those other primordial rocks could include sedimentary rocks (without fossils) as well as some non-Precambrian granites and metamorphic rocks, such as some which occur in New England. While Day 1 includes the preeminent geological event of earth history, the geologic occurrences of Day 3 may also have been quite significant. Specifically, the appearance of dry land out of a watery environment on Day 3 may have been accompanied by the rapid formation of certain sedimentary rocks, in particular those that geologists classify as Precambrian. (Initially, of course, these "creation-week" sedimentary rocks would have been free of fossils.) The events of Day 3 might have included vulcanism and the formation/creation of some intrusive rocks as well. Conceivably, there may have been limited mixing of the different created-rock types during creation week.

My creation model of the global flood envisions tremendous upheavals of the earth's crust and many opportunities for the deposition, intrusion, mixing, erosion, and reorientation of different rock types. Here are some of the possibilities: Although the flood itself lasted just a year, long-term geological effects may have lasted for hundreds of years thereafter. For example, while the sedimentary rock formations observed in the Grand Canyon are ascribed to the period of the flood itself, the erosional processes that cut through the freshly deposited sediments may well have continued for a number of years after the flood. In my model the bulk of fossil-bearing sedimentary rocks would have formed during the opening and closing stages of the flood, with lesser amounts being formed during the long period of subsidence and run-off after the flood.

Extensive vulcanism is envisioned as occurring during the same periods, which means that opportunities existed for the intrusion of volcanic magma into sedimentary formations. Vulcanism during and after the global flood provides a mechanism whereby the primordial and other rocks, created during creation week, could have mixed with flood-related volcanic and sedimentary material. To illustrate, consider that, as magma (hot fluid rock) formed deep in the earth passes upward toward the earth's surface, it may pass through and melt, or alternatively encapsulate, a variety of rocks, beginning with those created on Day 1 or Day 3, and extending through those formed by volcanic and sedimentary activity during the time of the flood. Thus, when that magma finally cools to a solid, it would be a composite of all the rocks just mentioned. If the magma temperature was not too high, then the composite rock would contain unmelted fragments of all the rocks through which the magma had passed.

This description of my creation model is by no means exhaustive; however, I trust it will provide an expanded framework for interpreting diverse geological data. To me the Genesis record of creation and the flood is the master key which unlocks all of Earth's geologic history.

15

Current Attacks on Creation Science

This closing chapter illustrates how some confirmed evolutionists continue to ignore, disparage, misconstrue, or suppress the scientific evidence for creation. Having said this, I nevertheless respect the right of anyone who chooses to accept evolution as his model of origins. This is what democracy is all about. Here we have the right to choose any philosophy—or scientific hypothesis—after having the opportunity to evaluate *all* the relevant data.

Survey of Creation-Science Literature Yields Questionable Results

In his capacity as Research News Editor of *Science* magazine, Roger Lewin again attacks creation science in his May 17, 1985, article "Evidence for Scientific Creationism?" (Lewin 1985). The reader will recall my attempt to correct his inaccurate account of my testimony at the Arkansas trial (see Chapter 12). At that time the Letters Editor, Christine Gilbert, replied that the editorial staff of *Science* had decided not to publish my technical response to his comments on my research. She added that Lewin was unable to include certain details because of his space limitations. My complaint to *Science* did not question the column space given to my testimony but his garbled and incomplete account. It is interesting to note, however, that *Science* printed the entire *Opinion* written by Judge Overton. Evidently, space limitations are no problem when the commentary supports evolution.

This 1985 article by Roger Lewin erroneously portrays to the scientific community that creation science is devoid of published material in the eminent scientific journals of the world. He uses information obtained from a computer survey by Eugenie C. Scott, an anthropologist at the University of Colorado, and Henry P. Cole, a professor in educational psychology at the University of Kentucky. Lewin refers to their article (Cole and Scott 1982) to back up his contention that "so-called creation science" is based on "putative pillars," not genuine evidence documented in the technical literature. He quotes Scott and Cole's conclusion, that "nothing resembling empirical or experimental evidence for scientific creationism was discovered" in their survey of the scientific literature. Lewin re-emphasizes this point in discussing their latest survey (Scott and Cole 1985), when he focuses on their central theme: "why don't the professional scientists among creationists publish empirical, experimental, or theoretical evidence for scientific creationism?"

As soon as I read Lewin's article, I tried unsuccessfully to contact Dr. Scott at the University of Colorado and left word for her to return my call. I was able to reach Dr. Cole, though, at the University of Kentucky. Over the telephone I went into much detail outlining the basic results of my research efforts since the mid-sixties and questioned the conclusions of their recent survey. He replied defensively that Dr. Scott was more familiar with the radiohalos than he and that he would ask her to call me. In particular I asked Cole about their report in *The Quarterly Review of Biology* and the following statements made therein relative to my research:

> . . .Probably the best anomaly in the scientific creationists' arsenal is the existence of polonium halos, a "minor mystery" in Judge Overton's words, of which the scientific creationists are quite proud. Gentry [Gentry 1982] claims that the existence of Po halos in granite, coalified wood, mica, and other substances indicate that such materials were formed suddenly, under cool conditions, an interpretation supporting special creation. These observations, however, have alternative explanations within normal physical science, and are therefore not unambiguous evidences for Special Creation [Dutch 1983 and Hashemi-Nezhad et al. 1979]. (Scott and Cole 1985, 26)

Scott and Cole show their unfamiliarity about my work when they include coalified wood in the category of substances which "formed suddenly." More unfamiliarity is evident by their claim that my observations have alternative explanations within normal science, a claim they support by citing Dutch and Hashemi-Nezhad et al. But these scientists did not do specific *research on polonium halos* (Gentry 1983b, Gentry 1984a); thus they had no alternative explanations based on *demonstrable* evidence, only hypothetical solutions. *Postulating a hypothetical origin for the polonium halos in granites is something that anyone can do. But for a scientist to truthfully claim he has found a conventional explanation of polonium halos in granites, he must provide demonstrable evidence that his explanation is correct.* As I have noted several times, this can be accomplished only by the artificial synthesis of polonium halos in granites (Gentry 1979, Gentry 1980, and Gentry 1984a; Appendix). Such proof of a conventional explanation for these polonium halos has not been demonstrated. I explained this to Dr. Cole, and again he indicated that Dr. Scott was largely responsible for the comments about my work on halos.

Soon after our conversation he wrote me a letter, stating that he had reread the article written by Dr. Scott and him along with the pertinent references to my work. He insisted there were, "indeed, other scientists who provide alternative explanations for the existence of Po halos." He ended the letter, assuring me that he would call Dr. Scott and ask her to contact me.

I have yet to hear from Dr. Scott! And obviously my conversation with Dr. Cole had not changed his mind. He was more convinced than ever to uphold what had been written in their article. He was content to let *plausibility arguments* serve as "alternative explanations [for Po halos] within normal physical science." I suggested that, if in fact he knew of scientists who had demonstrable experimental evidence to refute the results of my work on the halos, they should by all means submit such evidence to the review process in journals like *Science* or *Nature,* where it could be critically analyzed along with my response. Published *theoretical* statements about the origin of the halos, on the other hand, do not and never will constitute an alternative explanation derived by the scientific method.

In Scott and Cole's article in *The Quarterly Review of Biology,* they quote from my 1974 report in *Nature* and then comment on the statement as follows:

> . . .In an article in *Nature* [Gentry et al. 1974] he asks "Do Po halos imply that unknown processes were operative during the formative period of the earth?" He makes no statement about special creation here, however, and in fact goes on to posit another kind of explanation: "Is it possible that Po halos in Precambrian rocks represent extinct natural radioactivity and are therefore of cosmological significance?. . ." (Scott and Cole 1985, 27)

Scott and Cole, not being geophysical scientists, misinterpret my conclusions because they do not understand the terminology. They are not aware that connecting polonium halos with "extinct natural radioactivity" is just a technical way of saying the primeval earth formed very rapidly. One of my earliest reports was almost rejected because a referee understood this connection with creation (see Chapters 2 and 3). Thus, Scott and Cole are wrong when they say, I went on "to posit another kind of explanation" about the implications of polonium halos. The terms "special creation" or "creation" were not used in my reports to avoid rejection of the manuscripts.

Their concluding remark about my article is:

> . . .Later in the [Gentry's] article (p. 566) another hint is offered: "Just as important as the existence of a new type of lead is the question of whether Po halos which occur in a granitic or pegmatic [sic, pegmatitic] environment. . .can be explained by accepted models of Earth history." . . .Articles of this sort are likely what creationists refer to as "masked" literature. (Scott and Cole 1985, 27)

Scott and Cole imply that there is something masked about the above statement, but actually they easily noticed the implications for creation, which, as Chapter 3 showed, was my intent in putting this statement in my article. But of far more significance is something they did not say: that is, the implications for creation, expressed in my article, have never been rebutted. This fact was carefully "masked" in the report of their survey.

Scott and Cole's final declaration to the scientific community had the effect of a trumpet, sounding the call to battle against creation science:

> . . .science teachers are faced with community campaigns for the teaching of scientific creationism by influential persons, some with scientific credentials, who repeatedly claim there is as much, and equally as good, scientific evidence for scientific creationist concepts as there is for evolution. Teachers, school administrators, and lay persons on school boards are hard pressed to deal effectively with these claims. Support from university-level scholars is often crucial to these disputes, but it is not always offered. Objective documentation of the fallaciousness of the scientific creationist claim that their views are based upon scientific evidence provides "ammunition" for these people. We hope the results of our study will be useful for those who directly confront the creationists. (Scott and Cole 1985, 29)

Apparently Roger Lewin wanted to do his part in providing ammunition "for those who directly confront the creationists," for he concluded his May 17, 1985, article in *Science* by quoting those very words. Unquestionably inveterate evolutionists were inspired with new zeal as they prepared to use his article as a basis for renewed attacks on creation science. Doubtless they thought that Lewin had furnished them with all the relevant facts in his possession. Did he?

Another Response Denied

In Chapter 12 I recounted that Lewin was present throughout my entire four-hour testimony at the Arkansas trial, when published reports of my research on creation were presented in detail to the Court. Why, then, did he choose to remain silent about my publications for creation science instead of completely supporting Scott and Cole's claim that such published evidence was practically nonexistent? In Chapter 7 I showed examples of evolutionists who claim that creation scientists tend to twist the facts and resist unwanted information. I ask: Is Roger Lewin's refusal to report the whole story about published evidences for creation due to his resistance against unwanted information? His journalistic bias for evolution prompted me to send a response to Christine Gilbert, Letters Editor of *Science*; it was an attempt to present the other side of the story, effectively omitted by Lewin:

RESPONSE TO ROGER LEWIN'S MAY 17, 1985,
ARTICLE IN *SCIENCE*—"EVIDENCE FOR
SCIENTIFIC CREATIONISM?"

Roger Lewin (1) quotes Scott and Cole (2,3) to deny both the existence of recent published evidences for creation and the possibility of censorship. Despite these denials, all three of these evolutionists have omitted discussion of a critical test of the evolution and creation models. This test is derived from my published evidence which implies that polonium halos in Precambrian granites originated with primordial polonium (4). On this basis, these granites must be the primordial Genesis rocks of our planet, having been created rather than having crystallized naturally, as evolutionary geology supposes. If the Precambrian granites, with their polonium halos, are indeed the handiwork of the Creator, then, in my view, it is impossible to duplicate them. On the other hand, if the granites just formed naturally, as evolution assumes, then it should be possible to reproduce a hand-sized piece of granite in a modern scientific laboratory. My first opportunity to present this test to the scientific community came in 1979 (5). There was no response to this challenge; so on every available occasion I have repeated it (6) and focused attention on how clearly the issues are defined: Success in duplicating a granite containing just one [218]Po halo would confirm the evolutionary view that both these entities formed by natural processes, and this would falsify my creation model. Failure in this experiment would mean the opposite is true.

Now Scott and Cole (3) say, "It is the nature of scientists to study and debate any scientific fact or finding that challenges existing scientific theories and models. If even one of the creationists' basic assumptions or concepts were supported by empirical evidence from any of the fields of scientific inquiry, scores of scientists would flock to the sites of the evidence and work earnestly to undo or 'falsify' prevailing scientific theories in light of this new evidence.'' Thus, when these authors were confronted with the falsification test in one of my publications (7), why didn't they issue an urgent call for ''scores of scientists'' to begin working ''earnestly'' on it?

A more penetrating question is why Lewin has maintained a deafening silence about this matter for over three years. He was present at the Arkansas trial when I testified about the polonium halo evidence for creation and explained the falsification test in detail. Yet he neglected to mention this decisive test of the two models in his coverage of the trial (8). I attempted to have this glaring omission (and other inaccuracies about my testimony) corrected through a rebuttal letter to *Science,* but my response was denied publication. Subsequently, I lost my position as a Guest Scientist at a national laboratory, even though shortly before my dismissal some of my latest research efforts (9) came to the favorable attention of the U.S. Senate (10).

How much longer will the scientific basis for creation be suppressed? For six years I have waited for those scientists who oppose creation to publish their results on the experimental challenge described above. Why would they wait interminably to refute what I claim to be unambiguous evidence for creation—except that they face an impossible task!

Robert V. Gentry

References

1. R. Lewin, *Science* 228, 837 (1985).
2. H. P. Cole and E. C. Scott, *Phi Delta Kappan* (April 1982), p. 557.
3. E. C. Scott and H. P. Cole, *Quat. Rev. Biol.* 60, 21 (1985).
4. R. V. Gentry, et al., *Science 194,* 315 (1976).
   R. V. Gentry, et al., *Nature 252,* 564 (1974).
   R. V. Gentry, *Science 184,* 62 (1974).
   _____, *Annual Rev. Nucl. Sci.* 23, 347 (1973).
   R. V. Gentry, et al., *Nature 244,* 282 (1973).
   R. V. Gentry, *Science 173,* 727 (1971).
   _____, *Science 160,* 1228 (1968).
   _____, *Nature 213,* 487 (1967).
5. R. V. Gentry, EOS 60, 474 (1979).
6. R. V. Gentry, *Proceedings of the 63rd Annual Meeting, Pacific Division,* AAAS 1, 38 (1984).
   _____, *Physics Today* (December 1984), p. 92.
   _____, *Physics Today* (April 1984), p. 108.
   _____, *Physics Today* (April 1983), p. 13.
   _____, *EOS 61,* 514 (1980).
7. R. V. Gentry, *Physics Today* (October 1982), p. 13.

8. R. Lewin, *Science 215*, 33 (1982);
   Ibid., p. 142 (1982).
9. R. V. Gentry, et al., *Geophys. Res. Lett. 9*, 1129 (1982).
   R. V. Gentry, et al., *Science 216*, 296 (1982).
10. *Congressional Record - Senate 128*, 4306 (1982).

I was hoping *Science* would be more open to publishing this response than they were to the one I submitted in 1982. Unfortunately, this rebuttal to Lewin's view of creation science was also rejected with the excuse: "We wish we could print more letters, but space restrictions limit us to a very small fraction of those we receive." I was curious as to whether there may have been other reasons for their refusal to publish my remarks and telephoned the Letters Editor. She informed me that the decision not to publish my response was made by Daniel Koshland, Editor of *Science*. Subsequently, on June 22, 1985, I wrote to Dr. Koshland, asking for a re-evaluation:

Dear Dr. Koshland:

Today I received a letter from Christine Gilbert indicating that my response to Roger Lewin's write-up would not be published. Space limitations were given as the main reason for rejecting my response.

I have talked to Ms. Gilbert about this decision and have decided to appeal directly to you for publication of my response. I realize my letter contains some potentially embarrassing information about one of *Science*'s staff reporters; but it is information that is nonetheless true, and the scientific community deserves to know what has been going on behind the scenes.

Thanking you in advance for consideration of this appeal to publish my response, I am

Sincerely,

/s/ Robert V. Gentry.

I never received a reply from Koshland about this appeal.

In May 1985, Dr. Russell Humphreys of the Sandia National Laboratories also wrote a letter responding to the implications of Lewin's article on Scott and Cole's surveys. His letter was likewise turned down for publication, and on July 30, 1985, he appealed to Christine Gilbert for a second consideration:

Dear Ms. Gilbert:

Thank you for informing me of your decision not to publish my 28 May letter. It is the most courteous rejection I have ever received. I would like to ask you, however, for a few details on why it was rejected. I know that you have very limited space, but there must be some reasons why you filled that space with other letters than mine.

The reason I am asking is that I have a suspicion the letter was rejected because it supported creationism. My suspicion is based on the fact that in six years, I have seen only one letter in *Science* which was in favor of creationism. I'm sure you have received many more than that, mine among many others. Even your sister magazine across the Sea, *Nature,* has published a reasonable number.

I'm sure you can see how this is related to the subject of my letter, which concerns Roger Lewin's claim that creationists don't submit articles to mainstream science journals. If *Science* does indeed have a hidden policy of suppressing creationist letters, surely Mr. Lewin can see why creationist scientists don't spend the much greater effort of submitting articles. I would appreciate it if you would tell me frankly: Does your journal have such a policy? If it does not, the best way you could prove it is by publishing a competent creationist letter every now and then.

Yours very truly,

/s/ Russ Humphreys

D. Russell Humphreys, Ph.D.
Division 1252
Sandia National Laboratories

On August 30, 1985, she replied to him as follows:

Dear Dr. Humphreys:

Thank you for your letter of 30 July. It is true that we are not likely to publish letters supporting creationism. This is because we decide what to publish on the basis of scientific content.

The letters we received objecting to the study reported by Roger Lewin contained arguments that were largely conjectural or anecdotal. They were therefore not considered acceptable material for *Science.*

Yours sincerely,

/s/ Christine Gilbert

Christine Gilbert
Letters Editor
Science   (Gilbert 1985; Appendix)

Notice that the excuse given was that the negative comments were "largely conjectural or anecdotal." The readers can decide if my June 1985 response, reprinted earlier in this chapter, fits this description. Note also that Gilbert's admission that *Science* has a discriminatory policy against publishing creation-science letters seems to contradict its own editorial policy stated in every issue—the claim to include "the presentation of minority or conflicting points of view."

In summary, the first intent of my response was to especially focus the attention of the scientific community on Lewin's continued silence about the scientific evidences for creation and the falsification test. The second intent was to emphasize that, in the case of my research, there had been no attempt within the scientific community to "flock to the sites of the evidence and work earnestly to . . . 'falsify' . . . this new evidence" as Scott and Cole assured would be the case if "even one of the creationists' basic assumptions or concepts were supported by empirical evidence . . ."

By refusing to publish my response the editor of *Science* effectively allied himself with Lewin and decided to stonewall the entire matter. Perhaps he felt secure in believing that his decision would never become known to the scientific community, or if it did, that he would have their full support in taking action to suppress dissent about such an unpopular cause. Whatever the reason, both the editor of *Science* and Lewin have shown how confirmed evolutionists can use the power of the Establishment to prevent free and open discussion of the published evidences for creation, evidences that most clearly and directly falsify the basic premise of the general theory of evolution.

Part of the *Affirmation of Freedom of Inquiry* discussed in the Overview says "that the search for knowledge and understanding of the physical universe...should be conducted under conditions of intellectual freedom . . ." and "that freedom of inquiry and dissemination of ideas require that those so engaged be free to search where their inquiry leads, free to travel and free to publish their findings without political censorship and without fear of retribution in consequence of unpopularity of their conclusions." The reader may decide whether the editor of *Science* followed the principles of this *Affirmation*.

Response to the National Academy of Sciences

The ultimate battle against creation science since the Arkansas trial has been waged by the National Academy of Sciences. Much discussion concerning how this prestigious scientific organization has denied the evidence for creation was presented in the Overview. In this concluding chapter, since the reader may now have a different perspective of the evolution/creation controversy, I ask: Is the National Academy of Sciences correct in claiming that special creation is an invalidated hypothesis? In the Conclusion of its booklet, *Science and Creationism*, we find the Academy's final evaluation of creation science:

It is, therefore, our unequivocal conclusion that creationism, with its accounts of the origin of life by supernatural means, is not science. It subordinates evidence to statements based on authority and revelation. Its documentation is almost entirely limited to the special publications of its advocates. And its central hypothesis is not subject to change in light of new data or demonstration of error. Moreover, when the evidence for creationism has been subjected to the tests of the scientific method, it has been found invalid. (National Academy of Sciences 1984, 26)

This paragraph contains five accusations, each deserving special comment:

(1) The first sentence effectively hides the failure of evolutionists to confirm a basic prediction of their own theory—the spontaneous origin of life from inert matter. Instead of admitting that this failure invalidates the entire theory of evolution, the Academy attempts to exclude creation science from the scientific landscape by defining science to exclude supernatural power. It is somewhat of a paradox that the Academy would advance such a view because the theory of evolution is in desperate need of a supernatural power both for the origin of life and for the Big Bang. Generally these facts have not been understood by the public.

(2) In the second sentence the Academy claims that the idea of a supernatural origin of life is equivalent to subordinating scientific evidence to revelation. In truth, the abject failure of scientists to synthesize life from inert matter points to only one conclusion—that life originates only with the Creator—just as indicated by the biblical account.

(3) By claiming that the documentation for creation science lies almost entirely within the realm of special publications of its advocates, the Academy Committee members disregarded the scientific publications described in this book supporting creation. Readers should understand that the Academy cannot plead ignorance of those publications. Through my testimony at the Arkansas creation trial in 1981 and my presentation at the American Association for the Advancement of Science symposium in 1982, a significant number of prominent evolutionists became aware of the implications of my research.

(4) The claim that the central hypothesis of creation science is not subject to change in the light of new data is directly refuted by the falsification test that I have proposed to the scientific community for over six years. As noted before, the failure of evolutionists to respond to this critical test leads to only one conclusion—the fundamental *uniformitarian principle* is not now, nor has it ever has been a sufficient basis for granites to form. Without this *principle* the evolutionary mosaic disintegrates.

(5) In the last sentence the Academy asserts that evidence for creation has been subjected to the scientific method and found to be invalid. This statement is definite and unequivocal, with no qualifications. Thus far, to my knowledge, whenever my evidences for creation have been critically examined, they have successfully withstood those examinations. Nevertheless, due to the impeccable reputation of the Academy for scientific integrity, we must ask: *Is the Academy able to back up its all-inclusive claim? If so, it should immediately reveal what published scientific report negates my published evidences for creation.*

American taxpayers, especially those who question the evolutionary model, deserve to know whether such a report actually exists. If it does exist, the integrity of the Academy remains intact. If it doesn't exist, then the Academy's claim must in reality rank as only one of its greatest wishes. In the latter case, it seems that all open-minded evolutionists should query whether their faith in evolution has been misplaced. They might consider that the Creator left trillions of "tiny mysteries" in earth's Genesis rocks to establish substantive faith in the inspired record of creation.

Addendum — Challenge to the National Academy of Sciences

Shortly before this book was printed, I thought of a way to settle the question of whether the Academy actually had a report which invalidated my work. I was scheduled to speak three times at the *International Conference on Creationism* to be held on the campus of Duquesne University in Pittsburgh, Pennsylvania, August 4-9, 1986. This well-organized conference, sponsored by the Creation Science Fellowship, featured about thirty speakers in the technical, educational, and basic sections. The following letter, which is self-explanatory, was sent via overnight courier to where the President of the National Academy of Sciences was located during the week of the conference.

(August 4, 1986)

Dr. Frank Press, President
National Academy of Sciences
2101 Constitution Avenue
Washington, DC 20418

Dear Dr. Press:

This letter concerns the claims about creation science that were made by you and others in the booklet, *Science and Creationism: A View from the National Academy of Sciences,* published in 1984 by the National Academy Press. In my book, *Creation's Tiny Mystery,* which is soon to be published, I focus attention on several statements in the booklet:

...The hypothesis of special creation has, over nearly two centuries, been repeatedly and sympathetically considered and rejected on evidential grounds by qualified observers and experimentalists. In the forms given in the first two chapters of Genesis, it is now an invalidated hypothesis. (p. 7)

It is, therefore, our unequivocal conclusion that creationism, with its accounts of the origin of life by supernatural means, is not science. It subordinates evidence to statements based on authority and revelation. Its documentation is almost entirely limited to the special publications of its advocates. And its central hypothesis is not subject to change in light of new data or demonstration of error. Moreover, when the evidence for creationism has been subjected to the tests of the scientific method, it has been found invalid. (p. 26)

Tuesday, August 5, 1986, marks the beginning of the International Conference on Creationism, to be held on the campus of Duquesne University in Pittsburgh. My technical presentation at this Conference begins at 2 p.m. on Thursday, August 7, 1986. On this occasion I will review the accumulated evidences for creation which I have discovered and published in recognized scientific journals for a period of almost two decades.

These evidences were critically examined by renowned evolutionists when I testified at the 1981 Arkansas creation/evolution trial, and again in 1982 when I gave an invited paper at the symposium ''Evolutionists Confront Creationists,'' sponsored by the Pacific Division of the American Association for the Advancement of Science. These events must surely have come to the attention of the Academy because: (1) two of the authors of the Academy's booklet were involved in the Arkansas trial — Francisco Ayala as a witness for the ACLU, and Joseph Flom as a member of the New York law firm that assisted the ACLU — and (2) another one of the authors, Preston Cloud of the University of California at Santa Barbara, attended the AAAS symposium where my invited paper was given.

Despite this certain knowledge of my work — which was not refuted at Little Rock or at Santa Barbara — the above quotes show that the Academy claims evidence for creation has been tested and found to be invalid. This all-inclusive claim makes no exceptions; so it must be assumed it includes my work as well. However, the Academy's booklet does not even mention my published evidences for creation, much less show that they have been refuted.

I am requesting, therefore, that you make available for this Conference the scientific report which invalidates my results. You could have it sent to one of the Conference organizers, Mr. Robert E. Walsh, 9312 Old Perry Highway, Pittsburgh, PA 15237, or better still just have your office phone the information on this report to Mr. Walsh or Mr. Henry Jackson at 412/341-4908. This number will be answered on a 24-hour basis this week.

In case no such report exists, I invite you to come to my presentation on Thursday afternoon with all the evidence which you think invalidates my scientific results. At the end of my presentation, you will be given the opportunity to speak and show where my results are wrong. I urge you to bring as many evolutionists as you can persuade to come with you — especially any Academy members who have investigated my work. This invitation is extended to anyone you may choose to send in your stead.

If the Academy fails to respond to this challenge, it will be evident that the Academy's claim about invalidating the evidences for creation was only one of their greatest wishes. Part or all of this letter will be read during my presentation on Thursday.

Cordially,

/s/ Robert V. Gentry

No response was received from Dr. Press during the first two days of the conference. Thursday afternoon marked a critical moment in history. Before my presentation began, I asked if any representatives of the National Academy of Sciences were present. No one in the audience responded. About two hours later, near the end of my talk, I asked once more if any representatives of the Academy were present. The audience was silent. The National Academy of Sciences had failed to meet the challenge of creation! To exclude any uncertainty about this matter, about a month later I sent Dr. Press a duplicate of the above letter. Still there was no reply.

Though this book is soon to close, perhaps the reader can begin anew to contemplate the significance of the evidences for creation presented in its pages, considering that in the final analysis *Creation's Tiny Mystery* may represent the Achilles' heel of evolution.

Again, my intent is not to cast aspersion on those who continue to accept the evolutionary model of origins. To trace God's handiwork of creation, as I will continue to do, is an end in itself; to try to duplicate the handiwork of the Grand Design is in another realm altogether.

EPILOGUE

The Grand Design

In the Overview I indicated that by the end of the book the reader should have sufficient information to decide whether the scientific evidence favors evolution or creation. I have presented new evidences for creation and given the reactions of these evidences by prominent scientific organizations, both governmental and private, as well as media representatives. The scientific community, by and large, has not accepted even the *possibility* that these evidences could be fitted into a creation model of origins. Historical considerations have some bearing on this attitude.

During the early nineteenth century the *uniformitarian principle* and its corollary, geological uniformitarianism, were becoming accepted as the bases for reconstructing the history of our planet and solar system. With the publication of Darwin's *Origin of the Species* in 1859, it appeared that the unifying link between geology and biology had been found. Uniformitarian biologists and geologists agreed that one factor—a vast expanse of time—was an absolutely essential prerequisite for evolution. It could not be otherwise. Events which the Creator could accomplish in moments, days, or months would take eons of time if explained on the basis of natural processes observed today. The creation event was one of those special periods when the uniformity of physical law was superseded. Likewise, the fall of man and the worldwide flood marked other special periods, characterized by the miraculous intervention of the Creator.

With the exception of the Big Bang event, the theory of evolution excludes any deviation from the premise of complete uniformity of the fundamental laws of the universe throughout endless time—past, present, and future. This view has been accepted by more and more influential scholars in each succeeding generation. Today, the majority of society accepts that evolution is true, not by knowledge gained from independent study, but rather from books which have pictured evolution as the *only* scientifically credible explanation of earth history.

The challenge I have presented in this book to the *uniformitarian principle* includes evidences of an instantaneous creation and a young age of the earth. Thus the essential time element needed for the geological evolution of the earth as well as the biological evolution of life on it vanishes, and the entire evolutionary scenario is devastated.

These conclusions perplex many scientists, who for decades have been conditioned to accepting evidences for evolution based on the *uniformitarian principle.* They feel to depart from this cherished assumption would be equivalent to regressing in time to the period of the Dark Ages, when superstitions and traditions molded scientific theories. To avoid that extreme, they have presumed to shift their thinking 180° and have concluded all religious foundations are unscientific. Actually, their conclusions are based on a false premise. Instead of excluding all religious concepts from science, they are only assisting in the establishment of a new order, antithetical to biblical foundations. This new order—evolutionism—has spread to the Western world in the form of theistic evolution. Under the guise of science, it has found acceptance in academic institutions throughout civilized societies.

In view of these historical influences within academia, few scientists realize that the biblical record provides a broad, expansive framework of earth history, capable of incorporating an almost unlimited variety of geological data. Invariably, I have found that "arguments" and/or "problems" proposed against the biblical framework as a model of earth history are ultimately those which result from imposing unwarranted constraints. As mentioned before, the deliberate or unwitting acceptance of the *uniformitarian principle* is the most profound example of such constraints. There is no obstacle in correlating Earth's geologic history with the biblical record once it is understood that the Creator is not governed or restricted by that *principle.*

But those who accept uniformitarian concepts, such as a worldwide geologic column and its counterpart—radiometric dating—should never expect to find that correlation. Those holding such views often insist that they have found evidence contrary to the biblical record, yet at the same time they generally fail to mention that their evidence is based on uniformitarian assumptions. Thus, in the last analysis, they have only confirmed that the biblical record of creation and the flood cannot be reconciled with a uniformitarian geological framework. Perhaps they should reflect on the inspired words spoken to Job, "Where wast thou when I laid the foundations of the earth? Declare, if thou hast understanding." (Job 38:4)

Evidently, many scientists are willing to accommodate God into science, provided His presumed activities can be fitted into their evolutionary framework. However, when unambiguous scientific evidence is discovered, which is incompatible with evolution and can only be attributed to God's creative power, there is a different reaction within the scientific establishment. Now we have creation science—something the National Academy of Sciences says has been scientifically invalidated, and hence should not have a place in the science curriculum at any level. The Academy has a right to its opinion, but this book has shown that when the Academy was confronted with the opportunity to prove its claim about creation at the *International Conference on Creationism,* it signally failed to meet the challenge. Nothing could have more effectively unmasked the Academy's spurious claims about creation than did its deafening silence on this occasion. And nothing could have more clearly pinpointed its contradictory position on the *Affirmation of Freedom of Inquiry and Expression.* On one hand, the Academy uses the *Affirmation* to defend the academic and civil liberties of foreign dissidents. On the other hand, it promotes the exclusive teaching of evolution in public schools notwithstanding that, as definitely implied in Lane's letter (pp. 94-96), this practice has involved the persecution of some American "dissidents" — students who have the courage to stand for their religious convictions.

The Academy and others opposed to creation science should have realized long ago that for some Americans the imposed study of evolution is a moral issue. The philosophy of evolutionism directly contradicts their conviction that the literal six-day creation account given in Genesis, and explicitly reaffirmed in the Fourth of the Ten Commandments (partly quoted at the close of this Epilogue), represents the correct description of earth history. Again I say, this book has demonstrated that valid, scientific evidence exists to support this biblical creation model. Therefore to eliminate the present discriminatory practice in the classroom against those students opposed to evolution, why not allow all public school and state university students the option of studying either a creation or evolution-based model of origins in their science courses?

In my opinion, no one, evolutionist or creationist, should be forced into a course of study that violates his conscience. After all, the freedom to choose — as long as our choices do not infringe adversely on the rights of others — is the essence of our democracy. If we fail to uphold that freedom for public school students on this critical issue, we open the door for coercion — the unmistakable hallmark of totalitarian governments — to gain the ascendancy in all phases of American society. What is at stake is religious and academic freedom for all Americans. Should science education prohibit the teaching of certain evidence just because of its philosophical setting? Science is the knowledge obtained from a quest for truth and can be illustrated by the "Parable of the Grand Design":

> Long ago, a master artist conceived a mural which he wished to use as an illustration of the Grand Design of nature. Much time and effort were spent in completing this enormous task. Tragically, before it was unveiled,

an accident occurred, shattering the mirror-like mural into innumerable fragments throughout the face of the earth. Later philosophers became interested in reconstructing the Grand Design. Most were aware of the ancient outline left by the master artist, but many came to question the authenticity of the outline, choosing instead to construct their own version, based on the pieces they found in the earth. After many years, the consensus of wise men and philosophers was that they had developed the basic skeletal framework of the Grand Design, even though there were large gaps in the center and many pieces which could not be reconciled with the overall Design. Nevertheless they propagated this skeletal framework as absolute truth until governments and universities everywhere provided the funds they needed to continue their work. There were still a few remaining artists who believed that the ancient outline was the blueprint for the real Design used by the master artist. They carefully pointed out that all of the collected pieces would also fit into this ancient outline. And most importantly, the millions of recently discovered microscopic pieces, which did not fit into the skeletal framework, were found to perfectly fit into the ancient outline. Some were convinced that they should redirect their study and use the ancient outline as their model for the Grand Design, but the great majority never accepted its validity. Someday the truth would be evident to all, for the master artist promised to return and restore the magnificent Grand Design to its original beauty.

Until that day, which I believe is imminent, *Creation's Tiny Mystery* will stand as the Rock of Gibraltar against the tide of evolution.

Nearly 6,000 years ago the Ruler of the Universe engraved an indelible record of creation in the Genesis rocks of our planet just as He later inscribed the Ten Commandments on tables of stone at Mount Sinai, including the words,

> *"For in six days the Lord made heaven and earth, the sea, and all that in them is, and rested the seventh day . . ." (Exodus 20:11)*

In a single stroke, the Master Artist irrevocably blended the Genesis record of creation and the moral law into His Grand Design.

REFERENCES

Ager, D. 1981. *The Nature of the Stratigraphical Record.* New York: Wiley.

Aller, L.H. and McLaughlin, D.B. 1965. *Steller Structure.* Chicago: University of Chicago Press.

American Association for the Advancement of Science 1982. AAAS Resolution on Creation Science. *Science* 215, 1072.

American Geological Institute 1981. "AGI Statement on Organic Evolution." AGI News Release, November 5.

Battson, A. 1982. Videotape, *Confrontation: Creation/Evolution, Part IV.* Santa Barbara, CA: UCSB Television Services.

Breger, I.A. 1974. "Formation of Uranium Ore Deposits." *Proceedings of a Symposium,* Athens, May 6-10, p. 99. Vienna: International Atomic Energy Agency.

Clark, S. 1982. Letter to D. Bumpers, U.S. Senate, not dated. Little Rock: State of Arkansas, Office of the Attorney General.

Cochran, T. 1982. Senate Proceedings *Congressional Record* 128, S4307.

Cole, H.P., and Scott, E.C. 1982. "Creation-Science and Scientific Research." *Phi Delta Kappan,* April, 557.

Dalrymple, G.B. 1982. "Radiometric Dating and the Age of the Earth; A Reply to 'Scientific' Creationism." Talk presented at the AAAS Pacific Division meeting, June 22-23.

Dalrymple, G.B. 1984. "How Old Is the Earth? A Reply to 'Scientific' Creationism." *Proceedings of the 63rd Annual Meeting of the Pacific Division, American Association for the Advancement of Science* 1, 66.

Dalrymple, G.B. 1985. Letter to K.H. Wirth dated March 26, 1985. *Creation/Evolution Newsletter* 5, 12.

Damon, P.E. 1979. "Time: Measured Responses." *EOS Transactions of the American Geophysical Union* 60, 474.

Davies, P. 1981. *The Edge of Infinity.* New York: Simon & Schuster.

de Camp, S. 1968. *The Great Monkey Trial.* New York: Doubleday.

Dutch, S. 1983. Letters. *Physics Today* 36, No. 4, 11.

Eichelberger, J.D. et al. 1985. "Research Drilling at Inyo Domes, California; 1984 results." *EOS Transactions of the American Geophysical Union* 66, 186.

Feather, N. 1978. "The Unsolved Problem of the Po-haloes in Precambrian Biotite and Other Old Minerals." *Communications to the Royal Society of Edinburgh,* No. 11, 147.

Fezer, K.S. 1985. "Gentry's Pleochroic Halos." *Creation/Evolution Newsletter* 5, 12.

Frazier, K. 1978. "Superheavy Elements." *Science News* 113, 236.

Fremlin, J.H. 1975. "Spectacle Haloes." *Nature* 258, 269.

Geisler, N.L. 1982. *The Creator and the Courtroom*. Milford, MI: Mott Media.

Gentry, R.V. 1966a. "Abnormally Long Alpha-Particle Tracks in Biotite (Mica)." *Applied Physics Letters* 8, 65.

Gentry, R.V. 1966b. "Alpha Radioactivity of Unknown Origin and the Discovery of a New Pleochroic Halo." *Earth and Planetary Science Letters* 1, 453.

Gentry, R.V. 1966c. "Anti-matter Content of the Tunguska Meteor." *Nature* 211, 1071.

Gentry, R.V. 1967. "Extinct Radioactivity and the Discovery of a New Pleochroic Halo." *Nature* 213, 487.

Gentry, R.V. 1968. "Fossil Alpha-Recoil Analysis of Certain Variant Radioactive Halos." *Science* 160, 1228.

Gentry, R.V. 1970. "Giant Radioactive Halos: Indicators of Unknown Alpha-Radioactivity?" *Science* 169, 670.

Gentry, R.V. 1971a. "Radioactive Halos and the Lunar Environment." *Proceedings of the Second Lunar Science Conference* 1, 167. Cambridge: MIT Press.

Gentry, R.V. 1971b. "Radiohalos: Some Unique Pb Isotope Ratios and Unknown Alpha Radioactivity." *Science* 173, 727.

Gentry, R.V. 1973. "Radioactive Halos." *Annual Review of Nuclear Science* 23, 347.

Gentry, R.V. 1974. "Radiohalos in Radiochronological and Cosmological Perspective." *Science* 184, 62.

Gentry, R.V. 1975. Response to J.H. Fremlin's Comments on "Spectacle Haloes." *Nature* 258, 269.

Gentry, R.V. 1978a. "Are Any Unusual Radiohalos Evidence for SHE?" *International Symposium on Superheavy Elements, Lubbock, Texas*. New York: Pergamon Press.

Gentry, R.V. 1978b. "Implications on Unknown Radioactivity of Giant and Dwarf Haloes in Scandinavian Rocks." *Nature* 274, 457.

Gentry, R.V. 1979. "Time: Measured Responses." *EOS Transactions of the American Geophysical Union* 60, 474.

Gentry, R.V. 1980. "Polonium Halos." *EOS Transactions of the American Geophysical Union* 61, 514.

Gentry, R.V. 1982. Letters. *Physics Today* 35, No. 10, 13.

Gentry, R.V. 1983a. Letters. *Physics Today* 36, No. 4, 3.

Gentry, R.V. 1983b. Letters. *Physics Today* 36, No. 11, 124.

Gentry, R.V. 1984a. "Radioactive Halos in a Radiochronological and Cosmological Perspective." *Proceedings of the 63rd Annual Meeting of the Pacific Division. American Association for the Advancement of Science* 1, 38.

Gentry, R.V. 1984b. "Lead Retention in Zircons" (Technical Comment). *Science* 223, 835.

Gentry, R.V. 1984c. Letters. *Physics Today* 37, No. 4, 108.

Gentry, R.V. 1984d. Letters. *Physics Today* 37, No 12, 92.

Gentry, R.V. 1986. "Gentry Responds to Dalrymple's Letter to Kevin Wirth." See Appendix Contents of this book.

Gentry, R.V. et al. 1973. "Ion Microprobe Confirmation of Pb Isotope Ratios and Search for Isomer Precursors in Polonium Radiohalos." *Nature* 244, 282.

Gentry, R.V. et al. 1974. "'Spectacle' Array of ^{210}Po Halo Radiocentres in Biotite: A Nuclear Geophysical Enigma." *Nature* 252, 564.

Gentry, R.V. et al. 1976a. "Radiohalos and Coalified Wood: New Evidence Relating to the Time of Uranium Introduction and Coalification." *Science* 194, 315.

Gentry, R.V. et al. 1976b. "Evidence for Primordial Superheavy Elements." *Physical Review Letters* 37, 11.

Gentry, R.V. et al. 1982a. "Differential Lead Retention in Zircons: Implications for Nuclear Waste Containment." *Science* 216, 296.

Gentry, R.V. et al. 1982b. "Differential Helium Retention in Zircons: Implications for Nuclear Waste Containment." *Geophysical Research Letters* 9, 1129.

Gilbert, C. 1982. Letter to R.V. Gentry dated March 9, 1982. Washington, D.C.: *Science*.

Gilbert, C. 1985. Letter to D.R. Humphreys dated August 30, 1985. Washington, D.C.: *Science*.

Hammond, A. and Margulis, L. 1981. "Farewell to Newton, Einstein, Darwin . . ." *Science 81* 2, No. 10, 55.

Harwit, M. 1986. Book Review. *Science* 231, 1201.

Hashemi-Nezhad, S.R. et al. 1979. "Polonium Haloes in Mica." *Nature* 278, 333.

Heffelfinger, W.S. 1982. Letter to U.S. Senator J. Sasser dated June 14, 1982. Washington D.C.: Department of Energy.

Howe, R.A. 1982. House of Representatives Proceedings. *Congressional Record* 128, H1653.

Hower, J. 1977. Letter to R.V. Gentry dated July 11, 1977. Washington, D.C.: National Science Foundation.

Jedwab, J. 1966. "Significance and Use of Optical Phenomena in Uraniferous Caustobioliths." *Coal Science* (Editor, P. Given). Washington, D.C.: American Chemical Society.

Johnson, F.S. 1982. Letter to R.S. Walker, House of Representatives, dated June 17, 1982. Washington, D.C.: National Science Foundation.

Johnson, F.S. 1983. Letter to R.J. Lagomarsino, House of Representatives, dated February 14, 1983. Washington, D.C.: National Science Foundation.

Joly, J. 1923. *Proceedings of the Royal Society,* London, Series A 102, 682.

Kazmann, R.G. 1978. "It's About Time: 4.5 Billion Years." *Geotimes* 23, 18.

Kazmann, R.G. 1979. "Time: In Full Measure." *EOS Transactions of the American Geophysical Union 60*, 21.

Kitcher, P. 1982. *Abusing Science.* Cambridge: The MIT Press.

Lane, J.W. 1982. Letters. *Physics Today* 35, No. 10, 15.

La Grone, J. 1984. Letter to J.H. Quillen dated September 4, 1984. Oak Ridge: Department of Energy.

Larsen, J. 1985. "From Lignin to Coal in a Year." *Nature* 314, 316.

Lewin, R. 1981. "A Response to Creationism Evolves." *Science* 214, 635.

Lewin, R. 1982a. "Creationism on the Defensive in Arkansas." *Science* 215, 33.

Lewin, R. 1982b. "Where Is the Science in Creation Science?" *Science* 215, 142.

Lewin, R. 1982c. "Recent Advances in Our Understanding of the Mechanisms of Evolution." *Bulletin American Physical Society* 27, 464.

Lewin, R. 1982d. *Thread of Life, The Smithsonian Looks at Evolution.* Washington, D.C.: Smithsonian Books.

Lewin, R. 1985. "Evidence for Scientific Creationism?" *Science* 228, 837.

Meier, H., and Hecker, W. 1976. "Radioactive Halos as Possible Indicators of Geochemical Processes in Magmatites." *Geochemical Journal* 10, 185.

Melnick, J. 1981. "Polonium Radiohalos & the Case of Dr. Robert V. Gentry." *Christian Citizen* (August 1981) 5; Reprinted as "The Case of the Polonium Radiohalos," in *Origins Research* 5, No. 1 (1982).

Menton, D.N. 1985. "'Inherit the Wind': A Hollywood History of the Scopes Trial." *Bible-Science Newsletter* 23, No. 1.

Merkel, P. 1981. Audio Tape of Robert V. Gentry's Testimony. McLean vs. Arkansas State Board of Education. Little Rock: Official Court Reporter, U.S. District Court.

Moazed, C. et al. 1973. "Polonium Radiohalos: an Alternate Interpretation." *Science* 180, 1272.

National Academy of Sciences 1984. *Science and Creationism.* Washington, D.C.: National Academy Press.

Osmon, P. 1986. Commentary on "Gentry's Pleochroic Halos." *Creation/Evolution Newsletter* 6, 17.

Overton, W. 1982. *Memorandum Opinion.* Little Rock: U.S. District Court.

Raloff, J. 1982a. "They Call It Creation Science." *Science News* 121, 44.

Raloff, J. 1982b. "Radwaste Solutions Pivot on Politics." *Science News* 121, 296.

Roth, A. 1984. "Is Creation Scientific?" *Origins* 11, 64.

Sasser, J. 1982a. Letter to W.S. Heffelfinger, Department of Energy, dated May 18, 1982. Washington, D.C.: U.S. Senate.

Sasser, J. 1982b. Letter to R.V. Gentry dated June 16, 1982. Washington, D.C.: U.S. Senate.

Science News editorial 1981. "Evolution at the AAAS." *Science News* 119, 19.

Scopes, J.T., and Presley, J. 1967. *Center of the Storm.* New York: Holt, Rinehart and Winston.

Scott, E.C., and Cole, H.P. 1985. "The Elusive Scientific Basis of Creation 'Science.'" *The Quarterly Review of Biology* 60, 21.

Sinclair, R.M. 1981. Creation-evolution Letters. *Science News* 119, 67.

Skow, J. 1981. "The Genesis of Equal Time." *Science 81* 2, No. 10, 54.

Smith, S. 1982a. Testimony of Harold Morowitz. McLean vs. Arkansas State Board of Education. Little Rock: Official Court Reporter, U.S. District Court.

Smith, S. 1982b. Testimony of Gary B. Dalrymple. McLean vs. Arkansas State Board of Education. Little Rock: Official Court Reporter, U.S. District Court.

Sparks, C.J., Jr. et al. 1977. "Search with Synchrotron Radiation for Superheavy Elements in Giant-Halo Inclusions." *Physical Review Letters* 38, 205.

Sparks, C.J., Jr. et al. 1978. "Evidence against Superheavy Elements in Giant-Halo Inclusions Re-examined with Synchrotron Radiation." *Physical Review Letters* 40, 507.

Stieff, L.R. et al. 1953. "A Preliminary Determination of the Age of Some Uranium Ores of the Colorado Plateaus by the Lead-Uranium Method." *U.S. Geological Survey Circular 271.*

Stutzer, O. 1940. *Geology of Coal,* translated by A.C. Noe. Chicago: University of Chicago Press.

Talbott, S.L. 1977. "Mystery of the Radiohalos." *Research Communications Network,* Newsletter Number 2.

Todd, E.P. 1977. Letter to R.V. Gentry dated September 15, 1977. Washington, D.C.: National Science Foundation.

York, D. 1979. "Polonium Halos and Geochronology." *EOS Transactions of the American Geophysical Union* 60, 617.

CREDITS

· *American Association for the Advancement of Science* (AAAS). "Fossil Alpha-Recoil Analysis of Certain Variant Radioactive Halos," *Science* Vol. 160, pp. 1228-1230, June 14, 1968, by Robert V. Gentry. Copyright © 1968 by the AAAS. "Giant Radioactive Halos: Indicators of Unknown Radioactivity?", *Science* Vol. 169, pp. 670-673, August 14, 1970, by Robert V. Gentry. Copyright © 1970 by AAAS. "Radiohalos in a Radiochronological and Cosmological Perspective," *Science* Vol. 184, pp. 62-66, Photo and Table I, April 5, 1974, by Robert V. Gentry. Copyright © 1974 by AAAS. "Radiohalos in Coalified Wood: New Evidence Relating to the Time of Uranium Introduction and Coalification," *Science* Vol. 194, pp. 315-317, Photos, October 15, 1976, by Robert V. Gentry et al. Copyright 1976 by AAAS. "A Response to Creationism Evolves," *Science* Vol. 214, pp. 635-636, 638, November 6, 1981, by R. Lewin. Copyright © 1981 by AAAS. "Creationism on the Defensive in Arkansas," *Science* Vol. 215, pp. 33-34, January 1, 1982, by R. Lewin. Copyright © 1982 by AAAS. "Where Is the Science in Creation Science?," *Science* Vol. 215, pp. 142-144, 146, January 8, 1982, by R. Lewin. Copyright © 1982 by AAAS. "Differential Lead Retention in Zircons: Implications for Nuclear Waste Containment," *Science* Vol. 216, pp. 296-298, April 16, 1982, by Robert V. Gentry et al. Copyright © 1982 by AAAS. Review of "Naissance et Enfance des Etoiles," *Science* Vol. 231, pp. 1201, March 7, 1986, by Martin Harwit. Copyright © 1986 by AAAS. All of the above articles were excerpted or reprinted by permission of the AAAS.

American Association for the Advancement of Science. "Farewell to Newton, Einstein, Darwin . . ." by Allen Hammond and Lynn Margulis, and "The Genesis of Equal Time" by John Skow, *Science 81,* December, pp. 54-60. Copyright © 1981 by the AAAS. Excerpted by permission of *Science 84* Magazine.

Pacific Division of the American Association for the Advancement of Science. "Radioactive Halos in a Radiochronological and Cosmological Perspective," Proceedings of the 63rd Annual Meeting of the Pacific Division, AAAS, *Evolutionists Confront Creationists,* Vol. 1, Part 3, pp. 38-65, April 30, 1984, by Robert V. Gentry. Copyright © 1984 by the Pacific Division of the AAAS. Reprinted by permission of the Pacific Division of the AAAS.

American Geophysical Union. "Differential Helium Retention in Zircons: Implications for Nuclear Waste Containment," *Geophysical Research Letters,* Vol. 9, pp. 1129-1130, October 1982, by Robert V. Gentry et al. Copyright © 1982 by the American Geophysical Union. Used by permission of the American Geophysical Union.

American Institute of Physics. Letters, *Physics Today* Vol. 35, October 1982, pp. 15, 103 by J. Willits Lane. Copyright © 1982 by the American Institute of Physics. Reprinted by permission of J. Willits Lane and the American Institute of Physics.

Annual Reviews, Inc. "Radioactive Halos," *Annual Review of Nuclear Science* Vol. 23, pp. 347-362, by Robert V. Gentry. Copyright © 1973 by Annual Reviews, Inc. Reproduced, with permission, from the Annual Review of Nuclear Science, Annual Reviews, Inc.

Doubleday & Company, Inc. The Great Monkey Trial, p. 432, by L. Sprague de Camp. Copyright © 1968 by Doubleday & Co., Inc. Used by permission of Barthold Fles Literary Agency, New York.

Loma Linda University, University Relations, Loma Linda, CA. "Evolution Model" graphic adapted from drawing by Glenn Thomas. Used by permission of University Relations, Loma Linda University.

Macmillan Journals Limited. "Ion Microprobe Confirmation of Pb Isotope Ratios and Search for Isomer Precursors in Polonium Radiohalos," *Nature* Vol. 244, No. 5414, pp. 282-283, August 3, 1973, by Robert V. Gentry et al. Copyright © 1973 by Macmillan Journals Ltd. "'Spectacle' Array of ^{210}Po Halo Radiocentres in Biotite: A Nuclear Geophysical Enigma." *Nature* Vol. 252, No. 5484, pp. 564-566, December 13, 1974, by Robert V. Gentry et al. Copyright © 1974 by Macmillan Journals Ltd. These articles were reprinted by permission from *Nature.*

National Academy Press. Science and Creation: A View from the National Academy of Sciences by the Committee on Science and Creationism, National Academy of Sciences. Copyright © 1984 by the National Academy of Sciences. Used by permission of the National Academy Press.

Research Communications Network. "*Mystery of the Radiohalos,*" *Research Communications Network* Newsletter #2, February 10, 1977, by Stephen L. Talbott. Copyright (1977) by Research Communications Network. Used by permission of Stephen L. Talbott.

Science Service, Inc. "Superheavy Elements," *Science News* Vol. 113, pp. 126-238, April 15, 1978, by Kendrick Frazier. Copyright © 1978 by Science Service, Inc. "Evolution at the AAAS," *Science News* Vol. 119, p. 19, January 10, 1981. Copyright © 1981 by Science Service, Inc. "They Call It Creation Science," *Science News* Vol 121, No. 3, pp. 44-45, January 16, 1982, by Janet Raloff. Copyright © 1982 by Science Service, Inc. The above articles are reprinted with permission from *Science News,* the weekly news magazine of science.

Stony Brook Foundation, Inc. "The Elusive Scientific Basis of 'Creation' Science," *The Quarterly Review of Biology*, Vol. 60, No. 1, pp. 21-30, March 1985, by Eugenie C. Scott and Henry P. Cole. Copyright © 1985 by Stony Brook Foundation, Inc. Excerpted by permission of *The Quarterly Review of Biology.*

University of California—Santa Barbara, Television Services. "Creation Model" graphic adapted from videotape, *Confrontation: Creation/Evolution*, Part IV. Used by permission of Television Services, UCSB.

Radiohalo Catalogue

Contents

Approximate Magnification

	(a)	(b)	(c)	(d)	(e)	(f)	(g)	(h)
1	× 900	× 1015	× 1120	× 1015				
2	× 540	× 875	× 875	× 770				
3	× 800	× 800	× 800	× 865				
4	× 930	× 730	× 965	× 985				
5	× 500	× 500	× 500	× 500				
6	× 725	× 725	× 880	× 725	× 725	× 725	× 725	× 725
7	× 725	× 725	× 725	× 860	× 725	× 725	× 725	× 725
8	× 725	× 725	× 725	× 725	× 725	× 725	× 725	× 725
9	× 245	× 575	× 165	× 265				
10	× 420	× 250	× 500	× 250	× 275	× 140	× 275	× 260
11	*	*	× 300	*				

* Scale shown in figure

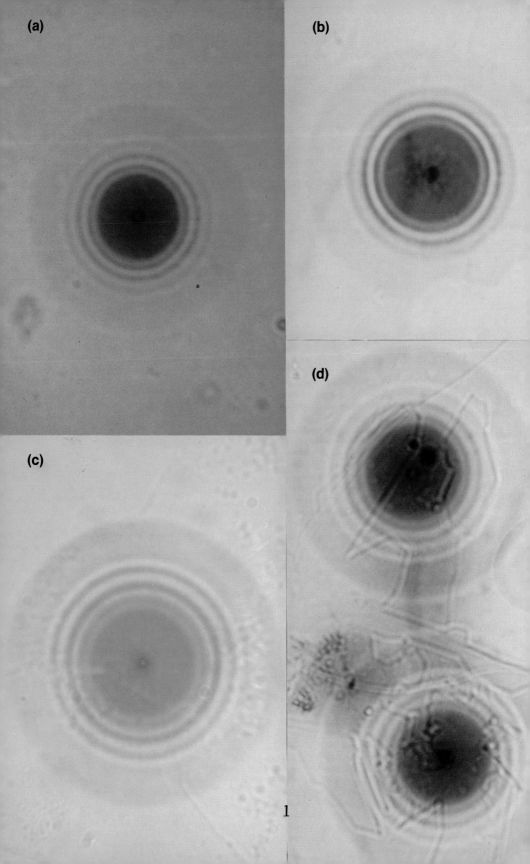

1

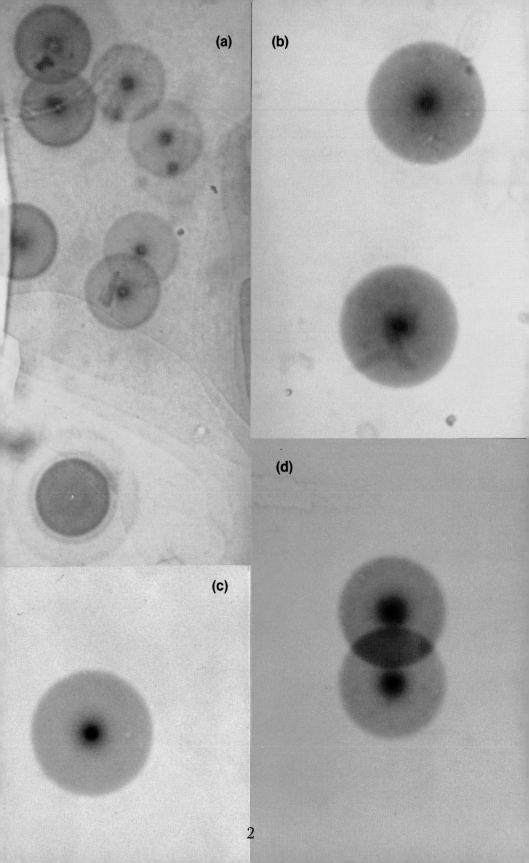

(a)

(b)

(d)

(c)

2

(a)

(b)

(c)

(d)

3

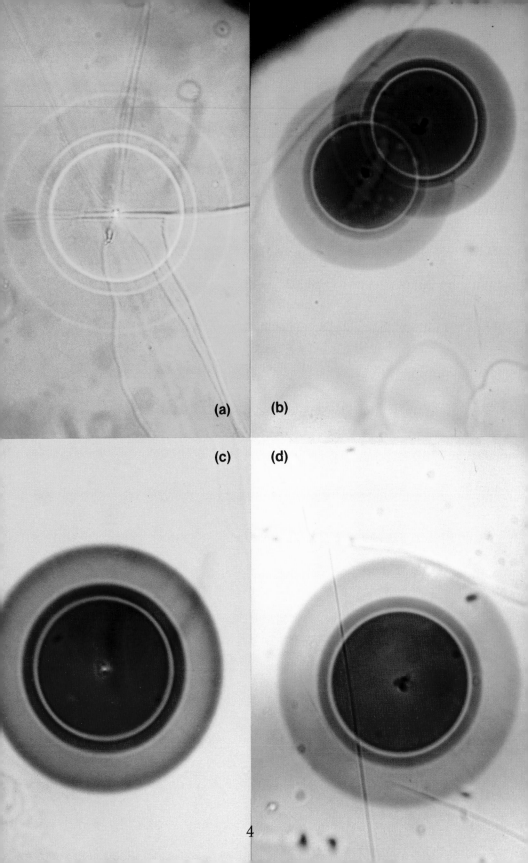

(a)　(b)

(c)　(d)

4

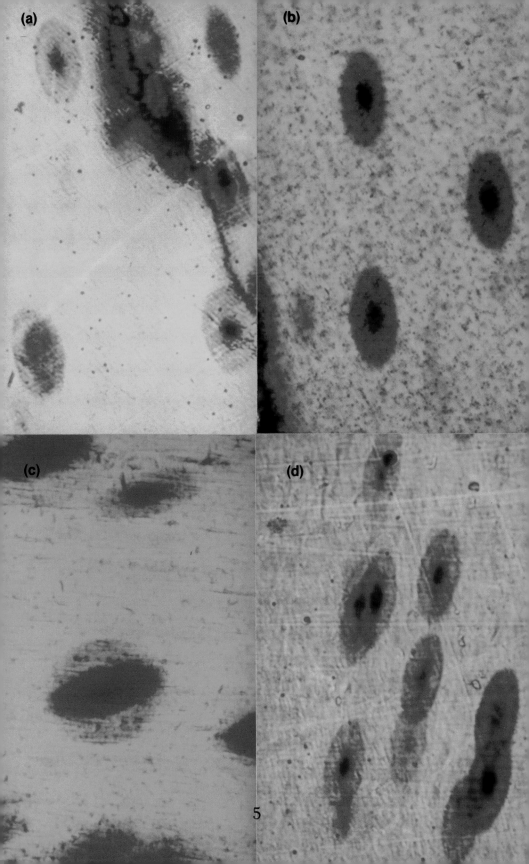

5

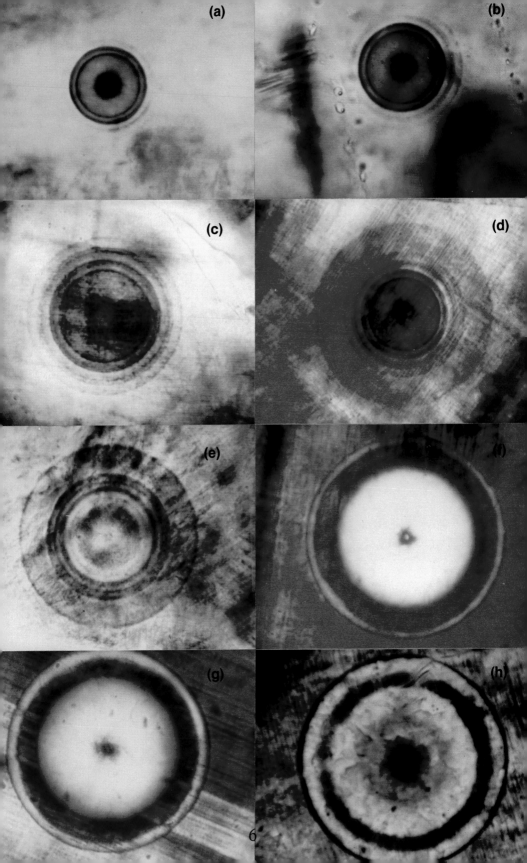

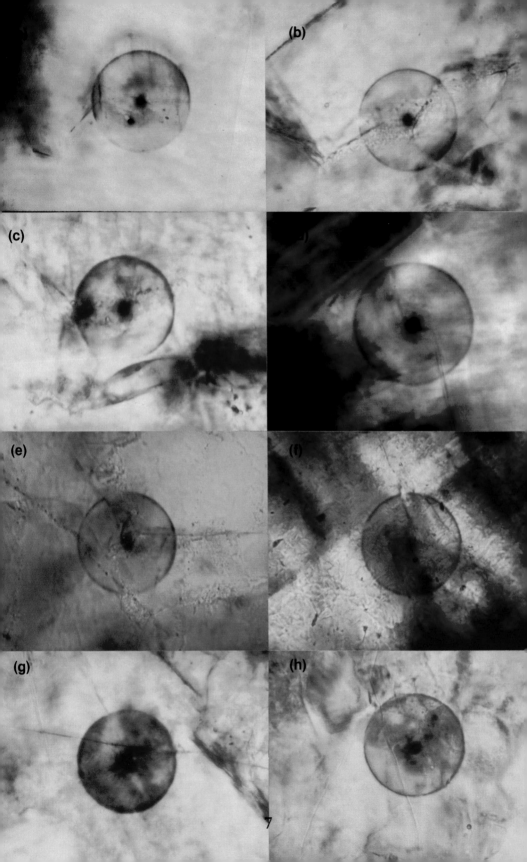

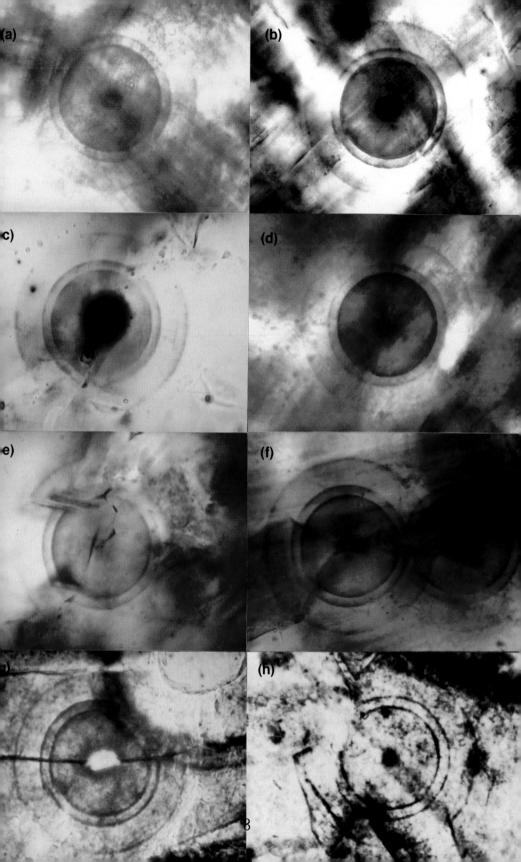

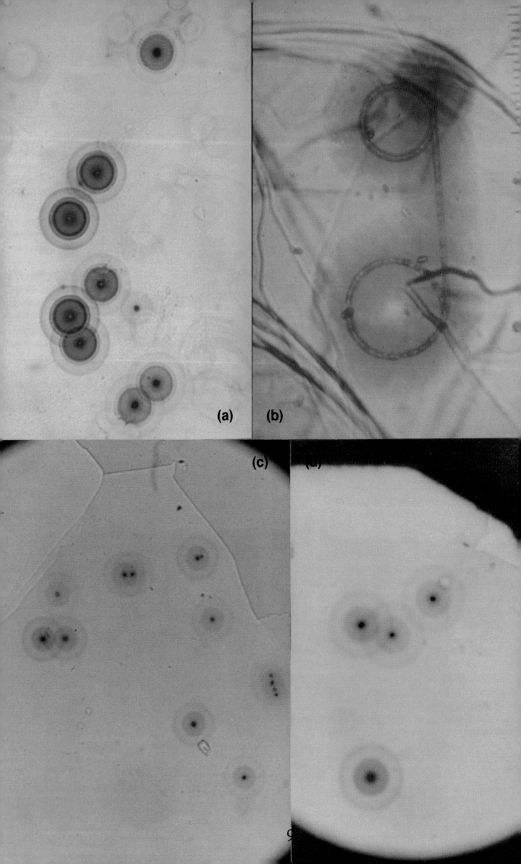

(a)

(b)

(c)

(d)

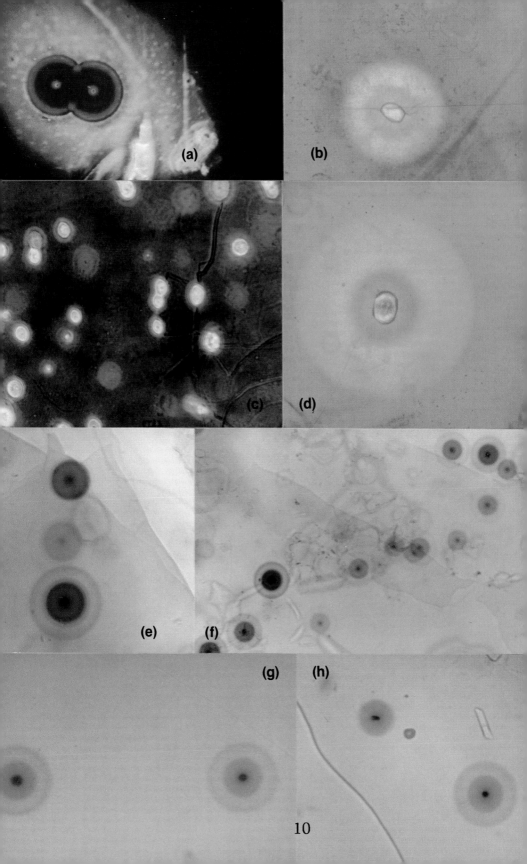

(a)　(b)

(c)　(d)

APPENDIX

Contents *

296   Dalrymple, G.B. 1985. Letter to K.H. Wirth dated March 26, 1985. *Creation/Evolution Newsletter 5, 12.*

299   Gentry, R.V. 1986. Gentry Responds to Dalrymple's Letter to Kevin Wirth.

304   Gilbert, C. 1985. Letter to D.R. Humphreys dated August 30, 1985. Washington, D.C.: *Science.*

305   Merkel, P. 1981. Transcription of Robert V. Gentry's Cross-Examination from Audio Tape. McLean vs. Arkansas State Board of Education. Little Rock: Tape recorded by P. Merkel, Official Court Reporter, U.S. District Court.

*Note: Official documents are listed according to first citation in book. Audio tape transcription listed at end.

Fossil Alpha-Recoil Analysis of Certain Variant Radioactive Halos

Abstract. *The distribution of alpha-radioactivity in the vicinity of uranium and of certain variant radioactive halos in biotite was investigated by the fossil alpha-recoil method. Within the limits of the method I could not confirm a previously proposed hydrothermal mechanism for the origin of certain variant halo types due to polonium isotopes.*

Microscopic examination of thin sections ($\approx 20\ \mu$) of certain minerals sometimes reveals a distinctive pattern of colored concentric rings surrounding a minute central inclusion about 0.5 to 1 μ in radius. Although these structures had long been observed by mineralogists, their origin was a mystery until almost simultaneously Joly (*1*) and Mugge (*2*) correctly attributed the phenomenon to the presence of radioactivity in the central inclusion. While in some instances the inclusions have been identified as zircon (*1, 3*), xenotime, or monazite (*4*), the halo nuclei are often too small for petrologic analysis.

In polarized light, the appearance of the varicolored ring patterns in such anisotropic minerals suggested the designation "pleochroic halos," although "radioactive halos" is clearly more appropriate. While the radioactivity in the central inclusion may consist of α-, β-, and γ-radiation, the development of a halo is basically due only to the proportionately much higher ionization effects of the α-particles. This is an extremely fortuitous situation because, since the α-particle has a rather precise range R in a mineral for a given initial energy E, one can often ascertain not only the elements responsible for a particular halo type but also the specific isotopes. If the halo nucleus contains uranium, the α-emission from the eight α-emitters in the decay chain produces a region of radiation damage surrounding the inclusion. In certain biotites this region becomes faintly visible when about 10^8 atoms of ^{238}U have decayed; with increased α-emission a series of colored, spherically concentric shells eventually appears, corresponding to the ranges of the respective α-emitters of the ^{238}U decay chain. The three-dimensional nature of the halo becomes strikingly apparent when a sample of biotite is prepared for microscopy. The leaves of a book of mica are easily cleaved

with transparent cellophane tape, and each successive layer of mica reveals a ring pattern of increasing size until the diametral section is obtained. Years ago there was great interest in the ring structure of uranium and thorium halos in investigation of the invariance of the radioactive transformation rate over geological time (*5*). It is in this connection that radioactive halos have again drawn interest (*6*).

Naturally ring sizes are always measured from diametral sections; results are best from specimens having exceptionally small nuclei. Use of a filar micrometer shows the ring radii for the uranium and thorium halos to agree very well with the calculated α-particle ranges of ^{238}U and ^{232}Th and their respective α-emitters. Thus an experimental range:energy relation for α-particles may be determined for any mineral containing well-defined uranium or thorium halos, with small central inclusions.

Certain types of halos (I call them variant halos) exist that cannot be identified with the ring structure of either the uranium or thorium halos. What is the nature of the α-emitters responsible for these variant halos? Several types of variant halos were discovered but were not claimed to be evidence of new α-emitters because radioactive-decay schemes of uranium and thorium were still being refined. Nevertheless Joly (*7*) reported three variant halo types: one he attributed to "emanation" (^{222}Rn), a dwarf having a very small radius; another was simply designated the X-halo. Others (*8–10*) have reported unusual halo sizes, and I have found halos having anomalous ring structure (*11, 12*). For greater clarification of the variant halos, I classify as class I those rather easily identifiable with known α-emitters; as class II those (such as Joly's X-halo) whose ring structure has not been correlated with known α-emitters. For example, Henderson reported four variant

halo types: A, B, C, and D. Types A, B, and C were correctly attributed to the polonium isotopes ^{210}Po, ^{214}Po, and ^{218}Po, respectively; thus they are of class I. But I have been unable to confirm Henderson's association of the D-halo with ^{226}Ra (*13*). I confine this report to investigation of class-I halos —in particular to analysis of Henderson's proposed origin of the polonium halos.

The polonium isotopes have relatively short half-lives; any mechanism proposed for their origin must be consistent with this fact. The ^{218}Po halo (Fig. 1, left), so-called because ^{218}Po is the initiating isotope, exhibits three rings arising from successive α-decay of ^{218}Po (E_1, 6.0 Mev; r_1, 23 μ), ^{214}Po (E_2, 7.68 Mev; r_2, 34 μ), and ^{210}Po (E_3, 5.3 Mev; r_3, 19 μ). E_i and r_i denote, respectively, the α-particle kinetic energy and the corresponding average halo-ring radius. By analogy the ^{214}Po and ^{210}Po halos (Fig. 1, right) are, respectively, dual and single ring patterns. I have observed the polonium halos in many Precambrian biotites, and the halos in Fig. 1 were found in biotites from the Baltic (Norway) and Canadian shields, respectively. Since these polonium isotopes are daughter products of ^{238}U, it was initially conceived (*10*) that they were preferentially fixed out of uranium-bearing solutions at localized deposition centers along small conduits or veins within the host mineral (mica, for example).

While coloration surrounding minute veins in the mica is an indication of the flow of radioactive solutions (very weak solutions may show no staining whatsoever), it does not follow that halos that formed around small nuclei in the conduits were necessarily derived from radioactivity in solution. For example, polonium, uranium, and thorium halos also form around very small inclusions, with no visible conduit or crack in the mica connecting the halo nuclei, and it is certainly not clear that these halos are of hydrothermal origin.

An attempt to determine whether the halo nuclei were capable of acting as selective fixation sites for certain radionuclides, by electron-microprobe analysis of the halo inclusions, failed because of the small size involved. However, refinement of techniques may lead to clarification of the nature of the inclusions (*14*). Thus a more sensitive technique is required for testing of the hypothesis regarding genesis of the polonium halos from a uranium-bearing solution.

Fission-track techniques (*15*) may serve this purpose. Uranium-238 fissions spontaneously, and the damaged regions in the host mineral, produced by the fission fragments, can be enlarged sufficiently by acid etching for visibility under an optical microscope. Immersion of biotite samples, containing the polonium and uranium halos in hydrofluoric acid for a few seconds and subsequent observation of the areas in the vicinity of the inclusions reveal a striking difference: the polonium halos are characterized by complete absence of fission tracks, whereas the uranium halos always show clusters of fission tracks.

To eliminate the possibility that fission tracks may have been annealed out of the sample, I have irradiated mica specimens containing the uranium and polonium halos with a neutron flux of 5×10^{17} neutrons per square centimeter and again etched the mica. The uranium halos, as expected, show marked increase in the number of fission tracks emanating from the central inclusion, due to neutron-induced ^{235}U fission, whereas the polonium halos are again completely devoid of tracks (*12*).

If a uranium solution had been in a conduit feeding the central inclusions of the polonium halos with daughter-product activity, about 70 fission tracks per centimeter of conduit would be expected by use of Henderson's model (*10*). This result depends on such parameters as the uranium concentration in the solution, the rate of flow (conservatively I have assumed that the solution ceased to flow when the polonium halos formed), and the total number of polonium atoms (5×10^8) necessary to form a well-developed ^{218}Po halo. This last value I determined by observing the degree of coloration in uranium halos as a function of the number of fission tracks emanating from the halo nucleus, the total number of α-particles required for production of a halo being computed as eight times the number of fission tracks times the ratio of the half-lives for spontaneous fission and alpha decay for ^{238}U. While fission tracks are observed along stained conduits, in general I cannot correlate the distribution of fission tracks along clear conduits with the presence of polonium halos.

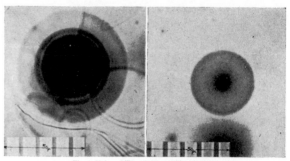

Fig. 1. Halos of ^{218}Po (left) and ^{210}Po (right).

Polonium halos are also found randomly distributed throughout the interior of large mica crystals far removed from any conduit. (A limited survey may indicate halos occurring within certain cleavage planes, but more extensive search shows this is not the case.) The question now arises of whether the source of the short-half-life radioactivity, characteristic of such polonium halos, was due to (i) the laminar flow of a non-uranium-bearing solution, containing disequilibrium amounts of daughter-product α-activity, through a thin cleft parallel to the cleavage plane, or (ii) the diffusion of gaseous radon through the mica. The latter case has been considered (*8*), but only recently has the discovery of α-recoil tracks in micas (*16*) enabled quantitative checking of either of these mechanisms. This technique is based on the fact that an atom recoiling from α-emission impinges on the host mineral and forms a damaged region large enough to produce a pit which is visible in phase contrast when etched with hydrofluoric acid.

The original experiment (*16*) determined that a series of multiple recoils, such as is expected in the sequential α-decay of ^{238}U and ^{232}Th, yields α-recoil .tracks. Two additional points necessary for a complete α-recoil analysis—(i) whether a single α-recoil produces a track, and (ii) whether α-recoil pits form in a sample placed in contact with an α-emitter—have now been resolved.

Several samples of mica were annealed for removal of background α-recoil pits; three different concentrations of dilute solutions of americium (5 percent ^{241}Am and 95 percent ^{243}Am) were evaporated on separate samples, and an α-count was taken. The daughter

products of the americium isotopes have very long half-lives, so that any α-recoil pits occurring reflect only single α-decay. The higher α-count samples yielded correspondingly higher α-recoil densities within the area of deposition, accompanied by almost complete absence of tracks outside the radioactive zone. Thus was established the existence of one α-track from a single α-recoil (*17*).

Corresponding α-recoil densities were also noted in annealed mica samples placed in contact with the americium-coated samples. It follows that any excess α-radioactivity in micas may be effectively determned by analysis of the samples by the α-recoil technique.

The procedure for ascertaining the extent of increased α-activity consists in measuring background fossil α-recoil track densities in areas far removed from the halos themselves, and in comparing these values with the densities near the halos for determination of the degree of excess α-activity. Samples of Precambrian mica from Canada and Ireland (*18*), containing uranium and polonium halos, were investigated by etching in 48 percent hydrofluoric acid for about 15 to 50 seconds. As in earlier experiments, ^{238}U halos revealed the presence of fission tracks emanating from the central inclusions, whereas no fission tracks were noted from the central inclusions of the polonium halos.

The experimental procedure was to photograph in phase contrast a given etched area, enlarge, and count anywhere from several hundred to 1000 α-recoil centers for each density measurement. The enlargement factor was determined by photographing the rulings of a stage micrometer, using each objective. Replicate measurements were made on several areas with different

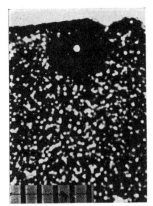

Fig. 2. Fossil α-recoil centers in the vicinity of a ²¹⁰Po halo (phase contrast).

halo types. The background fossil α-recoil density was measured before a count was made in the mica cleavage plane about 5 to 10 μ directly above the halo nucleus. The mica was then cleaved until the central inclusion appeared on the surface; the mica was etched again and another count was made to enable a density comparison of three separate regions.

The mean fossil α-recoil densities were 12.7 × 10⁶ and 11.6 × 10⁶ α/cm² for the Canadian and Irish micas, respectively, regardless of where the α-recoil count was taken. For a given etch period these results are reproducible within ± 10 percent. The fission-track density exhibited a random distribution in each piece of mica except (as expected) near the ²³⁸U halos. The α-recoil:fission-track ratios were about 2.5 × 10³ and 3.0 × 10³, respectively, for the Canadian and Irish micas. Huang and Walker (16) have shown that the background α-recoil density in micas is due to both uranium and thorium α-decay; by using 100 Å and 10 μ for the alpha-recoil and fission-track ranges, respectively, one can determine that uranium alone contributes an α-recoil:fission-track ratio of about 2.2 ×

10³, any excess being due to thorium. Figure 2 portrays a ²¹⁰Po halo (Irish mica) showing the distribution of α-radioactivity (fossil α-recoil centers) in the vicinity.

As far as the experimental analysis is concerned, there is no detectable difference in the microscopic distribution of α-radioactivity (with respect to background density) near either the uranium or the polonium halos. [I note that thin clefts, which usually result near the edges of the mica from weathering (but not within the bulk of the mica), are easily detected by an acid etch since α-recoil tracks appear throughout the extent of the cleft area.] This finding seems to imply that there was no gross transport of α-radioactivity to the polonium-halo inclusions (i) by way of laminar flow of solutions (through thin clefts) disequilibrated as to uranium daughter-product activity, or (ii) by diffusion of radon, since an increased α-recoil density, higher than background by several orders of magnitude, should be evident within a 10-μ radius of the halo inclusions in either case. This last value is a conservative estimate, for I have considered only the decay of ²¹⁸Po atoms en route to an inclusion. Furthermore, autoradiographic experiments on the samples of Canadian mica containing ²³⁸U, ²³²Th, and polonium halos showed only the normal background distribution of α-tracks, indicating that if excess activity now exists it is below the detection level of the method.

Thus, as far as the experimental analysis is concerned, I cannot confirm Henderson's model for the secondary origin of the polonium halos. To the question of what mode of origin is consistent with the relatively short half-lives of the polonium isotopes (or their β-decaying precursors), I can say only that other mechanisms are under study.

Whatever hypothesis is invoked, to explain the origin of the polonium halos, must also explain both the one found by Henderson (19) [due to a combination of isotopes from both the thorium series (²¹²Po and ²¹²Bi) and the uranium series (²¹⁰Po)] and a halo presumably due to ²¹¹Bi (12) from the ²³⁵U series.

Perhaps most interesting of all is the occurrence of 20,000 to 30,000 ²¹⁸Po and ²¹⁰Po halos per cubic centimeter in a Norwegian mica—without the ²¹⁴Po halos.

ROBERT V. GENTRY
*Institute of Planetary Science,
Columbia Union College,
Takoma Park, Maryland 20012*

References and Notes

1. J. Joly, *Phil. Mag.* **13**, 381 (1907).
2. O. Mugge, *Zentr. Mineral.* **1907**, 397 (1907) (see Oak Ridge National Laboratory *ORNL-tr-757*).
3. J. Joly, *Phil. Trans. Roy. Soc. London Ser. A* **217**, 51 (1917); P. Ramdohr, *Geol. Rundschau* **49**, 253 (1960) (see *ORNL-tr-758*).
4. C. O. Hutton, *Amer. J. Sci.* **245**, 154 (1947).
5. J. Joly, *Nature* **109**, 480 (1922); F. Lotze, *ibid.* **121**, 90 (1928).
6. G. Gamow, *Phys. Rev. Letters* **19**, 759 (1967).
7. J. Joly, *Proc. Roy. Soc. London Ser. A* **102**, 682 (1923).
8. S. Iimori and J. Yoshimura, *Sci. Papers Inst. Phys. Chem. Res.* **5**, 11 (1926); A. Schilling, *Neues Jahrb. Mineral. Abhandl.* **53A**, 241 (1926) (see *ORNL-tr-697*).
9. J. S. van der Lingen, *Zentr. Mineral. Abt. A* **1926**, 177 (1926) (see *ORNL-tr-699*); C. Mahadevan, *Indian J. Phys.* **1**, 445 (1927); H. Hirschi, *Vierteljahrsschr. Naturforsch. Ges. Zuerich* **65**, 209 (1920) (see *ORNL-tr-702*); E. Wiman, *Bull. Geol. Inst. Univ. Uppsala* **23**, 1 (1930); G. H. Henderson, *Proc. Roy. Soc. London Ser. A* **173**, 238 (1939).
10. G. H. Henderson, *Proc. Roy. Soc. London Ser. A* **173**, 250 (1939).
11. R. V. Gentry, *Appl. Phys. Letters* **8**, 65 (1966); *Earth Planetary Sci. Letters* **1**, 453 (1966).
12. ———, *Nature* **213**, 487 (1967).
13. Observations on this and other class-II halos will be reported.
14. I thank Larry Kobren, Goddard Space Flight Center, for the electron-microprobe analysis. Also I thank Truman Kohman, Carnegie-Mellon University, for suggesting the microprobe experiments and for valuable discussions concerning the origin of the halos.
15. R. L. Fleischer, P. B. Price, R. M. Walker, *Science* **149**, 383 (1965).
16. W. H. Huang and R. M. Walker, *ibid.* **155**, 1103 (1967).
17. J. Boyle and R. V. Gentry, in preparation.
18. G. H. Henderson, *Proc. Roy. Soc. London Ser. A* **145**, 591 (1934).
19. I thank G. C. Milligan and other members of the geology and physics departments of Dalhousie University, Halifax, for the loan of Henderson's halos and microphotographs. The halo referred to is in this collection.
20. I thank Paul Ramdohr, University of Heidelberg, for this particular specimen. Also I thank R. R. Gorbatschev (Uppsala), B. Loberg (Stockholm), D. E. Kerr-Lawson (Swastika, Ontario), J. H. J. Poole (Trinity College), and J. A. Mandarino (Royal Ontario Museum) for other mica specimens containing halos. I also thank H. L. Price for assisting in the α-recoil analysis and John Boyle, Oak Ridge National Laboratory, for the α-recoil experiments. For more extensive investigation it would appreciate contributions of samples of biotite from as many Precambrian localities as possible.

26 April 1968

Giant Radioactive Halos: Indicators of Unknown Radioactivity?

Abstract. *A new group of giant radioactive halos has been found with radii in excess of anything previously discovered. Since alternate explanations for these giant halos are inconclusive at present, the possibility is considered that they originate with unknown alpha radioactivity, either from isomers of known elements or from superheavy elements.*

A radioactive halo is generally defined as any type of discolored, radiation-damaged region within a mineral and usually results from either alpha or, more rarely, beta emission from a nearby radioactive inclusion containing either uranium or thorium. When the inclusions are very small (≈ 1 μm), the uranium and thorium daughter alpha emitters produce a series of discolored concentric spheres surrounding the inclusion, which in thin section appear microscopically as concentric rings whose radii correspond to the ranges of the respective alpha emitters (1). Although the radii of normal uranium and thorium halos vary from 12 to 42 μm in mica, possible evidence of unknown radioactivity exists in the scattered reports of unusual halos with anomalous ring radii (2, 3) varying from 5 to 10 μm in the dwarf halos to about 70 μm in the giant halos.

The very few previously reported occurrences of giant halos seem to have been largely ignored, perhaps because either definite information on the presence and size of the halo inclusion was absent (3) or because subsequent confirmation of the report was lacking. Hoppe (4), for example, was unable to confirm the existence of giant halos found by Wiman in certain Swedish granites, but this is not surprising in view of the large variability in the occurrence of particular halo types and the relatively small number of thin sections that Hoppe examined. Indeed, after a more extensive search in which I examined about 1000 thin sections from these granites, I find that giant halos in the 55-μm range do exist in the biotite along with ordinary uranium and thorium halos. These giant rings invariably occur only around very densely colored thorium halos, a result which implies a correlation of this ring with a high thorium content of the inclusion. Examination of the thorium decay scheme shows that the daughter alpha emitter, Po^{212}, emits a low-abundance (1 : 5500) alpha particle of slightly higher energy (10.55 Mev, compared to a normal 8.78 Mev), whose range may be correlated with the observed giant ring. Although there is some question whether the frequency of the low-abundance alpha particles in this energy range can produce a halo ring, I presently infer this association to be correct. The density of giant halos in these granites is quite low, however, and after a further search I have found a mica sample from Madagascar with uranium and thorium halos, in addition to an exceptionally fine collection of giant halos including all the sizes reported by Wiman as well as several much larger varieties of halos heretofore unreported.

The close proximity of occurrence of different halo types in the Madagascar mica provides an excellent range-energy relation which checks with coloration band widths produced experimentally in Van de Graaff helium ion irradiation of the mica matrix (5). Whereas the induced coloration bands are darker than the mica, the halos show reversal (bleaching) effects and are generally lighter than the surrounding matrix, except adjacent to the inclusion. Electron microprobe analyses indicate that the inclusions are monazites (6), and, since they are somewhat large (> 10 μm in diameter), they do not show ring structure as well as halos with point-like inclusions. Also, the high radioactive content of some of the inclusions leads to an overexposed condition which tends to further obliterate inner ring structure.

The visual appearance of the giant halos (Figs. 1–3) is similar to that of the combination uranium-thorium halos, and the question arises whether long-range alpha particles have produced the giant halos. The affirmative answer to this question cannot be accepted without a critical examination of other modes of origin, since the

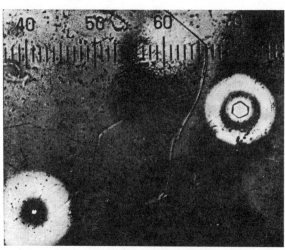

Fig 1. The halo on the right is a combination uranium and thorium halo, with the inner ring radius of 34 μm from the uranium daughter emitter Po^{214} ($E = 7.68$ Mev) and the outer ring radius of 40 μm from the thorium daughter emitter Po^{212} ($E = 8.78$ Mev). The halo on the left with a relatively small inclusion is a giant halo with about a 50-μm radius. One scale division = 10 μm.

magnitude of the giant halo radii involved implies the previous existence of naturally occurring alpha emitters with energies higher than any currently known.

Hence it is considered that the giant halos may have originated from:

1) *Variations in alpha particle range due to structural changes in mica.* Observations show that certain halo inclusions exhibit shapes or structural symmetry not exactly identical to the present outline of the inclusion in the mica matrix, and such deformations of the inclusion from radiation-damage effects might very well alter the structure of the matrix in the vicinity of the inclusion. However, there are numerous sites where uranium and thorium halos of normal size exist adjacent to and, in some cases, actually overlap giant halos (the inclusions of which show no evidence of any expansion or contraction). At least in these cases it would appear that the giant halos do not arise from normal-range alpha particles, which passed through a region of lower mica "density."

2) *Diffusion of a pigmenting agent from the inclusion into the matrix.* Although it is possible that some pigmenting substance may have been present, electron microprobe traverses across the region of the halo revealed no variations in elemental abundances of the matrix. Furthermore, in annealing experiments that were carried out at 450°C for 24 hours the yellowish tint of the halos either remained the same or in some cases became opaque; that is, there was no fading or otherwise any difference between the reaction of the uranium and thorium halos and that of the giant halos. In essence, if a purely chemical diffusion mechanism is operable, it is producing a type of coloration that is thus far indistinguishable from that initiated by radiation-damage effects. [Small crystalline structures (Liesegang patterns) often occur in mica, but these are easily distinguished from radioactive halos.]

3) *Diffusion of radioactivity from the inclusion to the matrix.* Electron microprobe analyses showed that uranium and thorium were confined to the inclusion; techniques by which fission tracks were induced indicated only a background uranium concentration surrounding the inclusion, and autoradiographic experiments with Kodak NTA emulsion showed alpha radioactivity restricted to the site of the inclusion. If diffusion of radioactivity has occurred, it is below the detection limit of these three methods.

Table 1. Frequency of halo sizes of radii 32 to 110 μm.

Group	Interval of halo radius (μm)	Maximum energy of alpha particles (Mev)	Total No. of halos
I	32–35	7.68	22
II	37–43	8.78	274
III	45–48	≈ 9.5	28
IV	50–58	≈ 10.6	130
V	60–67	≈ 11.7	69
VI	70–75	≈ 12.3	58
VII	80–85	≈ 13.2	30
VIII	90–95	≈ 14.1	10
IX	100–110	≈ 15.1	5

4) *Channeling.* Even though different optical properties in the region parallel to the cleavage plane make it difficult to observe a transverse halo section in any mica, the giant halos do exhibit a three-dimensional structure typical of radioactive halos when successive mica layers are cleaved. The idea that channeling of normal-range alpha particles parallel to the cleavage plane would be instrumental in the formation of giant halo rings is certainly correct in principle. Whether the relatively small number of alpha particles emitted along any given cleavage plane is sufficient to produce coloration is not clear. Furthermore, if channeling were the explanation, a series of successive outer bands corresponding to a given multiple of the ranges of the uranium or thorium daughter alpha emitters, or both, might be expected in a given giant halo. This situation is not observed.

5) *Beta radiation instead of alpha emission.* Laemmlein (7) found beta halos of rather diffuse boundaries with radii up to several thousand micrometers surrounding thorium-containing monazite inclusions in quartz. The fact that many of the perimeters of these giant halos in this mica are well-defined does not favor the association of these halos (Figs. 1–3) with the beta halos; neither do the radii correspond. In addition, Laemmlein noted a correlation between the radius of the beta halo and the volume of the halo inclusion (that is, the thorium content). This is understandable, since energetic beta rays producing coloration at maximum range would emanate throughout the volume of the inclusion. In contrast, no such effect is observed in this mica. Giant halos and uranium and thorium halos occur around relatively small inclusions as well as around larger ones.

6) *Long-range alpha particles from spontaneous fission.* Long-range alpha particles with a broad energy spectrum accompany normal spontaneous fission events from U^{238} in an abundance of about 1:400. Neither of these factors is favorable for the production of relatively sharp boundaries such as are seen in certain giant halos. Upon etching several giant halos with hydrofluoric acid to reveal fission tracks, I have found that fission tracks emanate from the inclusions of some, but not all, giant halos. The tracks emanating from some of the inclusions may be attributed to

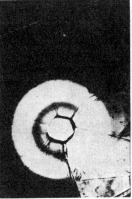

Fig. 2 (left). A giant halo approximately 57 μm in radius, presumably due to the long-range alpha particles from Po^{212} ($E = 10.55$ Mev). One scale division = 10 μm.
Fig. 3 (right). A giant halo approximately 84 μm in radius, whose origin is unknown. If the halo is due to long-range alpha particles, the energy would be about 13.1 Mev. One scale division = 10 μm.

the uranium content of the halo inclusions. The lack of fission tracks in other inclusions implies that at least in these cases long-range alpha particles from spontaneous fission are not instrumental in producing the giant halos.

7) *Alpha particles or protons from (n,α) or (α,p) reactions.* Mica sandwiches containing halo inclusions were irradiated with a total flux of 5×10^{18} neutron/cm^2. No induced coloration was noted in the mica section adjacent to the inclusion after irradiation. Since this integrated flux is several orders of magnitude higher than would be expected in naturally occurring inclusions, it appears that (n,α) reactions have not produced the giant halos. Calculations show that (α,p) reactions are also insufficient to produce coloration (see 8).

From the preceding comments it would appear that, although some of the above explanations cannot be definitely excluded, neither can any be presently confirmed as a factor responsible for the origin of the giant halos. Therefore, a few remarks may be made concerning the distribution of halos in this mica and the possibility that the giant halos may have originated with long-range alpha activity either from isomers of known elements or from superheavy elements.

The radii of several hundred halos that were measured with a precision of about ± 1.5 μm are given in Table 1. Greater accuracy was possible but seemed unnecessary, since for halos with large inclusions the actual radius of the halo as measured from the inclusion edge to the halo perimeter will vary up to around 5 to 6 μm with the variation dependent upon the stage of halo development (9). Other uncertainties in the radii measurements arise if the inclusion is inclined with respect to the cleavage plane. The intervals of halo radii were thus chosen to be rather broad; it may well be that certain of the groups listed are composites of subgroups of halos with slightly different maximum radii, but further subdivision did not seem justified at present. The maximum energy values of the alpha particles are recorded for purposes of comparison *only* and are *not* meant to necessarily imply that the respective halo groups originated with alpha particles of that energy. There were a few halos which did not fall into any of the above categories, but the number of this type was only a small percentage of the total (2 percent). Halos in groups I and II are the normal uranium and thorium halos,

whose maximum radii may be identified with the respective daughter alpha emitters Po^{214} ($E = 7.68$ Mev) and Po^{212} ($E = 8.78$ Mev) of these decay series. Halos in group IV may be associated with the low-abundance, long-range alpha particles from Po^{212} ($E = 10.55$ Mev) in the Th^{232} decay series.

An attempt to relate other groups of long-range alpha emitters of polonium isotopes in the uranium and thorium decay chain with the giant halo radii is more difficult. For example, the 9.5-Mev group of Po^{212}, which conceivably could produce a 48-μm halo, occurs in an abundance of only about 1:30,000; the 9-Mev group (1:45,000) of Po^{214} could produce a 45-μm halo; and there exist still other groups with energies up to 10.5 Mev, but these occur in an abundance of only about 1:10^6. If it is considered that these alpha particles were emitted in the same abundance as is presently observed, only the halos in group IV may reasonably be attributed to known low-abundance alpha particles of higher energy. G. N. Flerov has suggested that Po^{212m}, an isomer of polonium with a half-life of 47 seconds and an alpha-particle energy of 11.7 Mev, not known to occur naturally, may have been responsible for the halo group in the 62- to 67-μm range, since the energy correlates with the prescribed range (10). This identification, if correct, would, first, constitute another example of a rather peculiar phenomenon, namely, the occurrence of halos originating with polonium isotopes apparently unrelated to uranium and thorium daughter products (11), and, second, raise the interesting possibility that the other giant halo groups may be associated with unknown isomers emitting high-energy alpha particles in the 10- to 15-Mev range. Kohman has suggested that such alpha emitters, if they exist, may be shape isomers (12) of known nuclides.

Very recent mass spectrometric studies in which the Ion Microprobe Mass Analyzer (IMMA) (Applied Research Laboratories) was used revealed an isotope ratio for Pb^{207} to Pb^{206} of about 0.16 for the halo inclusions as contrasted with a value of about 0.35 for the bulk monazite crystals (13), which occur adjacent to the mica (both values were uncorrected for common Pb). If subsequent work shows that this difference cannot be attributed to common Pb, this result might suggest that a closer examination be made of possible high-energy isomers, namely, an isomer in a chain decaying to Pb.

The possibility that the giant halos originate with a postulated superheavy element (14) in the region from atomic numbers 110 to 114 seems remote, since these elements (i) would not be expected to occur in monazites and (ii) would be expected to exhibit spontaneous fission activity either directly or indirectly (that is, to decay by way of alpha emission to the known spontaneous fission region below atomic number $Z = 105$) (15). As noted earlier, some giant halo inclusions do not exhibit background fission tracks. However, of special interest in this context are very recent theoretical calculations by Bassichis and Kerman (16), which indicate an island of superheavy element stability at somewhat higher Z (around 120). If such an element exists, it might be expected to occur in a pegmatitic mica.

ROBERT V. GENTRY
Chemistry Division, Oak Ridge National Laboratory, Oak Ridge, Tennessee 37830

References and Notes

1. G. H. Henderson and S. Bateson, *Proc. Roy. Soc. London Ser. A Math. Phys. Sci.* **145**, 563 (1934).
2. E. Wiman, *Bull. Geol. Inst. Univ. Uppsala* **23**, 1 (1930); H. Hirschi, *Vierteljahresschr. Naturforsch. Ges. Zuerich* **65**, 209 (1920) (see Oak Ridge Nat. Lab. Rep. ORNL-tr-702); J. Joly, *Proc. Roy. Soc. London Ser. A Math. Phys. Sci.* **102**, 682 (1923); R. V. Gentry, *Earth Planet. Sci. Lett.* **1**, 453 (1966); J. S. van der Lingen, *Zentralbl. Mineral. Geol. Palaeontol. Abt. A* **1926**, 177 (1926) (see Oak Ridge Nat. Lab. Rep. ORNL-tr-699).
3. R. V. Gentry, *Appl. Phys. Lett.* **8**, 65 (1966).
4. G. Hoppe, *Geol. Foren. Stockholm Forhandl.* **81**, 485 (1959) (see Oak Ridge Nat. Lab. Rep. ORNL-tr-756).
5. I thank Dr. F. Young, Physics Department, University of Maryland, for cooperation in the Van de Graaff experiments.
6. L. Kobren, Goddard Space Flight Center, National Aeronautics and Space Administration, and C. Feldman, Oak Ridge National Laboratory, performed the electron microprobe analyses, which revealed typical monazite constituents.
7. G. G. Laemmlein, *Nature* **155**, 724 (1945).
8. I assumed that (i) (α,p) reactions in the inclusion occur mainly with phosphorus and have a cross section of 0.1 barn, (ii) weight fractions are 0.25, 0.05, and 0.2 for Th, U, and P, respectively, and (iii) 0.2 is the fraction of U atoms decayed (≈ 0.1 from fission-track analysis on the mica); calculations then show that the integrated proton flux from (α,p) reactions in an inclusion 25 μm in diameter is at least a factor of 10^2 below that required to produce threshold coloration ($\approx 10^{13}$ alpha particles per square centimeter from Van de Graaff irradiation) in a giant halo of radius 75 μm. Similar considerations hold for (α,p) reactions on nuclides of low Z in the surrounding matrix.
9. Variations of 1 μm in halo radii are noted even with point-like inclusions, possibly resulting from maximum ionization (coloration) occurring at slightly less than end-point range. In larger inclusions the halo radius appears to increase several micrometers as the halo develops because a greater fraction of the alpha particles are being emitted within the outermost micrometer of the inclusion. Nonuniform halo boundaries also occur and may result from a nonuniform distribution of radioactivity in the inclusion or may be related to one of the unusual development modes previously described herein.
10. I thank G. N. Flerov, Director, Laboratory for Nuclear Reactions, Joint Institute for

Nuclear Research, Dubna, U.S.S.R., for this suggestion.

11. R. V. Gentry, *Science* **160**, 1228 (1968).
12. T. P. Kohman, personal communication; *U.S. At. Energy Comm. Rep. No. NYO-844-76* (1969), p. 74.
13. This work was performed by C. Andersen, Applied Research Laboratories, Goleta, Calif. In the IMMA a finely focused (10 μm) O_2^- ion beam is used to sputter material directly into a mass spectrometer.
14. G. T. Seaborg, *Annu. Rev. Nucl. Sci.* **18**, 53 (1968).
15. Mass spectrometric analyses of mica containing the giant halo inclusions, performed at Ledoux & Company, Teaneck, N.J., and the GCA Corporation, Bedford, Mass., indicated that, if superheavy elements are present, their abundance must be less than 200 parts per million. The IMMA analysis of the monazite inclusions revealed the presence of what are almost certainly molecular ions with a mass of 303 (possibly CaThP), 310 (possibly La_2O_2) and somewhat higher values.
16. W. H. Bassichis and A. K. Kerman, *Phys. Rev.*, in press.
17. I thank P. Ramdohr, University of Heidelberg, and H. de la Roche, National Center for Scientific Research, Nancy, France, for the Madagascar mica samples; I thank J. Boyle, Oak Ridge National Laboratory, for valuable assistance with the experiments. The research performed at Oak Ridge National Laboratory was sponsored by the U.S. Atomic Energy Commission under contract with Union Carbide Corporation. Part of the research was performed at Columbia Union College, Takoma Park, Maryland, from which I am presently on leave of absence.

12 February 1970; revised 9 June 1970

Reprinted from SCIENCE
14 August 1970, Volume 169, pp. 670-673

Copyright© 1970 by the
American Association for the Advancement of Science

(*Reprinted from Nature*, Vol. 244, No. 5414, *pp.* 282–283,
August 3, 1973)

Ion Microprobe Confirmation of Pb Isotope Ratios and Search for Isomer Precursors in Polonium Radiohaloes

RADIOHALOES associated with decay of several Po α emitters[1,2] have been studied by optical microscopic techniques and more recently by mass spectrometric examination of the halo inclusion using ion microprobe techniques[3,4]. In such cases a large excess of ^{206}Pb compared with ^{207}Pb was found to be incompatible with the radiogenic decay of ^{238}U and ^{235}U, yet was explainable on the basis of polonium decay independent of uranium[3]. A straightforward attempt to account for the origin of these Po haloes by assuming that Po was incorporated into the halo inclusion at the time of host mineral crystallization meets with severe geological problems: the half-lives of the polonium isotopes ($t_{\frac{1}{2}} = 3$ min for ^{218}Po) are too short to permit anything but a rapid mineral crystallization, contrary to accepted theories of magmatic cooling rates.

This dilemma might be resolved (R. V. G., unpublished) if several long half-life high-spin or shape isomers of polonium (or the β-decaying precursors) were formed at nucleosynthesis and were subsequently incorporated into the halo inclusions during crystallization. This hypothesis eliminates the geological difficulties, and is open to experimental verification using several techniques such as charged particle reactions, though the long half-lives may present an obstacle. But long half-lives imply that some of the isomers may still exist, in which case a mass analysis of the polonium halo inclusions should reveal whether significant quantities are still present. We now report additional ion microprobe analyses of these Po inclusions as well as U inclusions in search of the isomers and for additional information on the Pb isotope ratios.

Mass scans were taken on areas of the biotite free from haloes All the normal elemental constituents as well as some trace elements were seen in these scans. The mass region from 150 to 300 is conspicuously free from any mass peaks. Generally Fe_2^+ at position 112 is the only high mass peak of significance observed from the biotite itself.

In the pure uranium, thorium, or uranium-thorium inclusions, ion microprobe analysis showed that the inclusions were either zircons or monazites; in many cases the ^{204}Pb ion current or signal was near background, so that it was difficult to make a common Pb correction; the $^{238}U/^{235}U$ ratio was normal in inclusions which contained uranium; and the $^{238}U/^{206}Pb$ signal ratio varied from 10 to 70 in the different inclusions analysed. The actual $^{238}U/^{206}Pb$ atom ratio is difficult to determine because of the uncertainty in the U and Pb secondary ion yield from different minerals. In general, U is detected with several times greater efficiency than Pb. The radiogenic $^{206}Pb/^{207}Pb$ ratio was difficult to evaluate in those inclusions where the ^{204}Pb signal was near background. In other cases it was found to vary within normal limits.

There is a wide spectrum in the U and Th halo types—some inclusions contain just U or Th without the other element, while other inclusions contain varying amounts of U and Th and in some cases exhibit rings from both decay series; it seems that the same situation prevails with Po and U type haloes in certain micas. In the analyses thus far it seems that the larger the Po halo inclusion the greater the U content tends to be; but more work is needed to verify this. Also the larger inclusions seem to be definite mineral types (usually rare earths but not specifically identified as yet), whereas some of the point-like Po halo inclusions consist of only elemental Pb (without ^{204}Pb) and Bi. Previously no detectable U was found in such cases as the latter type.

In contrast to the Pb ratios in the U and Th halo inclusions, we again report exceptionally high $^{206}Pb/^{207}Pb$ ratios which are characteristic of the ^{218}Po decay sequence type Po halo. The results may be summarized as follows: $^{206}Pb/^{207}Pb$ ratios of 10, 12, 18, 22, 25, 40, and 100 were observed. In four of these cases no ^{204}Pb was detected. In the other two cases ^{204}Pb was almost background, so that no common Pb correction was made on any of the ratios (any such correction would have produced a larger $^{206}Pb/^{207}Pb$ ratio). In three of the cases (10, 12, and 22) the small uranium signal seen was 10 to 100 times less than that required to support the Pb observed. These results confirm the earlier ion microprobe analyses of Po halo inclusions in which Pb ratios were found that were impossible to explain on the basis of U decay. They give confidence that we are indeed dealing with a class of haloes that is distinct from the ordinary U and Th types as the optical microscopic measurements invariably suggest. Otherwise, the most important aspect of the results is that the decay product of the polonium (Pb) still exists in these inclusions in measurable quantities (10^8–10^{10} atoms) and has not diffused away. On such a basis we then expect that any isomer precursor of Po, if the half-lives were sufficiently long, would also still exist and be detectable by ion microprobe techniques.

The only source of geochemical data about the postulated isomer is derived by inference from the type of halo inclusion. Some Po halo inclusions are of the rare earth variety while others contain only elemental Pb and Bi. The latter case might suggest the existence of an isomer geochemically similar to those elements, whereas the former case is rather non-specific. Fortunately ion probe mass analysis techniques do not depend on knowing the chemical identity of the postulated isomer.

To obtain these Pb ratios, we first cleaved the mica until the halo inclusion appeared on the surface. In some cases the sample was coated with a thin conducting layer of carbon, but it was better to overlay the sample with electron microscope-type Cu grids. In the latter case there was no extraneous material introduced anywhere near the region of interest. Before taking mass scans on the Po haloes the ion microprobe was optimized to obtain the best Pb signal from large U type halo inclusions that were mounted on the same sample but in a different area. In many cases the ion probe was peaked on mass 206 position and then moved to the area in the vicinity of the Po halo inclusion. The signal at this mass position remained at background (1 Hz) until the beam was shifted to the Po inclusion itself. In some cases several minutes elapsed before the signal reached maximum intensity. Generally mass positions 204, 207, 208, 218 and 238 were monitored, as well as the regions considerably below Pb, for possible interference from molecular ions. In other cases mass scans of the entire region from mass 1 to 250 were taken. It can be definitely stated that the exceptionally high 206 signal, compared with 207, occurs only in the Po halo inclusions and is not an artefact due to a molecular ion originating with the mica itself, the inclusion, or a combination of the mica and the elemental constituents of the inclusion. This is not to say the ion microprobe does not generate molecular ions, for in certain cases it does so very efficiently. But in the case of the Po haloes, we took care to monitor the various possibilities, which could have interfered with the results.

The search for the isomer consisted of carefully scanning the region around mass 218, for the Po haloes used in these experiments originated with ^{218}Po α decay. To be certain of the mass position, a small amount of Hg was placed on the sample holder to use as a mass marker at the 218 position (^{202}Hg^{16}O). In all Po inclusions except one no signal was observed at the 218 position. That one exception was due to interfering HgO ions from the presence of Hg in the inclusion itself.

A very rough estimate of what these results mean in terms of the present existence of the isomer in the inclusion may be obtained because the ^{206}Pb sputtered ion count rate was greater than 1,000 Hz in some Po inclusions. If it is assumed the isomer resembles Pb in sputtered ion efficiency (Pb has a relatively poor sputtered ion yield), then the present abundance of the isomer in the inclusion is $\leqslant 10^{-3}$ that of the ^{206}Pb. One interpretation of these results is the isomer has simply decayed to the point where it was not detected in these experiments. (These samples were from an early Precambrian pegmatite in Scandinavia.) It is yet to be determined whether this information is consistent with the half-lives of the proposed isomers that can be ascertained by determining the latest geological epoch in which such haloes occur.

This work was sponsored by the US Atomic Energy Commission under contract to the Union Carbide Corp., the General Electric Company and Columbia Union College with National Science Foundation grants.

ROBERT V. GENTRY

Chemistry Division,
Oak Ridge National Laboratory,
Oak Ridge, Tennessee 37830

S. S. CRISTY
J. F. McLAUGHLIN

Laboratory Development Department,
Oak Ridge Y-12 Plant,
Oak Ridge, Tennessee 37830

J. A. McHUGH

Knolls Atomic Power Laboratory,
Schenectady, New York 12301

Received April 13, 1973.

[1] Henderson, G. H., *Proc. R. Soc.*, A, **173**, 250 (1939).
[2] Gentry, R. V., *Science, N.Y.*, **160**, 1228 (1968).
[3] Gentry, R. V., *Science, N.Y.*, **173**, 727 (1971).
[4] Andersen, C. A., and Hinthorne, J. R., *Science, N.Y.*, **175**, 853 (1972).

Printed in Great Britain by Flarepath Printers Ltd., St. Albans, Herts.

Research Communications NETWORK
BREAKTHROUGH REPORT

MYSTERY OF THE RADIOHALOS

- Current physical laws may not have governed the past.

- Earth's primordial crustal rocks, rather than cooling and solidifying over millions or billions of years, crystallized almost instantaneously.

- Some geological formations thought to be one hundred million years old are in reality only several thousand years old.

Grant these propositions and—any researcher will tell you—the entire structure of the historical natural sciences would dissolve into formlessness. Few certainties would remain. Yet these very possibilities (and others equally disintegrative) have been suggested in a remarkable series of papers published over the past several years in the world's foremost scientific journals—*Nature, Science,* and *Annual Review of Nuclear Science,* among others. Nor has this assault upon orthodoxy elicited a vigorous counterattack: the research results published to date have been so cautiously and capably elaborated, and evidence so thoroughly piled upon evidence, as to forestall any outcry by those whose scientific sensibility may have been outraged. While some investigators appear finally to be arming themselves for combat, the issue has not yet been joined.

It was over a decade ago that Robert V. Gentry, puzzling over questions about the Earth's age, directed his attention to an obscure and neglected class of minute discolorations in certain minerals. He has since examined more than 100,000 of these "radiohalos," and without doubt stands as the world's leading authority on the subject. As an assistant professor of physics at Columbia Union College (Takoma Park, Maryland), he has brought to bear upon the halos an array of sophisticated instrumentation such as few researchers ever have the privilege to wield. As a result, he has converted the entire field of radiohalo research into an exact science, transmuting the microscopic spheres of mystery into rich mines of exciting and challenging information.

RADIOACTIVE HALO (or RADIOHALO): "In some thin samples of certain minerals, notably mica, there can be observed tiny aureoles of discoloration which, on microscopic examination, prove to be concentric dark and light circles with diameters between about 10 and 40μm [one micrometer is one-millionth of a meter] and centered on a tiny inclusion. The origin of these halos (first reported between 1880 and 1890) was a mystery until the discovery of radioactivity and its powers of coloration; in 1907 Joly and Mugge independently suggested that the central inclusion was radioactive and that the alpha-emissions from it produced the concentric shells of coloration. . . . halos command attention because they are an integral record of radioactive decay in minerals that constitute the most ancient rocks" (1).

Gentry's studies have led him to the following conclusions:
1) Some halos ("polonium" halos) imply a nearly instantaneous crystallization of Earth's primordial rocks: and this crystallization must have occurred simultaneously with the

This review is based upon a series of telephone interviews with Robert V. Gentry, as well as the available technical literature.

Glossary of Technical Terms

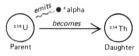

Parent   Daughter

A **parent** radioactive atom **decays** into a **daughter** atom in various ways, one of which is by the emission of an **alpha** particle from the parent atom's nucleus. Numerous types of radioactive atoms occur in nature, but only three are the initiators of a **decay series**: uranium-238 (^{238}U); uranium-235 (^{235}U); and thorium-232 (^{232}Th).

(The numerical superscript signifies how **heavy** the element is. **Isotopes** of the same element have different weights but nearly identical chemical behavior—as for example (^{238}U) and (^{235}U). An alpha particle has a weight of 4.)

Each of the three decay-series initiators decays, by a chain of steps, into lead. For example, the alpha-decay steps in the ^{238}U series are the following (steps not involving alpha-decay are not shown here):

^{238}U	$\Rightarrow$ ^{234}Th	^{222}Rn	$\Rightarrow$ ^{218}Po
^{234}U	$\Rightarrow$ ^{230}Th	^{218}Po	$\Rightarrow$ ^{214}Pb
^{230}Th	$\Rightarrow$ ^{226}Ra	^{214}Po	$\Rightarrow$ ^{210}Pb
^{226}Ra	$\Rightarrow$ ^{222}Rn	^{210}Po	$\Rightarrow$ ^{206}Pb

Similarly, ^{235}U decays by a different series of steps to ^{207}Pb, and ^{232}Th decays to ^{208}Pb. Note that while all the series end up with lead, each one results in a different isotope of lead.

The **half-life** of a given type of radioactive atom is the time during which half the atoms in any collection will decay. The half-life of ^{238}U is 4½ billion years. Half-life, **decay rate,** and **decay constant** are closely related quantities. If we assume that the decay rate has not changed over geologic time,* and if we measure 1) how much of a parent in a rock has decayed into its daughter; and 2) the current rate of this decay, then we can, it is generally believed, assess the date when the parent was incorporated into the rock—that is, the date when the rock was formed. In the case of Earth's oldest rocks, this date (some 3½ billion years ago) is thought to be the time when the molten Earth first cooled down sufficiently for rocks to solidify from the primordial magma.

*Numerous other assumptions and technicalities also come into play.

---

synthesis/creation of certain elements.
2) Some halos correspond to types of radioactivity which are unknown today.
3) Whereas radiohalos have been thought to afford the strongest evidence for unchanging radioactive decay rates

throughout geological time (and these rates enable scientists to determine rock ages), in actuality the overall evidence from halos requires us to question the entire radioactive dating procedure: something appears to have disrupted the radioactive clocks in the past.

4) Halos in coal-bearing formations that are conventionally thought to be 100 to 200 million years old suggest these strata to be only several thousand years old. Further, the time required for coal formation is much less than previously thought.

5) Taken together, these conclusions point to one or more great "singularities" in Earth's past—events or processes that are discontinuous with the rest of history, unique occurrences that critically affect the data we now have. If we attempt to interpret these data solely in terms of current processes, we go astray.

In this report we will discuss only those researches leading to conclusion (1), reserving the rest for a subsequent report.

January 29, 1975

You ask for my opinion of Dr. Robert Gentry's work on pleochroic polonium halos. I spent a number of hours reviewing this fascinating work with him some weeks ago. I was impressed with the clarity of the evidence for "anomalous halos"—that is, cases where there are rings indicating the presence of some members of the normal radioactive decay chain without the other members of the family tree that normally are present, and that normally do show up in rings of their own, and that have to be there on present views of the radioactive decay chains involved. If the evidence is impressive, the explanation for it is far from clear. I would look in normal geologic process of transfer of materials by heating and cooling; in isomeric nuclear transitions; and in every other standard physical phenomenon before I would even venture to consider cosmological explanations, let alone radical cosmological explanations. To explore all the avenues that need exploring would take months, not the few hours I was privileged to spend in Dr. Gentry's company. A few days ago I reviewed this work, all too briefly, with Dr. G. Wasserburg of Cal Tech, who is an expert in the radioactive dating of rocks, whose opinion would be much more to the point than mine, especially if he will give it to you in writing.*

JOHN A. WHEELER
(Professor of Physics,
Princeton University)

**Professor Wheeler requested that his letter be printed in full. Dr. Wasserburg's views have not been obtained.*

THE CONSERVATISM OF SCIENCE

Many have noted a conservatism in science essential to its orderly advance: skepticism toward radically new ideas enables scientific journals to retain focus, prevents anarchic descent into theoretical chaos, and makes it possible to extend currently reigning theories as far as they can bear before replacing them with other theories yet more embracive. A successfully modified, "tested" theory is preferable to a new "untried" theory. And so scientific knowledge advances in an orderly fashion, with as few wrong turns as possible.*

Gentry has so far avoided clashing with this conservatism,

**This conservatism—and its deceptive advantages—will receive continuing discussion in these newsletters.*

chiefly by concentrating his efforts on publication of data rather than discussion of their implications—and also by the good fortune that his work has been slow to draw widespread attention. That is beginning to change, however. But perhaps the reaction of a number of prominent physicists to Gentry's work on polonium halos (see insets on this and the following page) is the most significant gauge of what will be forthcoming. This reaction is noteworthy both for the confidence expressed in Gentry's work and for the almost uniformly conservative—albeit open—stance toward any extrapolations from the raw data that challenge accepted theory. Of those whose opinions we sampled, only one seemed to suggest (without wishing to be quoted) that we not publicize Gentry's work. He felt that the subject should be "left to the experts," while cautioning that it is too early to reject the conventional view of Earth's history.

In the end, it is, presumably, the evidence which will decide the issue. Let us look more closely at the radiohalos themselves.

THE NATURE OF HALOS

If a small grain (inclusion) containing radioactive atoms is embedded in certain rock minerals, the alpha particles emitted from the radioactive atoms travel outward from the inclusion and damage the crystalline structure of the mineral, in time producing the visible discoloration typifying halos. Since each type of radioactive atom emits alpha particles with a characteristic energy, and since this energy determines how far the particle will travel in the host mineral, the diameter of a halo's rings guides researchers in determining which radioactive element is responsible for the halo. If the radioactive element in an inclusion is the initiator of a decay series, then a group of concentric halo rings results, each ring corresponding to a step in the decay series, that is, to alpha particles of a particular energy. In the case of the ^{238}U series, with eight alpha-decay steps, there are five distinct halo rings (some of the alpha particles are so close together in energy that their rings are not distinguishable).

The conventional argument drawn from observed radiohalo sizes is summarized by Struve:

"There is excellent evidence that the rates of radioactive processes measured in the laboratory at the present time are valid also for the remote past. If a radioactive element and its decay products are embedded in a crystal, each alpha particle emitted during disintegration travels a certain distance that depends only on the rate of that particular decay step. The more rapid this rate, the greater the energy of the alpha particles, and the farther they go before being stopped and producing a color change in the crystal.

"Suppose a speck of ^{238}U has remained undisturbed since the formation of a mineral containing it. Then, because the rate of disintegration at each successive emission is different, eight concentric rings of mineral discoloration will be found surrounding the particle of uranium. These rings . . . have been found in many rocks of different geological ages, and the diameters of the respective rings are always the same.

"Thus it can be concluded that the rates of disintegration of uranium and thorium are constant. . . ." (2).

As we will learn in a subsequent review, the evidence from halos has led Gentry in a direction quite opposite from Struve's. But more than that, Gentry's halo research appears to strike at the roots of virtually all contemporary cosmologies, posing a fundamental problem which has so far resisted every effort to solve it in conventional terms. This is the problem of the polonium halos.

POLONIUM HALOS

The last three alpha decay steps in the uranium-238 decay series (see glossary above) involve the successive decay of polonium-218 (^{218}Po), polonium-214 (^{214}Po), and polonium-

A uranium-238 halo (left) and a polonium-210 halo in biotite. Scale is 1 cm equivalent to 45 μm.

210 (^{210}Po). In contrast to the decay of the parent uranium, these steps occur very quickly; the half-lives of the three forms of polonium are 3.05 minutes, 164 microseconds, and 140 days, respectively. Polonium, therefore, is not thought to be observed in nature except as a daughter product of uranium and thorium decay.

That is where the enigma begins. For Gentry has analyzed numerous polonium halos possessing, in some cases, the rings for all three polonium isotopes; in other cases the rings for ^{214}Po and ^{210}Po; and in other cases, the ring for ^{210}Po alone— *but none of these halos exhibits rings for the earlier uranium-238 daughters.* These halos are evidence for parentless polo-

nium, not derived from uranium.*

But the question then arises, How did the polonium inclusions ever become embedded in the host rocks (more specifically, in Earth's oldest—Precambrian—rocks)? On the conventional view, these rocks slowly cooled and crystallized out of the primordial magma (molten rock) over millions of years. Under such circumstances, any polonium (with its extremely short half life) that was incorporated into the solidifying rocks would have completely decayed long before the crystalline rock structure was established. No halos could have formed, for they consist precisely of radiation damage to this crystalline structure. Polonium rings should exist only *in conjunction with* the other uranium series rings. But since the actual halos were caused by parentless polonium, they require nearly instantaneous crystallization of the rocks, simultaneously with the synthesis or creation of the polonium atoms.

Gentry, well aware that this conclusion is unthinkable to most, has buttressed it with impressive experimentation: fission track and neutron flux techniques (3) reveal no uranium in the inclusions that could have given rise to the polonium—a conclusion more recently confirmed by electron microscope x-ray fluorescence spectra (4); fossil alpha recoil

---

Gentry has also found halos with rings from polonium-218, -214, or -210, combined with a ring from polonium-212 which is in the thorium decay series. This last form of polonium is also parentless— that is, there are no halo rings for thorium itself or its other daughters.

Comments by Leading Scientists

Before the demise of the journal, Pensée, *the editor—in preparation for a planned article on Gentry's work— approached a number of leading scientists for their assessment of polonium halos. The following responses were received during the first month or so of 1975.*

PROFESSOR TRUMAN P. KOHMAN, Department of Chemistry, Carnegie-Mellon University, Pittsburgh. "I do not believe that 'Gentry's contentions' can be regarded as of 'rather startling nature.' However, some of his experimental findings (like those of his predecessors) are quite difficult to understand, and the ultimate explanations could be interesting and even surprising. Many persons probably do not take them seriously, believing either that there is something wrong with the reported findings or that the explanations are to be found in simple phenomena which have been overlooked or discarded. . . . I believe it can be said that Gentry is honest and sincere, and that his scientific work is good and correctly reported. It would be very hard to believe that all, or any, of it could have been fabricated. . . ."

PROFESSOR EDWARD ANDERS, Enrico Fermi Institute, University of Chicago. "His [Gentry's] conclusions are startling and shake the very foundations of radiochemistry and geochemistry. Yet he has been so meticulous in his experimental work, and so restrained in his interpretations, that most people take his work seriously. . . . I think most people believe, as I do, that some unspectacular explanation will

eventually be found for the anomalous halos and that orthodoxy will turn out to be right after all. Meanwhile, Gentry should be encouraged to keep rattling this skeleton in our closet for all it is worth."

DR. EMILIO SEGRE, Istituto Di Fisica "Guglielmo Marconi," Università Degli Studi, Rome. "The photos [of radiohalos] are remarkable, but their interpretation is still uncertain."

PROFESSOR FREEMAN DYSON, Institute for Advanced Study, Princeton. "Supposing that the results of Gentry are confirmed, what will it mean for theory? I do not think it will mean any radical changes in geology or cosmology. It is much more likely that the explanation will be some tricky point in nuclear physics or nuclear chemistry that the experts have overlooked. That is of course only my personal opinion and I am accustomed to being proved wrong by events. (I just lost a $10 bet that Nixon would be in office till the end of 1974. I will be glad to lose this one too.)"

ACADEMICIAN G. N. FLEROV, Joint Institute for Nuclear Research, Moscow. "We made sure that [Gentry] carried out his investigations very thoroughly. . . . Therefore his data deserve serious attention. . . . It is not excluded that [polonium halos] have been formed as a result of the extremely rare combination of geochemical, geological and other conditions, and their existence does not contradict the logically grounded system of concepts involved in the history of Earth formation."

DR. PAUL RAMDOHR, Emeritus Professor of Mineralogy, Heidelberg University, Heidelberg. "The very careful and timetaking examinations of Dr. Gentry are indeed very interesting and extremely difficult to explain. But I think there is no need to doubt 'currently accepted cosmological models of Earth formation'. . . . Anyhow, there is a very interesting and essential question and you could discuss it, perhaps with cautious restrictions against so weighty statements like the one above in quotes. It would be interesting and good if more scientists would have more knowledge of the problems."

PROFESSOR EUGENE P. WIGNER, Department of Physics, Rockefeller University, New York. "Even though I know Dr. Gentry personally, I am not sufficiently familiar with his scientific results to be able to judge them. Personally, however, I have a very high regard for him."

DR. E. H. TAYLOR, Chemistry Division, Oak Ridge National Laboratory, Oak Ridge, Tennessee. "I can attest to the thoroughness, care and effort which Gentry puts into his work. . . . In a general way these puzzling pieces of information might result from unsuspected species or phenomena in nuclear physics, from unusual geological or geochemical processes, or even from cosmological phenomena. Or they (or one of them) might arise from some unsuspected, trivial and uninteresting cause. All that one can say is that they do present a puzzle (or several puzzles) and that there is some reasonable probability that the answer will be scientifically interesting."

analysis (3) demonstrates that neither polonium nor other daughter products migrated from neighboring uranium sources in the rock, which agrees with calculations based on diffusion rates (5); ion microprobe mass spectrometry yields extraordinarily high $^{206}Pb/^{207}Pb$ isotope ratios that are wholly inconsistent with normal decay modes (6), but which are exactly what one would expect as a result of polonium decay in the absence of uranium.

To date there has been only one effort (7) to dispute Gentry's *identification* of polonium halos. As it turned out (4), that effort might better never have been written, the authors having been impelled more by the worry that polonium halos "would cause apparently insuperable geological problems," than by a thorough grasp of the evidences. Challenges to Gentry's *interpretation* of the polonium halos have been more noteworthy. English physicist J. H. Fremlin wrote in *Nature* (November 20, 1975) that "The nuclear geophysical enigma of the ^{210}Po halos is quite fascinating, but the explanation put forward . . . is not easy either to understand or to believe." Fremlin proposed two possible explanations:

Geologic transfer. If there are uranium inclusions reasonably close to polonium halos, then it is possible that one or more of the uranium daughter products migrated from the uranium site to a new location, where subsequent decay gave rise to the polonium halo. Since the daughter products have much shorter half-lives than uranium, we would not expect to find any quantity of them remaining at the site of the halo. The polonium would therefore appear to be "parentless." The difficulty with this view is that transfer of uranium daughters in minerals occurs so slowly that the daughters would decay long before they could migrate any significant distance (3, 5).

If the sophisticated experimentation cited above proved telling against the transfer hypothesis, Gentry and several co-workers delivered a yet more conclusive blow in a very recent paper: polonium halos derived by geologic transfer from uranium sources have now actually been found in coalified wood deposits (8). Their presence here was to be expected: prior to coalification the wood was in a gel-like condition permeated by a uranium-bearing solution. Such a material "would exhibit a much higher transport rate as well as unusual geochemical conditions which might favor the accumulation of ^{210}Po"—quite different from the situation in mineral rocks. Further, of these uranium-derived polonium halos, none were found due to ^{218}Po, and only three could conceivably (but doubtfully) be attributed to ^{214}Po, in contrast to numerous ^{210}Po halos. The half-life of ^{210}Po, we will recall, is 140 days, whereas the half-life of those forms of polonium which failed to generate halos in the coalified wood is a few minutes or less. So even under the ideal conditions in this wood, the short-half-lived ^{218}Po and ^{214}Po were not able to migrate rapidly enough from the parent uranium to form "parentless" halos. Clearly, then, such migration could not account for the ^{218}Po and ^{214}Po halos Gentry has found in Precambrian minerals, where the diffusion rate is very much lower even than in wood (5).

Isomer precursors. Two atoms with identical nuclear composition but different radioactive behavior are termed "isomers." For example, ^{212}Po (in the thorium decay series) decays to ^{208}Pb by emission of an alpha particle with an energy of 8.78 MeV. However, about one out of every 5500 ^{212}Po atoms emits an alpha particle with a much higher energy of 10.55 MeV. These rarely occurring, higher-energy ^{212}Po atoms are isomers, and they are apparently explained by some variation in nuclear structure. The suggestion has been made,

therefore, that polonium halos may result from the presence of heretofore unknown isomers which are long-lived and which decay* into polonium. These isomers ("precursors" of polonium) would circumvent the cosmological problem caused by the short-half-life polonium.

However, not only are such isomers unknown, but a careful search has revealed the presence of no elements which might qualify as the required isomers (4, 5). "Experimental results have ruled out the isomer hypothesis" (5).

"SINGULARITIES"

And so we have Gentry's conclusion in his reply to Fremlin: "But if isomers and uranium-daughter diffusion do not produce polonium halos in rocks, we are left with the idea that polonium halos originate with primordial Po atoms just as U and Th halos originate with primordial ^{238}U and ^{232}Th atoms. . . . Carried to its ultimate conclusion, this means that polonium halos, of which there are estimated to be 10^{15} [one million billion] in the Earth's basement granitic rocks, represent evidence of extinct natural radioactivity, and thus imply only a brief period between 'nucleosynthesis' [creation of elements] and crystallization of the host rocks" (5). In plainer terms, these rocks must have formed almost instantaneously upon the synthesis of the elements comprising them.

Gentry believes the evidence points to one or more great "singularities" that have affected Earth in the past, representing physical processes which we do not now observe. If this is so, then attempts to define these processes in conventional terms will prove fruitless, and the span represented by geologic time is a wide open question. Further (as we will explore in a subsequent review), Gentry concludes that the most recent "singularity" may have occurred only several thousand years ago. And he finds compelling reasons to question the entire radioactive dating scheme which undergirds our concept of geological time.

Gentry realizes that he still must reckon with the conservatism of science. While his experimental work has been impressive, few would yet concede that it is impregnable, or that his explanations are the only possible ones. As Wheeler remarked:

"If the evidence [for the polonium halo] is impressive, the explanation for it is far from clear. I would look in normal geologic process of transfer of materials by heating and cooling; in isomeric nuclear transitions; and in every other standard physical phenomenon before I would even venture to consider cosmological explanations, let alone radical cosmological explanations."

While the evidence does not seem to favor the specific mechanisms Wheeler suggested in early 1975, Gentry can be sure that, in pressing his own decidedly radical explanations, the sound and fury lie yet before him.

---

*by beta-emission

REFERENCES

1. R.V. Gentry, *Annual Review of Nuclear Science 23* (1973), p. 347.
2. O. Struve, *Sky and Telescope 18* (June, 1959), pp. 433-5.
3. R.V. Gentry, *Science 160* (June 14, 1968), pp. 1228-30.
4. R.V. Gentry, *Science 184* (April 5, 1974), pp. 62-66.
5. R.V. Gentry, *Nature 258* (November 20, 1975), pp. 269-70.
6. R.V. Gentry, L.D. Hulett, S.S. Cristy, et al., *Nature 252* (December 13, 1974), pp. 564-66.
7. C. Moazed, R.M. Spector, and R.F. Ward, *Science 180* (June 22, 1973), pp. 1272-74.
8. R.V. Gentry, W.H. Christie, D.H. Smith, et al., *Science 194* (October 15, 1976), pp. 315-18.

Reprinted from
5 April 1974, Volume 184, pp. 62-66

Radiohalos in a Radiochronological and Cosmological Perspective

Robert V. Gentry

Copyright© 1974 by the American Association for the Advancement of Science

Radiohalos in a Radiochronological and Cosmological Perspective

Abstract. *New photographic evidence, data on halo ring sizes, and x-ray fluorescence analyses provide unambiguous evidence that polonium halos exist as a separate and distinct class apart from uranium halos. Because of the short half-lives of the polonium isotopes involved, it is not clear how polonium halos may be explained on the basis of currently accepted cosmological models of Earth formation.*

I have examined some 10^5 or more radiohalos, mainly from Precambrian granites and pegmatites located in several continents. In addition to U and Th halos, originally studied (*1, 2*) for information on the constancy of the α-decay energy E_α and the decay constant λ, I have discussed X halos (*2, 3*), dwarf halos (*3*), and giant halos (*4*), and explained how these remain prime candidates for identifying unknown α-radioactivity and, not impossibly, unknown elements as well.

I have also reported (*5*) on a class of halos which had been tentatively attributed (*6, 7*) to the α-decay of ^{210}Po, ^{214}Po, and ^{218}Po. Earlier investigators (*2, 7–10*), possessing only a sparse collection of Po halos, at times confused them with U halos or invented spurious types such as "emanation"

halos (*2*) or "actinium" halos (*8*) to account for them. (Figure 1, a to d, is a schematic comparison of U and Po halo types with ring radii drawn proportional to the respective ranges of α-particles in air.) To explain Po halos, Henderson (*7*) postulated a slow accumulation of Po isotopes (or their respective β-decay precursors) from U daughter product activity. I demonstrated that this secondary accumulation hypothesis was untenable and showed, using the ion microprobe (*3*), that Po halo radiocenters (or inclusions) exhibit anomalously high ^{206}Pb/^{207}Pb isotope ratios which are a necessary consequence of Po α-decay to ^{206}Pb.

Recently, these ion microprobe results have been questioned, Henderson's results misinterpreted, Po halos considered to be only U halos, and allusions made to the geological difficulties that Po halos would present if they were real (*11*) [see (*12*) for comments].

Admittedly, compared to ordinary Pb types, the Pb isotope ratios of Po halos are unusual, but new ion microprobe analyses have confirmed (*13*) my earlier results (*3*). It is also apparent that Po halos do pose contradictions to currently held views of Earth history.

For example, there is first the problem of how isotopic separation of several Po isotopes [or their β-decay precursors (*13*)] could have occurred naturally. Second, a straightforward explanation of ^{218}Po halos implies that the 1-μm radiocenters of very dark halos of this type initially contained as many as 5×10^9 atoms (a concentration of more than 50 percent) of the isotope ^{218}Po (half-life, 3 minutes), a problem that almost defies reason. A further necessary consequence, that such Po halos could have formed only if the host rocks underwent a rapid crystallization, renders exceedingly difficult, in my estimation, the prospect of explaining these halos by physical laws as presently understood. In brief, Po halos are an enigma, and their ring structure

Fig. 1. The scale for all photomicrographs is 1 cm $\simeq$ 25.0 μm, except for (h') and (r'), which are enlargements of (h) and (r). (a) Schematic drawing of ^{238}U halo with radii proportional to ranges of α-particles in air. (b) Schematic of ^{210}Po halo. (c) Schematic of ^{214}Po halo. (d) Schematic of ^{218}Po halo. (e) Coloration band formed in mica by 7.7-Mev ^{4}He ions. Arrow shows direction of beam penetration. (f) A ^{238}U halo in biotite formed by sequential α-decay of the ^{238}U decay series. (g) Embryonic ^{238}U halo in fluorite with only two rings developed. (h) Normally developed ^{238}U halo in fluorite with nearly all rings visible. (h') Same halo as in (h) but at higher magnification. (i) Well-developed ^{238}U halo in fluorite with slightly blurred rings. (j) Overexposed ^{238}U halo in fluorite, showing inner ring diminution. (k) Two overexposed ^{238}U halos in fluorite showing inner ring diminution in one halo and obliteration of inner rings in the other. (l) More overexposed ^{238}U halo in fluorite, showing outer ring reversal effects. (m) Second-stage reversal in a ^{238}U halo in fluorite. The ring sizes are unrelated to ^{238}U α-particle ranges. (n) Three ^{210}Po halos of light, medium, and very dark coloration in biotite. Note the differences in radius. (o) Three ^{210}Po halos of varying degrees of coloration in fluorite. (p) A ^{214}Po halo in biotite. (q) Two ^{218}Po halos in biotite. (r) Two ^{218}Po halos in fluorite. (r') Same halo as in (r) but at higher magnification.

as well as other distinguishing characteristics need to be made abundantly clear.

In order to ascertain the E_α corresponding to a specific halo radius, I have produced a new series of standard sizes against which halo radii may be compared without relying on estimates derived from ranges of α-particles in air. Standard sizes may be prepared by irradiation of halo-bearing mineral samples with ^{4}He ions (*4*); the coloration bands thus produced show varying sizes (as measured from edge to coloration extinction) which are dependent on energy, total dose, and dose rate, the latter two factors not being ac-

counted for in other comparative methods.

I made more than 350 irradiations 1 to 10^4 seconds in duration using ^{4}He ions with energies ranging from 1 to 15 Mev, on over 40 samples of biotite, fluorite, and cordierite (*14*). Selecting the band sizes which correspond to the energies of the ^{238}U α-emitters (see Table 1) permits a direct comparison with new as well as previous (*1, 9, 10, 15*) U halo measurements in biotite, fluorite, and cordierite. Figure 1e shows a coloration band in biotite produced by 7.7-Mev ^{4}He ions, and Fig. 2a shows a densitometer profile of Fig. 1e.

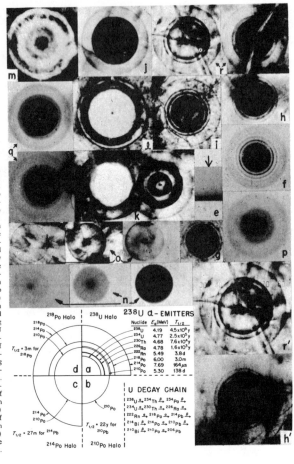

Nuclide	E_α(Mev)	$T_{1/2}$
^{238}U	4.19	4.5×10^9 y
^{234}U	4.77	2.5×10^5 y
^{230}Th	4.68	7.6×10^4 y
^{226}Ra	4.78	1.6×10^3 y
^{222}Rn	5.49	3.8 d
^{218}Po	6.00	3.0 m
^{214}Po	7.69	164 μs
^{210}Po	5.30	138 d

238U α - EMITTERS

U DECAY CHAIN

^{238}U $\xrightarrow{\alpha}$ ^{234}Th $\xrightarrow{\beta}$ ^{234}Pa $\xrightarrow{\beta}$
^{234}U $\xrightarrow{\alpha}$ ^{230}Th $\xrightarrow{\alpha}$ ^{226}Ra $\xrightarrow{\alpha}$
^{222}Rn $\xrightarrow{\alpha}$ ^{218}Po $\xrightarrow{\alpha}$ ^{214}Pb $\xrightarrow{\beta}$
^{214}Bi $\xrightarrow{\beta}$ ^{214}Po $\xrightarrow{\alpha}$ ^{210}Pb $\xrightarrow{\beta}$
^{210}Bi $\xrightarrow{\beta}$ ^{210}Po $\xrightarrow{\alpha}$ ^{206}Pb

Table 1. Comparison of sizes of induced bands (columns 1 to 5) with halo radii (columns 8 to 21). Column 6 gives the ^{4}He ion energies at which the induced bands were formed, or the α-particle energies corresponding to the nuclides in column 7. Thus, the nuclide or α-particle energy that produced any halo ring in columns 8 to 21 can be found from column 6 or 7. The letters K-L, H, S, M, and G represent halo measurements by Kerr-Lawson (15), Henderson (1, 6, 7), Schilling (9), Mahadevan (10), and Gentry. Subscripts L, M, and D indicate: medium (dose 10 to 20 times coloration threshold), and dark (dose about 50 times coloration threshold) induced bands; L→D and L→M indicate light to dark and light to medium; these were visually determined. Gentry's measurements were made with a filar micrometer readable to 0.07 μm. The estimated overall uncertainty was ±0.3 μm. Other abbreviations: N.M., not measured; N.R., not resolved; N.P., not present.

Coloration band size (μm)					E (Mev)	Nuclide	U halo radius (μm)						Po halo radius (μm) in							
Biotite			Fluorite	Cordierite			Biotite			Fluorite		Cordierite	Biotite						Fluorite	
1. G_L	2. G_M	3. G_D	4. G	5. G	6.	7.	8. K-L	9. H	10. G	11. S	12. G	13. M	14. H	15. $G_{L\to D}$	16. H	17. $G_{L\to M}$	18. H	19. G_M	20. G	21. G
13.4	13.8	14.2	14.1	16.2	→4.2	^{238}U→	12.3	12.7	12.2→13.0	14.0	14.2	16	N.P.	N.P.	N.P.	N.P.	N.P.	N.P.	N.P.	N.P.
N.M.	16.7	N.M.	17.3	19.2	→4.77	^{226}Ra→	15.4	15.3	14.8→15.8	16.9	17.1	19	N.P.	N.P.	N.P.	N.P.	N.P.	N.P.	N.P.	N.P.
N.M.	N.M.	N.M.	N.M.	N.M.	→4.66	^{230}Th→	N.R.	N.R.	N.R.	15.8	N.R.	N.R.	N.P.	N.P.	N.P.	N.P.	N.P.	N.P.	N.P.	N.P.
N.M.	16.7	N.M.	17.3	19.2	→4.78	^{234}U→	15.4	15.3	14.8→15.8	16.9	17.1	19	N.P.	N.P.	N.P.	N.P.	N.P.	N.P.	N.P.	N.P.
N.M.	19.3	20.0	19.6	22.5	→5.3	^{210}Po→	18.6	19.2	18.1→19.0	19.3	19.5	N.R.	19.8	18.3→19.9	20.0	18.1→19.1	19.9	19.3	19.8	19.8
N.M.	20.5	21.1	21.1	N.M.	→5.49	^{222}Rn→	22.0	23.0	21.5→22.7	20.5	20.5	23.5	N.P.	N.P.	N.P.	N.P.	N.P.	N.P.	N.P.	N.P.
N.M.	23.0	23.9	23.6	26.7	→6.0	^{218}Po→	N.R.	N.R.	N.R.	23.5	23.5	26.5	N.P.	N.P.	N.P.	N.P.	24.0	23.3	23.7	23.7
33.1	33.9	34.4	34.6	38.7	→7.69	^{214}Po→	33.0	34.1	30.8→33.0	34.5	34.7	38.5	N.P.	N.P.	34.5	32.5→33.8	34.0	34.0	23.9	23.9

The coloration extinction boundary is poorly defined near threshold coloration; only a few very light bands in biotite could be reliably measured. Reproducible measurements were obtained in the plateau region (14), where variations in band size are minimal. Darker halos in biotite generally have slightly larger radii than lighter halos (3, 4). Also, reversal effects in some biotites immediately exterior to the terminus of a halo ring cause apparent diminution of the radius. Therefore, while there are differences between the sizes of medium coloration bands (Table 1, column 2) and the radii of U halos in biotite (Table 1, columns 8, 9, and 10) that could be interpreted in terms of an actual change in E_α and λ (16), such differences more likely arise from a combination of dose and reversal effects (15, 17), producing slightly diminished radii. Diminution of U halo radii may also result from attenuation of α-particles within the small but relatively dense zircon radiocenters. Even though slight differences between band sizes and U halo radii do exist in biotite, the idealized U halo ring structure (Fig. 1a) compares very well with an actual U halo in biotite (Fig. 1f).

Biotite and fluorite are good halo detectors, but fluorite is superior because the halo rings exhibit more detail, often have smaller radiocenter diameters (< 1 μm), and have almost negligible size variations due to dose effects in the embryonic to normal stages of development. Figure 1g shows an embryonic U halo in fluorite with only the first two rings fully developed; the other rings are barely visible because, due to the inverse square effect, threshold coloration has not been reached. Figure 1h shows a U halo in fluorite in the normal stage of development, when nearly all the rings are visible. This halo closely approximates the idealized U halo in Fig. 1a. Under high magnification even separation of the ^{210}Po and ^{222}Rn rings may be seen. Figure 1i shows another U halo in fluorite, with a ring structure that is clearly visible but not adequate for accurate radius measurements.

In Table 1, columns 4, 11, and 12, the fluorite band sizes agree very well with the U halo radii measured in this mineral by myself and Schilling (9). This suggests that the differences between U halo radii and band sizes in biotite are not due to a change in E_α. However, experimental uncertainties in measuring U halo radii preclude establishing the constancy of λ to within 35 percent, and under certain assumptions U halos provide no information at all in this respect (16).

While halos with point-like nuclei which show well-defined, normally developed rings (as in Fig. 1h) can be used to determine the E_α's of the radionuclides in the inclusion, there are pitfalls in ascertaining what constitutes a normally developed ring. In contrast to the easily recognizable U halos in fluorite in Fig. 1, g to i, the overexposed fluorite U halo in Fig. 1j shows a diminutive ghost inner ring, which could be mistaken for an actual ^{238}U ring. Figure 1k shows two other partially reversed U halos, one of which shows the diminutive inner ring, while in the other all the inner rings are obliterated. The U halo in Fig. 1l is even more overexposed, and encroaching reversal effects have given rise to another ghost ring just inside the periphery. Figure 1m shows a still more overexposed U halo; in which second-stage reversal effects have produced spurious ghost rings that are unrelated to the terminal α-particle ranges.

Since this association of the halos in Fig. 1, l and m, with U α-decay cannot be easily proved by ring structure analysis alone, I have utilized electron-induced x-ray fluorescence to confirm this identification. Figure 3a shows the prominent Ca x-ray lines of the fluorite matrix (the F lines are below detection threshold) along with some background Ag and Rh lines which are not from the sample, but are produced when backscattered electrons strike a Ag-Rh alloy pole piece in the sample chamber. Figure 3b, the x-ray spectrum of a halo radiocenter typical of the halos in Fig. 1, l and m, clearly shows the x-ray lines due to U (as well as a small amount of Si) in addition to the matrix and background peaks. A more detailed analysis (18) reveals that the Uζ line masks a small amount of Pb probably generated by in situ U decay.

The variety of U halos shown in Fig. 1, g to m, establishes two points: (i) only a thorough search will reveal the numerous variations in appearance of U halos, and (ii) unless such a search is made, the existence of halos originating with α-emitters other than ^{238}U or ^{232}Th could easily be overlooked.

So far, three criteria have been used to establish the identity of U halos: (i) close resemblance of actual halos in biotite (Fig. 1f) and fluorite (Fig. 1h) to the idealized ring structure

Fig. 2. Densitometer profiles of the photographic negatives of (a) Fig. 1e, (b) Fig. 1f, (c) the light ²¹⁰Po halo in Fig. 1n, (d) the medium ²¹⁰Po halo in Fig. 1n, (e) the dark ²¹⁰Po halo in Fig. 1n, and (f) Fig. 1p.

(Fig. 1a), (ii) identification of lines in x-ray fluorescence spectra, and (iii) agreement between U halo radii and equivalent band sizes (very good in fluorite and fair in biotite and corderite). Using the third criterion (either band sizes or U halo radii), I can determine E_α for a normally developed fluorite halo ring to within ± 0.1 Mev. For biotite halos, U halo radii may form a suitable standard for determining E_α for rings that show reversal or other effects characteristic of U halos in the same sample. If good U halos are not available, and if the halos with variant sizes show well-developed rings without reversal effects, then the band sizes form a suitable standard for E_α determination when coloration intensities of variant halos and band sizes are matched.

Therefore, if halos result from the α-decay of ²¹⁰Po to ²⁰⁶Pb, their appearance should resemble the idealized schematic (Fig. 1b), and the light and dark halos of this type in biotite should exhibit radius variations consistent with the differences between lower and higher coloration band sizes (Table 1, columns 2, 3, 6, 14, and 15). Further, such halos, whether very light or very dark, should appear without any outer ring structure, as illustrated in Fig. 1n. Compare also the densitometer profiles of the halo negatives of Fig. 1f (the U halo) and Fig. 1n, shown in Fig. 2b and Fig. 2, c to e, respectively. Fig. 1o shows three similar halos in fluorite; here, irrespective of coloration differences, the halo radii are the same and correspond to the E_α of ²¹⁰Po (Table 1, columns 4, 6, and 20). Accordingly, the halos in Fig. 1, n and o, are designated ²¹⁰Po halos. (Actually I should emphasize that since not all biotites exhibit the same coloration responses, the radius measurements in Table 1 are strictly valid only for the particular micas I used. I did try to illustrate a range of responses by utilizing different biotites for the U halo and the three Po halo types.)

By analogy, the moderately developed biotite halo in Fig. 1p shows a marked resemblance to the idealized halo that would form from the sequential α-decay of ²¹⁴Po and ²¹⁰Po (see Fig. 1c). Table 1, columns 2, 3, 6, 7, 16, and 17, shows the corre-

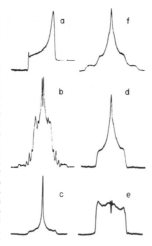

spondence of the radii with band sizes. The prominent unmistakable feature of the ²¹⁴Po halo is the broad annulus separating the inner and outer rings [see the densitometer profile of Fig. 1p shown in Fig. 2f and figures 7 to 9 in (6)]. With respect to comments in (11) it should be noted that the ²¹⁴Po halo can easily be distinguished from a U halo.

The last correspondence to be established is the resemblance of the two three-ring halos in biotite (Fig. 1q) and two similar halos in fluorite (Fig.

1r) to the idealized ²¹⁸Po halo (Fig. 1d) showing the ring structure from the sequential α-decay of ²¹⁸Po, ²¹⁴Po, and ²¹⁰Po. In biotite such halos may appear very light to very dark with radii correspondingly slightly lower and higher (excluding reversal effects) than those measured for medium coloration bands (compare Table 1, columns 2, 3, 18, and 19). Cursory examination of inferior specimens of this halo type could lead to confusion with the U halo, especially in biotite, where ring sizes vary slightly because of dose and other effects. However, good specimens of this type are easily distinguished from U halos, even in biotite. In fluorite, where the ring detail is better, a most important difference between ²³⁸U and ²¹⁸Po halos is delineated, that is, the presence of the ²²²Rn ring in the U halo (Fig. 1a) in contrast to its absence in the ²¹⁸Po halo (Fig. 1d). For example, note the slightly wider annulus (3.9 μm) between the ²¹⁰Po and ²¹⁸Po rings of the ²¹⁸Po halo compared to the equivalent annulus (3.0 μm) in the ²³⁸U halo (Fig. 1, a, d, h, h', r, and r'). This is evidence that the ²¹⁸Po halo indeed initiated with ²¹⁸Po rather than with ²²²Rn or any other α-decay precursor in the U chain. As further proof, Table 1 (columns 4, 11, 12, and 21) shows that the ²¹⁸Po halo radii agree very well with equivalent band sizes and U halo radii in this mineral. Additional Po halo types also exist (3) but are quite rare. [As yet I have found no halos at all in meteorites or lunar rocks (19)].

The preceding discussion has shown

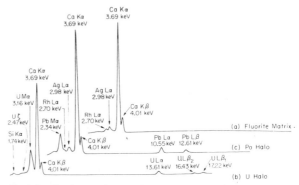

Fig. 3. Scanning electron microscope–x-ray fluorescence spectra of (a) the fluorite (CaF₂) matrix, (b) a U halo radiocenter in fluorite characteristic of Fig. 1, l and m, and (c) a ²¹⁸Po halo radiocenter in fluorite characteristic of Fig. 1r.

that Po halos can be positively identified by ring structure studies alone. That x-ray fluorescence analyses also provide quite convincing evidence is seen in Fig. 3c, where I show for the first time the x-ray spectra of a Po halo radiocenter (specifically, a ^{218}Po halo). Comparison of Fig. 3, b and c, reveals that the Pb in the Po halo radiocenter in fluorite did not arise from in situ decay of U. [Longer runs have shown small amounts as Se as well as U in some Po halo radiocenters (*18*).] On the other hand, the presence of Pb is to be expected in a ^{218}Po halo radiocenter because the decay product is ^{206}Pb. That the parent nuclide was ^{218}Po and not a β-decaying isomer precursor (*13, 20*) follows from half-life considerations of the U halo U/Pb ratio (> 10); the proposed isomer, if formed at nucleosynthesis, should now be detectable in Po halo radiocenters. No trace of this isomer has yet been found, and I thus view the isomer hypothesis as untenable.

The x-ray data in Fig. 3c are unambiguous and should remove any doubt that previously reported ^{206}Pb/^{207}Pb mass ratios (*3, 13*) actually are Pb isotope ratios, and in fact represent a new type of Pb derived specifically from Po α-decay. In summary, the combined results of ring structure studies, mass spectrometric analyses, and electron induced x-ray fluorescence present a compelling case for the independent existence of Po halos. The question is, can they be explained by presently accepted cosmological and geological concepts relating to the origin and development of Earth?

ROBERT V. GENTRY

Chemistry Division,
Oak Ridge National Laboratory,
Oak Ridge, Tennessee 37830

References and Notes

1. G. H. Henderson, C. M. Mushkat, D. P. Crawford, *Proc. R. Soc. Lond. Ser. A Math. Phys. Sci.* **158**, 199 (1934); G. H. Henderson and L. G. Turnbull, *ibid.* **145**, 582 (1934); G. H. Henderson and S. Bateson, *ibid.*, p. 573.
2. J. Joly, *ibid.* **217**, 51 (1917); *Nature (Lond.)* **109**, 480 (1920). I have examined Joly's collection and found that he associated certain Po halos with U halos and incorrectly associated the ^{210}Po halos as originating with Rn α-decay.
3. R. V. Gentry, *Science* **173**, 727 (1971).
4. ———, *ibid.* **169**, 670 (1970).
5. ———, *ibid.* **160**, 1228 (1968).
6. G. H. Henderson and F. W. Sparks, *Proc. R. Soc. Lond. Ser. A Math. Phys. Sci.* **173**, 238 (1939).
7. G. H. Henderson, *ibid.*, p. 250. A fourth type attributed to ^{226}Ra α-decay is in error.
8. S. Iimori and J. Yoshimura, *Sci. Pap. Inst. Phys. Chem. Res. Tokyo* **5**, 11 (1926).
9. A. Schilling, *Neues Jahrb. Mineral. Abh.* **53A**,

241 (1926). See translation, *Oak Ridge Natl. Lab. Rep. ORNL-tr-697*. Schilling, as did Joly, erroneously designated ^{210}Po halos as emanation halos. As for explanation of the 14.0-μm, 14.4-μm, and 15.8-μm rings which Schilling attributed to UI, UII, and Io, I can state that one of the rings at 14.0 μm and 14.4 μm is a ghost ring. I also rarely observe a light ring at about 16 μm, but do not presently associate this ring with ^{230}Th (Io) α-decay.
10. C. Mahadevan, *Indian J. Phys.* **1**, 445 (1927).
11. C. Moazed, R. M. Spector, R. F. Ward, *Science* **180**, 1272 (1973).
12. Moazed *et al.* (*11*) stated that because they could not find halos with dimensions matching those of Henderson's type B halo (the ^{214}Po halo in my terminology) such halos do not exist; however, Henderson gave both measurements and photographic evidence (6, figure 4, facing p. 242). They then inferred that a different halo (a U halo) must be the equivalent of the type B halo, although the radii of the inner ring of Henderson's type B halo and the outer second ring of their halo were significantly different (20 compared to 22.3 μm). They concluded that all Po halos are only U halos, without having U halos with normal ring structure available for comparison. I showed (*5*) that Po halos and U halos are distinguished by the number of fossil fission tracks after etching; that is, few, if any, compared to a cluster of 20 to 100 tracks. I also showed that the threshold coloration dose is directly obtainable by converting a U halo fossil fission-track count (20 to 100) to the number of emitted α-particles by using the ^{238}U branching ratio, λ_α/λ_f; this contradicts the supposition that such data are unknown to two orders of magnitude. Ion probe analyses of U halos show that a high U isotopic ratio can not be responsible for a small induced fission-track count. Furthermore, contrary to a statement by Moazed *et al.*, Henderson was able to distinguish reliably between his type B and type C halos (6, pp. 246–248).
13. R. V. Gentry, S. S. Cristy, J. F. McLaughlin, J. A. McHugh, *Nature (Lond.)* **244**, 282 (1973).
14. The irradiated biotite samples were cleaved in about 5-μm sections for microscopic examination. The coloration threshold (CT) for 30-μm biotite sections varied from 3×10^{13} to 6×10^{13} ^{4}He ions per square centimeter. Band sizes monotonically increased with dose to about 100 CT but were reproducible in a plateau region around 10 to 20 CT. Because band sizes were unpredictable at high beam intensities it was necessary to use beams of only about 10 na/mm^2.
15. D. E. Kerr-Lawson, *Univ. Toronto Stud. Geol. Ser. No. 27* (1928), p. 15.
16. From α-decay theory, $d\lambda/\lambda \simeq (3/2)(ZR)^{1/2}(dR/R) + (2Z/E^{1/2})(dE/E)$, where Z is the atomic number, R is the nuclear radius in 10^{-15} m, and E ($= E_\alpha$) is the α-decay energy in million electron volts. A particle of mass m and charge z has a range r (halo radius), given by the expression $r = \text{constant} \times E^2/mz^2$. Then $d\lambda/\lambda \simeq 43(dR/R) + 46(dr/r)$. If the difference between the halo radius and the coloration band size at 4.2 Mev is real, then $\Delta r = -0.4$ μm and $d\lambda/\lambda \simeq 46(-0.4/13) = -1.4$. Since the minimum uncertainty in making comparative range measurements is $\Delta r = 0.1$ μm, it is actually impossible to establish the constancy of λ (for ^{238}U) from radiohalo data any better than $d\lambda/\lambda \simeq 46(0.1/13) = 0.35$. Also, if $dE/E = 0$ while $dR/R \neq 0$, then $d\lambda/\lambda \neq 0$. In such a case, halos furnish no proof that λ is constant.
17. Some inner ring coloration in Fig. 1f results from other α-emitters in the U decay chain. Fission track analysis shows that the dose of α-particles from ^{238}U is only about 10^{13} per square centimeter, about ten times less that the ^{4}He ion dose for medium coloration.
18. R. V. Gentry, in preparation.
19. ———, in *Proceedings of the Second Lunar Science Conference* (MIT Press, Cambridge, 1971), vol. 1, pp. 167–168.
20. ———, *Annu. Rev. Nucl. Sci.* **23**, 347 (1973).
21. This work was sponsored by the Atomic Energy Commission under contract with Union Carbide Corporation, and by NSF grant GP-29510 to Columbia Union College, Takoma Park, Maryland.

2 July 1973; revised 26 December 1973

(Reprinted from Nature, Vol. 252, No. 5484, pp. 564–566, December 13, 1974)

'Spectacle' array of [210]Po halo radiocentres in biotite: a nuclear geophysical enigma

POLONIUM radiohaloes occur widely and not infrequently (total about $10^{15}-10^{20}$) in Precambrian rocks but their existence has so far defied satisfactory explanation based on accepted nucleo-cosmogeochemical theories[1]. Do Po haloes imply that unknown processes were operative during the formative period of the Earth? Is it possible that Po halos in Precambrian rocks represent extinct natural radioactivity[2] and are therefore of cosmological significance? A detailed comparison between an unusual array of Po halo radiocentres and U–Th halo radiocentres is presented here as bearing on the above questions.

Generally, radiohaloes occur in one of several mineralogical contexts[1,3,4]. First, as single haloes around discrete inclusions well isolated from other mineral defects and haloes; second, as single haloes around discrete inclusions lodged in conduits or cleavage cracks; third, as single haloes randomly spaced in clusters (sometimes overlapping); fourth, as vein haloes which formed from a continuous distribution of radioactivity (apparently deposited from hydrothermal solutions) along a conduit; and fifth, as line haloes, which surround, not conduits or cracks, but genuine single inclusions which are long (for example, 25 μm) compared with their width (perhaps 1 μm). Large, amorphous, coloured regions without discrete inclusions are not haloes.

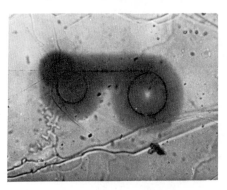

Fig. 1 'Spectacle' array of [210]Po haloes in biotite. Halo radius, 18.5 μm.

A striking exception[1] to this classification is the 'spectacle' coloration pattern (Fig. 1), which exhibits two almost circular rings of inclusions joined by a linear array of inclusions. As far as we know this is unlike any group of haloes previously seen. This geometrical arrangement of halo radiocentres, found in a Precambrian biotite from Silver Crater Mine, Faraday Township, Ontario, exhibits true radiohalo characteristics.

First, the coloration is identical to that of normal haloes found about 300 μm away in the same mica specimen. Second, the three-dimensional nature of the halo pattern was demonstrated when the specimen (initially about 50 μm thick) was cleaved; both halves revealed matching 'spectacle' coloration patterns, the only difference being the presence of the inclusion array in one half and its absence in the other half. Third, the radius of the coloration band (18.5 μm) implied an origin from [210]Po α decay. Mass spectrometric and X-ray fluorescence methods were used to ascertain whether this was indeed a Po halo array.

Before applying these techniques to the 'spectacle' halo, we established that ion-microprobe mass analyses and scanning electron microscope X-ray fluorescence (SEMXRF) studies of 'normal' or 'standard' halo radiocentres (those formed from both U and Th α decay) yielded data consistent with the visual means of identification. Several U–Th haloes (see, for example, photo insert, Fig. 2) found in a Precambrian pegmatitic mica from Rossi, New York, were analysed by X-ray and ion-probe techniques. Several U–Th halo radiocentres were chosen which contained only U, Th and Pb in any significant abundance, thereby virtually eliminating any molecular ion interference in the Pb–Th–U region ($m/e = 204-238$) in the ion probe.

That the mica matrix[5] yielded insignificant molecular ion currents in the region $m/e = 160-320$ is evident from the data in the lower portion of Fig. 2. In contrast, the recorded spectra of a U–Th inclusion (upper left portion of Fig. 2) revealed a significant number of ion counts accumulated in 12 passes of the regions $m/e = 204-209$ and (with a different scale) $m/e = 232-240$. Total ion counts are tabulated just above the two spectra. The scans on the Pb–Bi region ($m/e = 204-209$) lasted for several minutes and were taken before the scans (equal time) on the U–Th region.

Exact $^{206}Pb/^{238}U$ and $^{208}Pb/^{232}Th$ ratios are not obtainable from the ion count data in Fig. 2 because variable U and Th concentrations were observed as the ion probe beam sputtered away the inclusion; accurate ratios could be obtained by simultaneously accumulating counts in the region 204–238 provided that the greater secondary ion yield of U and Th as compared with Pb is taken into account. On the other hand, the separate Pb and U isotope ratios are meaningful. Note, for example, that after subtraction of background counts at $m/e = 240$ from the total counts at $m/e = 235$ and 238, the 235/238 value (0.76) satisfactorily approximates (considering the relatively small number of counts collected) the natural U isotopic ratio, $^{235}U/^{238}U = 0.72$. The absence of a peak at 204 shows there is little or no common lead in the inclusion and therefore, that the 206/207 ratio is that of $^{206}Pb/^{207}Pb$ as derived from *in situ* U decay.

Also shown in Fig. 2 are the SEMXRF spectra of the mica matrix and the U–Th halo radiocentre, both of which correlate well (with the exception of the low Z and low abundance elements in the former) with the respective ion-probe spectra. Only U, Th and Pb are exclusively in the inclusion.

The ion-microprobe mass spectrum of the mica matrix surrounding the 'spectacle' halo was nearly identical to the mica spectrum shown in Fig. 2 and is not repeated in Fig. 3. Figure 3 (top centre) shows the portion $m/e = 160-264$ of the ion-microprobe spectrum (verical log scale) of several of the inclusions. Also shown is the actual ion-probe trace of the important region from $m/e = 204-210$ using a linear vertical scale and an expanded horizontal scale. There is no significant ion current above $m/e = 209$; that is, no significant ion signals were detected at any of the prominent U and Th peaks: 238(U+), 254(UO+), 232(Th+) and 248(ThO+). No $m/e = 204$ was detected above background (1 c.p.s.), and the 206/207 mass ratio was ≃20 (206 signal ≃ 2,000 c.p.s.).

Figure 3 also shows SEMXRF spectra of the surrounding mica and of one of the Po halo radiocentres. Lead is the only element detectable in this radiocentre exclusive of the mica; some adjacent radiocentres revealed Bi as well. The use of two different instruments, and longer counting times, account for the slightly different X-ray spectra in Figs 2 and 3. The excellent resolution of the SEM showed the Pb-rich areas to coincide exactly with the Po halo radiocentres which are visible both in ordinary transmitted (Fig 1) and reflected light microscopy. Regions as close as 1 μm to the radiocentres showed virtually no Pb or Bi, implying little if any diffusion loss from the inclusions.

As the X-ray data definitely show Pb (and sometimes Bi) in the 'spectacle' halo radiocentres, and as there is no evidence for any molecular ion contribution in the region from $m/e =$

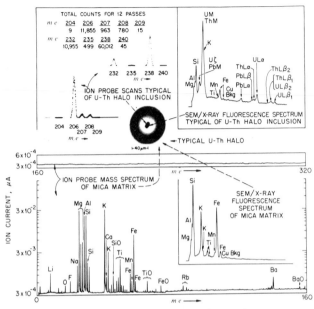

Fig. 2 Ion microprobe and XRF comparison between mica matrix and U-Th halo inclusion.

204–238, the 206, 207 and 208 peaks are interpreted as Pb isotopes and 209 as [209]Bi. [204]Pb, a constituent of both common and primordial Pb, is missing (no 204 peak), implying that the 'spectacle' halo inclusions analysed contained no detectable Pb of either of these types. Absence of the 232, 235 and 238 peaks is interpreted as showing the inclusions contain virtually no [232]Th, [235]U or [238]U and, therefore, no radiogenic [208]Pb, [207]Pb or [206]Pb derived from the *in situ* decay of these isotopes. The 207 and 208

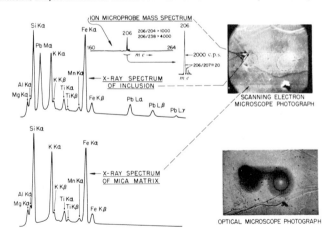

Fig. 3 Ion microprobe and SEMXRF spectra of mica matrix and [210]Po halo inclusions.

peaks are therefore attributed to ^{207}Pb and ^{208}Pb, perhaps arising from the decay of minute amounts of ^{211}Bi and ^{212}Bi within the inclusions[5,6]. The ^{209}Bi is considered to be primordial.

The outstanding feature of the mass analysis is the prominent 206 signal which, when attributed to the presence of ^{206}Pb in the inclusions, fits perfectly with the prediction based on ring structure measurements, that is, that the ^{206}Pb is radiogenically derived, not from U or Th, but directly from ^{210}Po α decay. In this respect, the large difference in the 206/238 (^{206}Pb/^{238}U) ratio between the 'spectacle' halo and the U–Th halo (Figs 2 and 3) is especially significant. Clearly the 'spectacle' halo resulted from ^{210}Po α decay; an explanation for its geometry is still under study.

Because the Pb isotope in these inclusions is not explicable as any combination of common, primordial, or from *in situ* Pb derived radiogenically *in situ* from U or Th, we conclude that a different type of Pb, derived from Po α decay, exists in nature. Supportive evidence comes from electron-probe and ion-probe analyses of a ^{218}Po halo radiocentre found in a mica from the Iveland District, Norway, which yielded a ^{206}Pb/^{207}Pb ratio of 23. This is consistent with that expected from ^{218}Po α decay to ^{206}Pb. Such a Pb ratio is impossibly high based on normal isotopic ^{238}U/^{235}U decay, the theoretical maximum being 21.8.

Other investigations have shown varying mixtures of U-derived and Po-derived Pb may occur in the same radiocentre, for there exists an almost continuous halo spectrum stretching from 'pure' U to 'pure' Po haloes. Only a few (< 0.01) Po haloes in biotite may survive the delicate sectioning process necessary for SEM X-ray analysis.

Just as important as the existence of a new type of lead is the question of whether Po haloes which occur in a granitic or peg-matitic environment (for example, in mica, fluorite or cordierite) can be explained by accepted models of Earth history[1]. (R. V. G. has found other ^{210}Po haloes that differ essentially from those in granites—unpublished information.)

This research has been sponsored by the United States Atomic Energy Commission under contract with Union Carbide Corp. and by Columbia Union College with an assistance grant from the National Science Foundation. Thanks are due to R. I. Gait and J. A. Mandarino, Royal Ontario Museum, Louis Moyd, National Museum of Canada, and G. Switzer, United States National Museum, for providing specimens.

ROBERT V. GENTRY

Chemistry Division,
Oak Ridge National Laboratory,

L. D. HULETT

Analytical Chemistry Division,
Oak Ridge National Laboratory

S. S. CRISTY
J. F. MCLAUGHLIN

Laboratory Development Department,
Oak Ridge Y-12 Plant,
Oak Ridge, Tennessee 37830

J. A. MCHUGH

Knolls Atomic Power Laboratory,
Schenectady, New York 12301

MICHAEL BAYARD

McCrone Associates,
2820 Michigan Street,
Chicago, Illinois 60616

Received July 31, 1974.

[1] Gentry, R. V., *Science*, **184**, 62 (1974).
[2] Kohman, T. P., *Ann. N. Y. Acad Sci.*, **62**, 503 (1956).
[3] Henderson, G. H., Mushkat, C. M., and Crawford, D. P., *Proc. R. Soc.*, **A158**, 199 (1934); Henderson, G. H., and Turnbull, L. G., *ibid.*, **145**, 582 (1934); Henderson, G. H., and Bateson, S., *ibid.*, 573 (1934).
[4] Gentry, R. V., *Ann. Rev. Nucl. Sci.*, **23**, 347 (1973).
[5] Gentry, R. V., Cristy, S. S., McLaughlin, J. F., and McHugh, J. A., *Nature*, **244**, 282 (1973).
[6] Gentry, R. V., *Science*, **173**, 727 (1971).

Reprinted from
15 October 1976, Volume 194, pp. 315-318

SCIENCE

Radiohalos in Coalified Wood: New Evidence Relating to the Time of Uranium Introduction and Coalification

Robert V. Gentry, Warner H. Christie, David H. Smith, J. F. Emery
S. A. Reynolds, Raymond Walker, S. S. Cristy and P. A. Gentry

Copyright© 1976 by the American Association for the Advancement of Science

Radiohalos in Coalified Wood: New Evidence Relating to the Time of Uranium Introduction and Coalification

Abstract. *The discovery of embryonic halos around uranium-rich sites that exhibit very high $^{238}U/^{206}Pb$ ratios suggests that uranium introduction may have occurred far more recently than previously supposed. The discovery of ^{210}Po halos derived from uranium daughters, some elliptical in shape, further suggests that uranium-daughter infiltration occurred prior to coalification when the radionuclide transport rate was relatively high and the matrix still plastically deformable.*

Even though the biological fossil record has been extensively documented, the rather abundant fossil record of radiohalos that exists in the coalified wood from the Colorado Plateau has remained virtually undeciphered. Jedwab (*1*) and Breger (*2*) have determined some important characteristics of such halos; in fact, earlier (*1, 2*) as well as present investigations on these samples (*3*) agree that: (i) the microscopic-size radiocenters responsible for halos (Fig. 1a) in coalified wood are actually secondary sites that preferentially accumulated α-radioactivity during an earlier period of earth history when uranium-bearing solutions infiltrated the logs after they had been uprooted; (ii) although autoradiography shows some α-activity dispersed throughout the matrix (*1, 2*), most of it is still concentrated in the discrete halo radiocenters; (iii) variations in coloration among radiohalos cannot necessarily be attributed solely to differences in the α-dose because there is evidence that the coalified wood was earlier far more sensitive to α-radiation than at present (*1*); (iv) halos that appear most intensely colored in unpolarized transmitted light also show evidence of induration; that is, when polished thin sections of coalified wood are viewed with reflected light (Fig. 1b), such high α-dose halos exhibit high reflectivity and pronounced relief; and (v) some areas of coloration are of chemical rather than radioactive origin (*1*).

In addition to the above verifications, the studies reported here mark the first time that (i) radii measurements have been made to determine the type and stage of development of halos in coalified substances and (ii) the radiocenters of such halos have been analyzed by modern analytical techniques. The discoveries reported herein raise questions

relative to when U was introduced into the wood, the duration required for coalification, and the age of the geological formations.

Specifically, it was discovered that the halos (Fig. 1a) surrounding the α-active sites are typically embryonic, that is, they do not generally exhibit the outer ^{214}Po ring characteristic of fully developed U halos in minerals (*4*). Such underdeveloped halos generally imply a low U concentration in the radiocenter. However, electron microprobe x-ray fluorescence (EMXRF) analyses (Fig. 2a) show many such radiocenters contain a large amount of U with the amount of daughter product Pb being generally too small to detect by EMXRF techniques (Fig. 2a). Although we discuss below the application of ion microprobe mass spectrometer (IMMA) techniques (*5*) to the prob-

Fig. 1. (a) Coalified wood halos with U radiocenters in transmitted light (× 125) [see (*7*)]. (b) The same halos in reflected light. The bright central spot in each halo is the radiocenter (× 125)

lem of quantitatively determining the $^{238}U/^{206}Pb$ ratios, two important points deserve mention here: (i) if there was only a one-time introduction of U into the wood (*2*), these radiocenters date from that event unless subsequent mobilization of U occurred, and (ii) if U was introduced prior to coalification (*1*), then the $^{238}U/^{206}Pb$ ratios in these radiocenters also relate to the time of coalification.

Another class of more sharply defined halos was discovered possessing smaller inclusions (≈ 1 to 4 μm in diameter) than the α-active sites. These inclusions exhibit a distinct metallic-like reflectance when viewed with reflected light. Three different varieties of this halo exist: one with a circular cross section, another with an elliptical cross section with variable major and minor axes, and a third most unusual one that is actually a dual halo, being a composite of a circular and an elliptical halo around exactly the same radiocenter (see Fig. 3, a to c).

Although the elliptical halos differ radically from the circular halos in minerals (*6*), the circular type resembles the ^{210}Po halo in minerals and variations in the radii of circular halos approximate the calculated penetration distances (≈ 26 to 31 μm) of the ^{210}Po α-particle (energy E_α = 5.3 Mev) in this coalified wood (*7*). Henderson (*8*) theorized that Po halos might form in minerals when U-daughter Po isotopes or their β-precursors were preferentially accumulated into small inclusions from some nearby U source. Although this hypothesis was not confirmed for U-poor minerals (*9*), it did seem a plausible hypothesis in this U-rich matrix.

The EMXRF analyses (Fig. 2b) showed that the halo inclusions were mainly Pb and Se. This composition fits well into the secondary accumulation hypothesis for both of the U-daughters, ^{210}Po (half-life, $t_{1/2}$ = 138 days) and its β-precursor ^{210}Pb ($t_{1/2}$ = 22 years), possess the two characteristics that are vitally essential for the hypothesis: (i) chemical similarity with the elements in the inclusion and (ii) half-lives sufficiently long to permit accumulation prior to decay. This latter requirement is dependent on the radionuclide transport rate. In minerals the diffusion coefficients are so low that there is a negligible probability that ^{210}Po or ^{210}Pb atoms would migrate even 1 μm before decaying, and thus the ori-

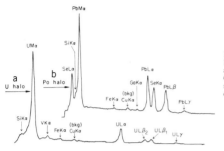

Fig. 2. Curve a,
EMXRF spectrum of
a U-rich radiocenter.
Curve b, EMXRF
spectrum of the radio-
center of a ^{210}Po halo.

gin of Po halos in minerals is still being argued (6, 10).

However, in this matrix the situation is quite different. A solution-permeated wood in a gel-like condition would exhibit a much higher transport rate as well as unusual geochemical conditions which might favor the accumulation of ^{210}Po and ^{210}Pb nuclides. Evidence that this accumulation was essentially prior to complete coalification comes from the fact that most Po halos are plastically deformed; furthermore, after coalification it is much more difficult to account for such rapid and widespread migration of the radionuclides (that is, within the ^{210}Po half-life). For example, a hundred or more ^{210}Po halos are sometimes evident in a single thin section (2 cm by 2 cm) of coalified wood, and they occurred quite generally in the thin sections examined (11). Of the thousands of Po halos seen in this matrix, only three show any trace of a ring that could possibly be attributed to ^{214}Po α-decay [that is, from the accumulation of the U-daughters ^{214}Pb ($t_{1/2} = 27$ minutes), ^{214}Bi ($t_{1/2} = 20$ minutes), or ^{214}Po ($t_{1/2} = 164$ μsec)], and none has been seen with a ring from ^{218}Po α-decay [that is, from the accumulation of short-lived ^{218}Po ($t_{1/2} = 3$ minutes)]. (Possibly these faint outer rings are of chemical rather than radioactive origin.)

Positive identification for the ^{210}Po halos comes from the IMMA analyses. Compared to a ^{238}U halo radiocenter, a ^{210}Po halo inclusion should contain much less ^{238}U (perhaps none at all) and much more of the ^{210}Po decay product ^{206}Pb. The IMMA analyses of Po halo inclusions showed that the ^{238}U content was low, the ^{238}U/^{206}Pb ratios varying from 0.001 to 2.0. [These values were corrected for the different ionization efficiencies ($\sim 2:1$) of Pb$^+$ and U$^+$ in this matrix.] This small ^{238}U content implies that only an extremely small amount of Pb could have been generated by in situ U decay. There are certainly three other

possible sources for the Pb in these inclusions: (i) common Pb, (ii) Po-derived radiogenic Pb generated by in situ decay of ^{210}Po, or (iii) U-derived "old" radiogenic Pb that had accumulated in the hypothesized (12) Precambrian U ore deposit (which is one possible source of the U now in the Colorado Plateau) prior to the time it was carried with the U in solution into the wood. Since the ^{204}Pb count rates, which are unique indicators of common Pb, ranged from undetectable to a few counts per second above background when ^{206}Pb count rates were several thousand counts per second, it was evident that relatively little common Pb was present. Thus only ^{206}Pb/^{207}Pb ratios had to be measured to obtain evidence of ^{206}Pb originating from the decay of ^{210}Po; the results were indeed confirmatory.

The ratios obtained were as follows: ^{206}Pb/^{207}Pb = 8 ± 0.5, 11.6 ± 0.3, 11.7 ± 0.4, 13.3 ± 0.7, 13.4 ± 1.0, 13.7 ± 0.6, 13.9 ± 0.6, 14.8 ± 0.9, 15.8 ± 1.1, and 16.4 ± 0.5. The variation in this ratio can easily be understood to have resulted from the addition of an increment of ^{206}Pb (generated by in situ ^{210}Po decay) to the isotopic composition of the "old" radiogenic Pb. The lowest Pb ratio, obtained from a very lightly colored ^{210}Po halo, differs slightly from the lowest Pb isotope ratio previously determined on bulk samples of Colorado Plateau U ore specimens (12).

What is the meaning of these Po halos? Clearly, the variations in shape can be attributed to plastic deformation which occurred prior to coalification. Since the model for ^{210}Po formation thus envisions that both ^{210}Po and ^{210}Pb were accumulating simultaneously in the Pb-Se inclusion, a spherical ^{210}Po halo could develop in 0.5 to 1 year from the ^{210}Po atoms initially present and a second similar ^{210}Po halo could develop in 25 to 50 years as the ^{210}Pb atoms more slowly β-decayed to produce another crop of ^{210}Po

atoms. If there was no deformation of the matrix between these periods, the two ^{210}Po halos would simply coincide. If, however, the matrix was deformed between the two periods of halo formation, then the first halo would have been compressed into an ellipsoid and the second halo would be a normal sphere. The result would be a dual "halo" (Fig. 3c). The widespread occurrence of these dual halos in both Triassic and Jurassic specimens (13) can actually be considered corroborative evidence for a one-time introduction of U into these formations (1, 2), because it is then possible to account for their structure on the basis of a single specifically timed tectonic event. The fact that dual halos occur in only about 1 out of 100 single Po halos is of special significance (14).

In halos with U radiocenters, the low Pb abundance made it generally quite difficult to measure U/Pb ratios with EMXRF (Fig. 2a) techniques. More sensitive IMMA measurements on these U radiocenters revealed ^{238}U/^{206}Pb ratios (15) of approximately 2230; 2520; 8150; 8300; 8750; 18,700; 19,500; 21,000; 21,900; and 27,300 (again corrected for different ionization efficiencies). Typically, the U$^+$ ion signals from which these ratios were derived were greater than 3×10^4 counts per seconds (cps); for example, the 19,500 value was obtained from a halo with a U$^+$ signal of 10^6 cps (± 5 percent) with background ≈ 3 cps. We checked the ^{238}U/^{235}U ratio independently (and found it normal) by excising several radiocenters and analyzing them directly on the filament of a high-sensitivity thermal ionization mass spectrometer (16).

Even without attempting to subtract out the ^{206}Pb component of the common and "old" radiogenic Pb (15), these ^{238}U/^{206}Pb ratios raise some questions. For example, if the ^{238}U/^{206}Pb = 27,300 value is indicative of the formation time of the radiocenter, this is more recent by at least a factor of 270 than the minimum (Cretaceous) and more recent by a factor of 760 than the maximum (Triassic) geological age estimated for the introduction of U into the logs (12, 17, 18). To obtain ^{238}U/^{206}Pb ratios that more accurately reflect the amount of Pb from in situ U decay, a search was made for sites with even higher ratios, for such areas possibly contained negligible amounts of extraneous Pb. Two halo radiocenters were found that exhibited ^{238}U$^+$ signals of 4×10^4 and 6.4×10^4 cps, respectively, while the ^{206}Pb$^+$ signals were indistinguishable from background (≳ 3 cps) in both cases (^{207}Pb also absent).

Such extraordinary values admit the

possibility that both the initial U infiltration and coalification could possibly have occurred within the past several thousand years. At the same time it may be argued that this view is quite improbable for there exists another explanation that could invalidate the association of the U/Pb ratios with the initial introduction of U. This explanation would admit that, although Po halos constitute evidence that U infiltration and hence U radiocenter formation occurred prior to coalification, some U may have been added or Pb may have been selectively removed, or both, by groundwater circulation after coalification. Hence variable U/Pb ratios would be expected, and the highest ratio would simply reflect the last time when U remobilization or Pb remobilization, or both, occurred. Although this hypothesis has been used to account for U disequilibrium (*18, 19*) in bulk specimens of U-impregnated Colorado Plateau material, there are some questions about its applicability here.

For example, if Pb was removed from the U sites, it must have been a very selective removal for both the EMXRF and IMMA results show that considerable quantities of Pb still remain in the nearby (within ≈ 50 μm of the U sites) Po halo Pb-Se inclusions. If Pb loss was minimal, then to explain the high $^{238}U/^{206}Pb$ ratios by remobilization requires that significant quantities of U were introduced into the U radiocenters quite recently. In any event, whether the hypothesis is U addition or Pb removal, the crucial point that seems quite difficult to explain under either assumption is the fact that, in general, the halos around U sites are embryonic (*20*). That is, since it seems clear that the U radiocenters formed during the initial introduction of U and if this were as long ago as the Triassic or Jurassic are generally thought to be, then there should be evident not only fully developed, but overexposed U halos as well (*21*).

Clearly, it was important to determine whether these phenomena were characteristic only of the U-rich Colorado Plateau coalified wood (*2, 3*). We therefore initiated studies on coalified wood fragments which are occasionally found in the Chattanooga shale (*3, 11, 22*). Thus far only embryonic halos have been seen, and the $^{238}U/^{206}Pb$ ratios are much too high (>10³) to correlate with the geological age of the formation (Devonian). The low U content of the Chattanooga shale (1 to 50 parts per million) makes it quite difficult to see how U remobilization could account for these very high isotope ratios. Thus the evidence does not appear to support the remobilization

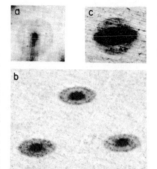

Fig. 3. (a) Circular ^{210}Po halo (× 250). (b) Compressed ^{210}Po halos (× 250). (c) Circular and compressed ^{210}Po halo (× 250).

hypothesis as a general explanation of these unusual $^{238}U/^{206}Pb$ ratios in either the Colorado Plateau or Chattanooga shale specimens.

If remobilization is not the explanation, then these ratios raise some crucial questions about the validity of present concepts regarding the antiquity of these geological formations and about the time required for coalification. Finally, in addition to again focusing attention on the question of the origin of Po halos in minerals (*6, 10*), the existence of U-derived single and dual Po halos in different formations suggests that the original source of U may have been a Precambrian ore deposit that was geographically not far removed from the present Colorado Plateau. Thus, in view of America's energy requirements, it might be profitable to search for such an ore deposit by deep drilling into selected areas around and within the Colorado Plateau.

ROBERT V. GENTRY
Chemistry Division,
Oak Ridge National Laboratory,
Oak Ridge, Tennessee 37830
WARNER H. CHRISTIE
DAVID H. SMITH
J. F. EMERY
S. A. REYNOLDS
RAYMOND WALKER
Analytical Chemistry Division,
Oak Ridge National Laboratory
S. S. CRISTY
Laboratory Development Division,
Y-12 Plant,
Oak Ridge National Laboratory
P. A. GENTRY
Columbia Union College,
Takoma Park, Maryland 20012

References and Notes

1. J. Jedwab, *Coal Science* (American Chemical Society, Washington, D.C., 1966).
2. I. A. Breger, in *Formation of Uranium Ore Deposits, Proceedings of a Symposium, Athens, 6–10 May 1974* (International Atomic Energy Agency, Vienna, 1974), pp. 99–124.
3. I. A. Breger donated Colorado Plateau coalified wood specimens from the following mines: (i) Jurassic—Peanut and Virgin No. 3, Colorado; Corvusite, Utah; and Poison Canyon, New Mexico; (ii) Triassic—Lucky Strike No. 2, Dirty Devil No. 2, Adams, and North Mesa No. 9, all in Utah; and (iii) Eocene—Docamour, Colorado. J. S. Levinthal provided 16 other specimens. However, only those from the Rajah 49 mine [Salt Wash member of the Morrison Formation (Jurassic)] were sufficiently well preserved to exhibit halos. The Chattanooga shale coalified wood (Devonian), which came from near Nashville, Tennessee, was donated by I. A. Breger and V. E. Swanson. Breger's analysis of this coalified wood yielded 0.001 to 16 percent U, 54 to 84 percent C, 3 to 7.5 percent H, 0.3 to 1.8 percent N, 6 to 38 percent O, and 0.6 to 14.5 percent S. Except where stated, all experimental results refer to work on Colorado Plateau coalified wood (Triassic and Jurassic formations). A thin section of a coalified wood specimen (earlier obtained from I. A. Breger) was provided by J. Jedwab and was used along with Breger's other specimens. Although personal communications with Breger and Jedwab proved of great value, this in no way implies that either Jedwab or Breger necessarily concurs with the results presented here.
4. R. V. Gentry, *Annu. Rev. Nucl. Sci.* **23**, 347 (1973). The halo in Fig. 1a would extend another 20 μm if fully developed.
5. C. A. Andersen and J. R. Hinthorne, *Science* **175**, 853 (1972).
6. R. V. Gentry, *ibid.* **184**, 62 (1974).
7. If the appropriate formulas [G. Friedlander, J. W. Kennedy, J. M. Miller, *Nuclear and Radiochemistry* (Wiley, New York, ed. 2, 1964), pp. 95–98] are used for computing α-ranges in various solids, the ranges of a 5.3-Mev α-particle in coalified wood [see (*3*)] of density 1.3 and 1.6 g/cm³ would be 31 and 25 μm, respectively. Uniform shrinkage of the matrix could also reduce the radius.
8. G. H. Henderson, *Proc. R. Soc. London Ser. A* **173**, 250 (1930).
9. R. V. Gentry, *Science* **160**, 1228 (1968).
10. ———, *Nature (London)* **252**, 564 (1974); *ibid.* **258**, 269 (1975).
11. This occurrence of Po halos refers to the Colorado Plateau coalified wood.
12. L. R. Stieff, T. W. Stern, R. G. Milkey, *U.S. Geol. Surv. Circ.* 271 (1953).
13. Dual halos have thus far been found in specimens from the North Mesa No. 9 mine in Utah and the Virgin No. 3 and Rajah 49 mines [see (*3*)].
14. The coloration pattern of the dual halos provides the key to understanding its rarity. If U with its daughters were concurrently flushed out of some Precambrian ore deposit, even with a relatively short transit time from the ore deposit to the wood, equilibrium conditions still require that more than 50 times as much ^{210}Pb as ^{210}Po be available for accumulation. If the wood exhibited constant sensitivity to α-induced coloration, then the outer circular halo resulting from ^{210}Pb accumulation would be expected to be much darker than the elliptical halo resulting from ^{210}Po accumulation. The fact that just the opposite is true is in good agreement with the evidence found by Jedwab [(*1*) and private communication] indicating that during the U infiltration the gel-like wood exhibited much higher sensitivity to α-induced coloration as compared to the later stages of coalification. Possibly then, a relatively dark halo could have formed rather quickly from as few as 10⁴ to 10⁵ ^{210}Po atoms, whereas some 20 to 50 years later the change in the coloration sensitivity of the matrix might require an α-dose 50 to several hundred times higher from the ^{210}Pb decay sequence to produce even a light halo. Thus possibly only in rare cases would the Pb-Se inclusions accumulate large enough quantities of ^{210}Pb to subsequently generate the outer circular halo.
15. The variation in the $^{238}U/^{206}Pb$ ratios may be attributed primarily to the "old" radiogenic Pb component and secondarily to ^{226}Ra and ^{210}Pb, which, in varying amounts, were also incorporated into the U-rich radiocenters. Evidence for this "old" radiogenic Pb was also found in larger, millimeter-size U-rich regions which also contained varying amounts of Na, Al, K, Ca, Ti, V, Fe, Y, Zr, Ba, and the rare earths. Such regions exhibit variable (but not very high) U/Pb ratios and very little common Pb.

16. D. H. Smith, W. H. Christie, H. S. McKown, R. L. Walker, G. R. Hertel, *Int. J. Mass Spectrom. Ion Phys.* **10**, 343 (1972–1973).

17. R. P. Fischer, in *Proceedings of the International Conference on the Peaceful Uses of Atomic Energy, Geneva, August 1955* (United Nations, New York, 1956), vol. 6, p. 605; *Econ. Geol.* **65**, 778 (1970).

18. S. C. Lind and C. F. Whittemore, *U.S. Bur. Mines Tech. Pap. 88* (1915), p. 1; T. W. Stern and L. R. Stieff, *U.S. Geol. Surv. Prof. Pap. 320* (1959), p. 151; J. N. Rosholt, in *Proceedings of the Second U.N. International Conference on the Peaceful Uses of Atomic Energy, Geneva, September 1958* (United Nations, New York, 1958), vol. 2, p. 231.

19. Nondestructive γ-ray spectrometry was utilized to check on U disequilibrium in gram-size specimens of the Colorado Plateau coalified wood. We found significant differences in the γ-spectra that could reasonably be attributed to U disequilibrium. By removing microportions of U-rich areas and physically smearing the material onto steel planchets for α-counting, we observed one α-spectra that unambiguously indicated U disequilibrium between ^{234}U and ^{230}Th, or ^{230}Th and ^{226}Ra, or both. Excess α-activity in the $\sim$ 4.7-Mev region was not attributed to excess ^{234}U because mass spectrometry measurements on a separate specimen showed an equilibrium ^{238}U/^{234}U value.

20. Less than 2.5 percent of the halos with U radio-centers have any trace of an outer ring. It is difficult to associate these with sequential α-decay from ^{238}U because such weak rings do not correlate with the U content. These weak rings may have resulted from diffusion of α-radioactivity out of the radiocenter prior to induration of the halo region by the α-radioactivity. Alternatively, these weak rings may have resulted from the accumulation of small amounts of ^{222}Rn, ^{214}Pb, or ^{226}Ra. In fact, the size of the dark halo region around the U-rich sites admits of the possibility that the inner halos may have formed from the accumulation of minute amounts of ^{226}Ra or ^{210}Pb, or both. Their more diffuse radiocenters, however, would prevent the formation of well-defined boundaries as in the case of the Pb-Se inclusions.

21. This would be true even if coalified wood is only 1/10 as sensitive to α-coloration as biotite.

22. I. A. Breger and J. M. Schopf, *Geochim. Cosmochim. Acta* **7**, 387 (1955); V. E. Swanson, *U.S. Geol. Surv. Prof. Pap. 300* (1956), p. 451. J. Jedwab informed me of halos in this material.

23. I thank I. A. Breger, J. S. Levinthal, V. E. Swanson, and J. Jedwab for supplying coalified wood specimens. Research sponsored by the Energy Research and Development Administration under contract with Union Carbide Corporation, and by Columbia Union College under NSF research grant DES 74-23451.

15 September 1975, revised 30 June 1976

NATIONAL SCIENCE FOUNDATION
WASHINGTON, D.C. 20550

Division of Earth Sciences

July 11, 1977

Dr. Robert V. Gentry

Dear Dr. Gentry:

This is in answer to your letter of June 27, requesting panel review comments on your proposal (EAR7713496). The panel review comments were not included in the declination letter because according to our rules that letter must go out first and then be followed by the comments, if requested. What I have done below is to give you the general nature of the panel discussion, based on my notes at the time, my memory of the discussion, and a short (two sentence) recommendation put on tape by the Chairman of the Geochemistry Panel. We do not tape the whole discussion. Here it is:

Much of the panel discussion centered on the general significance of the occurrence of "radioactive haloes" (both giant and dwarf) and the techniques the principal investigator has used to investigate them. The panel considers the occurrence of haloes of interest, but not of prime importance to geochemistry. One aspect of the past research was to try to detect "superheavy" elements in the mineral nucleii of giant haloes, and a tentative identification of superheavies was shown to be incorrect. The panel felt that the principal investigator and his colleagues handled the release of information concerning superheavy element detection judiciously - i.e., in an objective and straightforward manner with no sensationalism (which could have happened considering the potential scientific importance of the discovery). However, the panel did fault the principal investigator and his colleagues for the techniques used to try to detect superheavy elements. The initial method of an X-ray fluorescence attachment on a scanning electron microscope should have been known not to have the sensitivity. The tentative identification of elements with an atomic number near 26 resulted from using the proton induced X-ray emission method. The signal that resulted in the tentative identification of these elements has now been attributed with some confidence to a $Ce(p, n\gamma)$ nuclear reaction rather than X-ray fluorescence from elements with an atomic number of around 126. The panel felt that the principal investigator and his colleagues should have checked out all such possible reactions before publication because monazite (the mineral inclusion of the center of the halo) is a mineral known to contain large

Dr. Robert V. Gentry   - 2 -   July 11, 1977

amounts of cerium and other rare earth elements. The principal
investigator proposes a continuation of the search for superheavy
elements. The panel felt that there is little possibility of their
detection by the proposed techniques.

The most important criticism of the proposal did not, however, have
to do with superheavy element detection. The criticism stemmed from
the general nature of the proposed research on haloes. The principal
investigator has been collecting specimens, examining them petrographically,
and reporting their morphology and mineral occurrence for a number of years.
The panel considered that these descriptive contributions have been of
some value, but felt that more of the same approach had little potential
to contribute something new. The main difficulty with the proposal is
that (aside from the superheavy element search) there was no hypothesis
concerning the origin of the haloes that the principal investigator
proposed to test. He has already looked at and described a number of
occurrences. The panel felt that it was not justified in recommending
funding of a research project that merely proposed to make additional
observations of the phenomenon. There seems little possibility that the
principal investigator could arrive at a hypothesis by looking at
additional haloes since he has not been able to propose one at this time.

In summary, the panel considers giant and dwarf haloes to be of some geochemical interest, but feels that the proposed research was not likely to
make significant additional contributions to our knowledge of their origin.

I hope that this outline of the panel discussion will be of use to you in
your consideration of any future proposal you may want to submit.

                              Sincerely yours,

                              John Hower
                              Program Director
                                for Geochemistry

Copy to: Mr. Gordon E. Bullock
          Business Manager
          Columbia Union College

NATIONAL SCIENCE FOUNDATION
WASHINGTON, D.C. 20550

September 15, 1977

OFFICE OF THE
ASSISTANT DIRECTOR
FOR ASTRONOMICAL,
ATMOSPHERIC, EARTH,
AND OCEAN SCIENCES

Dr. Robert V. Gentry

Dear Dr. Gentry:

I am writing to respond to your letter of August 26, 1977, requesting reconsideration of the Foundation's earlier decision to decline your proposal, "Nuclear Geochemistry of Radiohalos," (EAR 77-13496), as provided for in the Foundation's Important Notice No. 61.

I requested a Section Head who is not a part of the Division of Earth Sciences to analyze the proposal jacket, to study the reviews, and to discuss the decision with me. Following his report to me, I have reviewed the documents in the case myself. It is my conclusion that your proposal received a thorough and fair peer review through the Geochemistry Program Office, a review that included a conscientious and careful consideration of six ad hoc mail reviews. As part of the reconsideration process your rebuttal to those reviews has been considered also.

It is my opinion that your proposal was fairly reviewed and that the decision to decline was justified. I would point out to you that under the terms of Important Notice No. 61 you are entitled to make a further appeal to the Deputy Director. You are also invited to revise or extend your thoughts for the submission of a new proposal at any time.

Sincerely yours,

Edward P. Todd
Acting Assistant Director

copy to: Business Manager
Columbia Union College

NATIONAL SCIENCE FOUNDATION
WASHINGTON, D.C.   20550

June 17, 1982

OFFICE OF THE
ASSISTANT DIRECTOR
FOR ASTRONOMICAL,
ATMOSPHERIC, EARTH,
AND OCEAN SCIENCES

Honorable Robert S. Walker
House of Representatives
Washington, D.C. 20515

Dear Mr. Walker:

Thank you for your memo to Mr. Raymond Bye, Jr. of May 12 and the enclosed letter from Mr. Leroy Anderson of Denver, Pennsylvania. Both of these are enclosed with this reply.

Mr. Anderson is correct when he states in his letter that Dr. Robert Gentry is the world's leading authority on the observation and measurement of anomolous radio-active haloes. Because of his recognized capabilities, Dr. Gentry's research was funded by the Foundation during the early 1970's. In 1977, however, a proposal presented by Dr. Gentry was declined. A copy of the pertinent correspondence is enclosed. That action was based upon the recommendations of six of his peer scientists, who found that the proposal did not measure up to either Dr. Gentry's earlier standards, as evidenced by his previously successful proposals, or to the standards of the Foundation. Following the declination, Dr. Gentry requested a formal reconsideration of his proposal. The procedures to be followed in such a situation are set forth in the enclosed NSF Circular No. 127 (revised August 1980). The outcome of the reconsideration was that the decision to decline was sustained. A copy of the letter informing Dr. Gentry of that decision is enclosed. The proposal to which your constituent, Mr. Anderson, refers was submitted in 1979. It was reviewed by mail by six of Dr. Gentry's peer scientists and by a panel of six additional scientists. Based upon the recommendations of these twelve knowledgeable persons, the proposed research was declined in April 1980. A copy of the correspondence is enclosed.

Please note that in each letter to Dr. Gentry he has been invited to resubmit his proposed research ideas. The funding process within the NSF is competitive for each submission of a proposal. The fact that a proposer has a grant in force does not bear upon whether he will be awarded a new grant. Each offering must stand on its own merit. We will be pleased to review and evaluate a proposal from Dr. Gentry at any time. I assure you that any submission will be given a fair, honest and open appraisal by his peers and that if they judge his ideas as worthy of support, he will be funded.

We appreciate your interest in the Foundation's programs and will be pleased to supply you with further information if you wish.

Sincerely yours,

Francis S. Johnson
Assistant Director

Enclosures

NATIONAL SCIENCE FOUNDATION

WASHINGTON, D.C.   20550

February 14, 1983

OFFICE OF THE
ASSISTANT DIRECTOR
FOR ASTRONOMICAL,
ATMOSPHERIC, EARTH,
AND OCEAN SCIENCES

Honorable Robert J. Lagomarsino
House of Representatives
Washington, D.C.   20515

Dear Mr. Lagomarsino:

I am replying to your letter of January 27, 1983, in which you asked about
NSF's handling of a research proposal from Dr. Robert Gentry of Columbia
Union College.

All NSF funding decisions are based on a process of peer review which
involves mail reviews by several experts in the field and, in many cases,
further consideration by a panel of scientists from outside NSF. Only
about half of the proposals we receive can be funded. Criteria used are
stated in our booklet "Grants for Scientific and Engineering Research"
(NSF 81-79, copy of relevant page enclosed). The holding of unorthodox
scientific views is not a barrier to the receipt of NSF support, and the
best evidence for this is the fact that during the 1970's NSF funded
several of Dr. Gentry's proposals including one for $54,900 for the study
of "Nuclear Geophysics of Radiohalos."

Please reassure your constituent that NSF funding decisions are based on
well identified criteria and that Dr. Gentry's views have not been a
barrier to his receiving NSF support.

Sincerely yours,

Francis S. Johnson
Assistant Director

Enclosure

Reprint Series
16 April 1982, Volume 216, pp. 296–298

Differential Lead Retention in Zircons:
Implications for Nuclear Waste Containment

Robert V. Gentry, Thomas J. Sworski, Henry S. McKown, David H. Smith, R. E. Eby, and W. H. Christie

Copyright © 1982 by the American Association for the Advancement of Science

Differential Lead Retention in Zircons:
Implications for Nuclear Waste Containment

Abstract. *An innovative ultrasensitive technique was used for lead isotopic analysis of individual zircons extracted from granite core samples at depths of 960, 2170, 2900, 3930, and 4310 meters. The results show that lead, a relatively mobile element compared to the nuclear waste–related actinides uranium and thorium, has been highly retained at elevated temperatures (105° to 313°C) under conditions relevant to the burial of synthetic rock waste containers in deep granite holes.*

We report here the measurement of Pb isotope ratios of whole, undissolved zircons, which were loaded directly onto the rhenium filament of a thermal ionization mass spectrometer. This innovation eliminates the Pb contamination introduced in standard chemical dissolution procedures. By using this technique, we were able to measure contamination-free Pb isotope ratios on single, microscopic ($\sim$ 50 to 75 μm) zircon crystals, which we estimate contained only $\sim$ 0.2 to 0.5 ng of Pb. We applied this ultralow-level detection method to study the differential retention of Pb in zircons ($ZrSiO_4$) extracted from Precambrian granite core samples (*1*) taken from depths of 960, 2170, 2900, 3930, and 4310 m. These depths correspond to presently recorded temperatures of 105°, 151°, 197°, 277°, and 313°C, respectively (*2*). We measured about the same $^{206}Pb/^{207}Pb$ ratio for zircons from all five depths, and we found that the total number of Pb counts measured per individual zircon was, to the limit of our experimental procedures, independent of depth. Taken together, these results strongly suggest that there has been little or no differential Pb loss which can be attributed to the higher temperatures existing at greater depths. As discussed below, this evidence for high Pb retention under adverse environmental conditions appears to have immediate and practical application to the question of long-term containment of hazardous nuclear wastes.

Samples of granite (*2*) from Los Alamos National Laboratory drill holes GT-2 and EE-2 from all five depths were individually crushed and then passed through different heavy liquid (methylene iodide) separatory funnels to obtain the high-density fraction containing the zircons. This procedure was repeated several times with different samples from each depth. The high-density fraction was then washed thoroughly with acetone to eliminate the methylene iodide residue before being placed on a standard 1 by 3 inch glass microscope slide. Under a polarizing microscope, the zircons were picked out of the high-density fraction with a fine-tipped needle and then loaded either onto pyrolytic graph-

ite disks for ion microprobe analysis or onto V-shaped rhenium filaments, which were mechanically compressed before mass spectrometric measurements. Surficial residues on the zircons burned off at temperatures well below that used to measure Pb from within the zircons.) Some zircons were analyzed by x-ray fluorescence before mass analysis.

Our efforts to measure lead isotope ratios in zircons with an Applied Research Laboratory ion microprobe failed because of molecular ion interferences. We then concentrated on determining relative abundances of U, Th, and Zr, using mostly an $^{16}O^-$ primary ion beam. Ion count rates were obtained on the $^{90}Zr^+$, $^{232}ThO^+$, and $^{238}UO^+$ peaks. The data were then quantified with sensitivity factors obtained from six different National Bureau of Standards glass standards containing Zr, Th, and U. Two or three zircons from three depths were analyzed, and usually four determinations were made from each zircon. Frequently, there were significant differences in the U and Th concentrations from two different locations on the same zircon. The results are given in Table 1 as a range of values obtained from each zircon.

The most important results came from the thermal ionization experiments. The thermal ionization mass spectrometer used in this work is similar to others described previously (*3*). It has a single magnet with 90° deflection and a 30-cm

Table 1. Ion microprobe determinations of U and Th concentration ranges in atomic parts per million on separate zircons from 960, 3930, and 4310 m. Calculations were based on a comparison of $^{238}UO^+$, $^{232}ThO^+$, and $^{90}Zr^+$ peak sizes and on the assumption that zircons were pure $ZrSiO_4$.

Zircon depth (m)	Th (ppm, atomic)	U (ppm, atomic)
4310	40–85	125–210
4310	63–175	110–550
3930	63–120	83–220
3930	60–90	90–110
960	220–750	465–1130
960	100–275	1250–3300
960	800–2000	240–5300

central radius of curvature. It is equipped with a pulse-counting detection system to allow complete isotopic analyses to be made on small quantities ($<$1 ng) of suitable elements ionized from a single filament. The filaments, made of V-shaped rhenium foil 0.64 cm long and 0.08 cm deep (*4*), were baked out at 2000°C before loading the zircons. Ions are formed by resistive heating of the filament; typical temperatures for this work were 1400° to 1470°C (uncorrected pyrometer readings).

Previous work done to develop a technique for analyzing small lead samples led to the use of silica gel to enhance ionization efficiency (*5*). Because individual zircons are chemically somewhat similar to silica, we decided to try to analyze lead from individual zircons loaded directly on the rhenium filament. Such a technique would have several advantages over traditional methods: contamination would be essentially eliminated because no chemical separation would be required and, since the zircons are small ($\sim$ 50 μm in diameter), they would provide an approximate point source of ions, which is known to optimize ion-optical conditions in the mass spectrometer (*6*).

Test experiments with zircons from other localities (*7*) were uniformly successful; ion signals were observed at masses (*m*) 206, 207, and 208 which could definitely be ascribed to Pb isotopes. To help ensure that we were at the correct ion lens conditions, we focused on the $^{138}BaO^+$ peak (the zircons contained some Ba), which was reasonably intense at 1200°C. Surficial residues left on the zircons after the acetone wash burned off before the operating temperature of 1450°C, where the lead signal was measured. Great care had to be exercised to avoid making the temperature too high; very rapid evaporation of the lead occurred only a little above the operating temperature. Typical count rates were 100 to 3000 counts per second for $^{206}Pb^+$. Traces of thallium (m = 203 and 205) were sometimes observed, but burned out more rapidly than the lead. Other than thallium, lead gave the only substantive peaks in the range m = 202 to 210. There was, however, a general background generated by the sample; chemically unseparated samples such as these zircons almost always yield such backgrounds. This background has little effect on the 206, 207, and 208 peaks, but made precise measurement of the ^{204}Pb signal, which was very small, impossible. For example, in an analysis typical of these experiments, 1.6 × 10⁵ counts from ^{206}Pb were collected; the background correction was about 40 counts and, after correction, 18 counts remained at mass 204. Although these counts are listed as ^{204}Pb counts in Table 2, more work is needed to determine how much may be uncompensated background.

Table 2 shows the results of our mass analyses of filaments loaded with single and multiple zircons from five granite cores. The range of ^{206}Pb/^{208}Pb values reflects the fact that this ratio varied from one group of zircons to another, and sometimes varied during measurements on a single zircon. These variations are not surprising in view of the ion microprobe analyses, which showed significant U/Th variations at different points on a single zircon (^{232}Th decays to ^{208}Pb and ^{238}U decays to ^{206}Pb). These variable ^{206}Pb/^{208}Pb ratios do not furnish any direct information on differential Pb retention in these zircons. For that purpose, it is generally accepted that the radiogenic ^{206}Pb/^{207}Pb ratios derived from ^{238}U/^{235}U decay are more specific. We note that Zartman's (8) isotopic measurements of Pb, which was chemically extracted from zircons taken from the GT-2 core at 2900 m, yield an adjusted ^{206}Pb/^{207}Pb ratio (9) that approximates our ratios.

In a conventional chemical extraction of lead from zircons, the lead measured in the mass analysis is considered to be a combination of radiogenic lead (from U and Th decay) and nonradiogenic lead (from common lead contamination and from some initial lead in the zircon). The radiogenic component is obtained by subtracting out a nonradiogenic component proportional to the amount of ^{204}Pb. In our experiments, however, the direct loading procedure virtually eliminated the common lead contamination, and we circumvented the need to make adjustments for initial lead in the zircons by accepting only analyses (10) showing a ratio of ^{204}Pb to total Pb of less than 2×10^{-3}. Thus the ^{206}Pb/^{207}Pb ratios shown in Table 2 represent highly radiogenic lead and hence are potential indicators of Pb retention.

We consider that the most important observations on the data in Table 2 are: (i) the fact that the ^{206}Pb/^{207}Pb ratios on single zircons closely approximate the ratio obtained when a group of similar zircons was loaded simultaneously on a single filament, (ii) the relative uniformity of the ^{206}Pb/^{207}Pb ratios for zircons from all depths, and (iii) the fact that the total number of Pb counts per zircon (the counts in column 4 of Table 2 divided by the product of columns 2 and 3) shows no systematic decrease with depth, as

Table 2. Results of thermal ionization mass measurements for zircons with a ^{204}Pb/total Pb ratio of less than 2×10^{-3}. The background correction was taken from the 208.5 mass position; it was applied to the raw data to obtain the isotopic abundances, which were used to compute the isotopic ratios. Standard deviations are listed with the Pb isotopic ratios.

Zircon depth (m)	Filaments analyzed	Average zircons per filament	Total Pb counts	Counts of ^{204}Pb	$\dfrac{^{204}\text{Pb}}{\text{total Pb}}$	Average ^{206}Pb/ ^{207}Pb	Range ^{206}Pb/ ^{208}Pb
960	4	~ 10	1.2×10^6	235	2×10^{-4}	9.6 ± 0.3	6.5–9.2
960	4	1	1.3×10^5	35	2.7×10^{-4}	9.9 ± 0.4	5.8–14
2170	3	~ 5	8.9×10^5	269	3×10^{-4}	10.0 ± 0.4	6.4–12.4
2900	3	~ 4	4.1×10^5	114	2.8×10^{-4}	11.2 ± 0.3	4–11.4
3930	2	~ 10	6.5×10^5	132	2×10^{-4}	11.0 ± 0.4	5.9–8.7
3930	2	1	8×10^4	46	5.8×10^{-4}	10.4 ± 0.1	3.1–6.9
4310	7	~ 10	5.6×10^6	1400	2.5×10^{-4}	9.7 ± 0.6	3.4–9.8
4310	2	1	1.6×10^5	100	6×10^{-4}	9.8 ± 0.4	4.5–10.7

would be expected if differential Pb loss had occurred at higher temperatures. Taken together, items (ii) and (iii) provide strong evidence for high Pb retention in zircons even for a prolonged period in an environment at an elevated temperature. These results have possible implications for long-term nuclear waste disposal.

For example, Ringwood (11, 12) has suggested that highly radiation-damaged minerals that have successfully retained U, Th, and Pb (13) over a significant fraction of earth history might also serve to immobilize high-level nuclear waste in synthetic rock (SYNROC) containers, which could be buried in deep granite holes. Even though zircons are not envisioned as part of Ringwood's special type of synthetic rock waste container, our results are relevant since they show that Pb, which is much more mobile in zircons than U and Th (12, 14), has been highly retained at depths (960 to 4310 m) which more than span the proposed burial depths (1000 to 3000 m) for synthetic rock containers in granite (11). The inclusion of this elevated temperature effect in our samples means that our results provide data which have heretofore been unavailable in support of nuclear waste containment in deep granite. In addition, the contamination-free method we used to analyze the zircons for radiogenic Pb may prove valuable in searching for other minerals suitable for synthetic rock waste containment.

Because it has been suggested that temperatures in the granite formation are rising (15), we do not know precisely how long the zircons have been exposed to the present temperatures. However, by using diffusion theory and the measured diffusion coefficient of Pb in zircon (16), we can estimate future loss of Pb by diffusion in synthetic rock–encapsulated zircons buried at the proposed depths of 1000 to 3000 m (11) if we assume a temperature profile similar to that in the

drill holes. At a burial depth of 3000 m (~ 200°C), we calculate that it would take 5×10^{10} years for 1 percent of the Pb to diffuse out of a 50-μm crystal. At 2200 m (~ 150°C) it would take 7.4×10^{13} years, and at 1000 m (~ 100°C) it would take 7.7×10^{17} years for 1 percent loss to occur (16). Since all these values greatly exceed the 10^5 to 10^6 years estimated for waste activity to be reduced to a safe level (11), and since, as noted earlier, U and Th are bound even more tightly than Pb in zircons (12, 14), our results appear to lend considerable support to the synthetic rock concept of nuclear waste containment in deep granite holes.

ROBERT V. GENTRY*
THOMAS J. SWORSKI
Chemistry Division,
Oak Ridge National Laboratory,
Oak Ridge, Tennessee 37830
HENRY S. McKOWN
DAVID H. SMITH
R. E. EBY
W. H. CHRISTIE
Analytical Chemistry Division,
Oak Ridge National Laboratory

References and Notes

1. A. W. Laughlin and A. Eddy, *Los Alamos Sci. Lab. Rep. LA-6930-MS* (1977). A. W. Laughlin provided the core samples used in this work.
2. R. Laney and A. W. Laughlin, *Geophys. Res. Lett.* **8**, 501 (1981).
3. D. H. Smith, W. H. Christie, H. S. McKown, R. L. Walker, G. R. Hertel, *Int. J. Mass Spectrom. Ion Phys.* **10**, 343 (1972).
4. W. H. Christie and A. E. Cameron, *Rev. Sci. Instrum.* **37**, 336 (1966).
5. A. E. Cameron, D. H. Smith, R. L. Walker, *Anal. Chem* **41**, 525 (1969).
6. D. H. Smith, W. H. Christie, R. E. Eby, *Int. J. Mass Spectrom. Ion Phys.* **36**, 301 (1980).
7. O. Kopp and H. McSween of the Department of Geological Sciences, University of Tennessee, Knoxville, provided zircons from Gjerstad, Norway; Oaxaca, Mexico; and Henderson County, North Carolina.
8. R. E. Zartman, *Los Alamos Sci. Lab. Rep. LA-7923-MS* (1979).
9. If the ^{204}Pb in Zartman's (8) Pb isotopic abundances in his zircons is attributed to common lead, the corrected ^{206}Pb/^{207}Pb ratio for the zircons from 2900 m is 11.03.
10. This criterion resulted in the rejection of four single zircon analyses whose average ^{206}Pb/^{207}Pb ratio was 8.8 ± 1.3. These lower ratios imply that some zircons contain more initial Pb than others, as noted in some other runs.

11. A. E. Ringwood, *Safe Disposal of High Level Nuclear Reactor Wastes: A New Strategy* (Australian National Univ. Press, Canberra, 1978).
12. A. E. Ringwood, K. D. Reeve, J. D. Tewhey, *in Scientific Basis for Nuclear Waste Management*, J. G. Moore, Ed. (Plenum, New York, 1981), vol. 3, p. 147.
13. V. M. Oversby and A. E. Ringwood, *J. Waste Manage.*, in press. See also A. E. Ringwood, *Lawrence Livermore Natl. Lab. Rep. UCRL-15347* (1981).
14. R. T. Pidgeon, J. O'Neill, L. Silver, *Science* **154**, 1538 (1966); *Fortschr. Mineral.* **50**, 118 (1973).
15. D. G. Brookins, R. B. Forbes, D. L. Turner, A. W. Laughlin, C. W. Naeser, *Los Alamos Sci. Lab. Rep. LA-6829-MS* (1977).
16. In general, if R is the gas constant, T is the absolute temperature, and D and Q are, respectively, the diffusion coefficient and activation energy of a certain nuclide in a given diffusing medium, then $D = D_0\, e^{-Q/RT}$, where D_0 is a temperature-independent parameter. In particular, if C_0 is the initial concentration of that nuclide within a sphere of radius a, then the average nuclide concentration $\overline{C}$ within that sphere at some later time t is given by

$$\overline{C}/C_0 = \frac{6}{\pi^2} \sum_{1}^{\infty} \frac{e^{-(n^2\pi^2 Dt/a^2)}}{n^2}$$

[see L. O. Nicolaysen, *Geochim. Cosmochim. Acta* **11**, 41 (1957)]. We used measured values of $D_0 = 2.2 \times 10^{-2}$ and $Q = 58$ kcal/mole for diffusion of Pb in zircon [see Sh. A. Magomedov, *Geokhimiya* **2**, 263 (1970)] and a computer program to calculate the times when $\overline{C}/C_0 = 0.99$ for $T = 100°$, $150°$, and $200°$C.
17. Research sponsored by the U.S. Department of Energy, Division of Basic Energy Sciences, under contract W-7405-eng-26 with the Union Carbide Corporation.
* Visiting scientist from Columbia Union College, Takoma Park, Md. 20112.

3 November 1981; revised 22 January 1982

Appendix 261

JIM SASSER
TENNESSEE

United States Senate

WASHINGTON, D.C. 20510

COMMITTEES:
APPROPRIATIONS
BUDGET
GOVERNMENTAL AFFAIRS
SELECT COMMITTEE ON
SMALL BUSINESS

May 18, 1982

Mr. William S. Heffelfinger
Assistant Secretary for Management
    and Administration
Department of Energy
James Forrestal Building
1000 Independence Avenue, S.W.
Washington, D.C. 20585

Dear Mr. Heffelfinger:

    This letter is written on behalf of Robert V. Gentry, Associate Professor of Physics at Columbia Union College and currently Guest Scientist at Oak Ridge National Laboratory.

    Mr. Gentry has been a Guest Scientist at ORNL for the past 13 years. During this time, he has published nearly 20 scientific reports, some of which have received national recognition. I have enclosed two published commentaries concerning Mr. Gentry's work which testify to the depth and importance of the research he has been able to conduct while at ORNL.

    In addition, Robert Gentry has been particularly helpful to me and my staff on energy-related matters, particularly nuclear waste site selection issues. He has provided valuable evaluations and technical expertise, which has assisted us in ascertaining the full implications of various energy policies.

    It is my understanding that Mr. Gentry has been notified that his current dollar-a-year consultant contract will be terminated on June 30, 1982. I also understand that he has recently discovered new evidence relating to nuclear waste containment about which he would like to conduct experiments and further research. However, he will be unable to do this if his contract is terminated on schedule.

    I wanted to take this opportunity to bring my interest in Mr. Gentry to your attention and to request that he be allowed to continue his work at Oak Ridge National Laboratory, if at all possible. I am sure that an extension of his contact would allow him to finish his research and prepare conclusions based on those experiments.

    I would greatly appreciate any assistance you can offer Mr. Gentry in this regard, and I look forward to hearing from you at your convenience.

        Sincerely,

        Jim Sasser
        United States Senator

Department of Energy
Washington, D.C. 20585   JUN 1 4 1982

Honorable Jim Sasser
United States Senate
Washington, DC 20510

Dear Senator Sasser:

This is in reference to your letter dated May 18, 1982, on behalf of
Robert V. Gentry, a guest scientist at the Oak Ridge National Laboratory
(ORNL) operated by Union Carbide Corporation for the Department of Energy.

At the time of his assignment at ORNL 13 years ago, Mr. Gentry's supporting
sponsor was Columbia Union College. The original purpose of his research was
to study pleochroic halos, an area of interest to ORNL at that time, but a
field of less significance to the Laboratory's mission in recent years.

Mr. Gentry's more recent efforts in nuclear waste containment referenced in
your letter are quite peripheral to the primary thrust of ORNL's ongoing
waste isolation programs.

When ORNL entered into its current subcontract with Mr. Gentry, effective
July 1, 1981, it was for him to continue his own research on halos, using
Laboratory facilities. It was anticipated that he could finish his work
during the year; no other work was authorized under the subcontract. He was
advised in June 1981 that he should seek other arrangements under which to
pursue his research interests beyond June 30, 1982.

Diminishing ORNL budgets require marked cutbacks in activities not directly
related to its priority program areas. Unfortunately, Mr. Gentry's work
does not fall in that category. Accordingly, we cannot be encouraging about
an extension of his agreement at ORNL.

Thank you for your continuing interest in Department of Energy programs.

                    Sincerely,

                    William S. Heffelfinger
                    William S. Heffelfinger
                    Assistant Secretary
                    Management and Administration

GEOPHYSICAL RESEARCH LETTERS, VOL. 9, NO. 10, PAGES 1129-1130, OCTOBER 1982

DIFFERENTIAL HELIUM RETENTION IN ZIRCONS: IMPLICATIONS
FOR NUCLEAR WASTE CONTAINMENT

Robert V. Gentry,[1]* Gary L. Glish,[2] and Eddy H. McBay[2]

[1]Physics Department, Columbia Union College, Takoma Park, MD 20012

[2]Analytical Chemistry Division, Oak Ridge National Laboratory, Oak Ridge, TN 37830

Abstract. A very sensitive helium leak detector was utilized to measure the helium liberated from groups of zircons extracted from six deep granite cores. The observed low differential loss of gaseous helium down to 2900 m (197°C) in these ancient Precambrian rocks is easily attributable to the greater diffusion of He at higher temperatures rather than losses due to corrosion of the zircons. This fact strongly suggests that deep granite burial should be a very safe corrosion-resistant containment procedure for long-term waste encapsulation.

Recent mass spectrometric studies (Gentry, et al. 1982) have revealed that lead has been retained in zircons extracted from deep (960 m to 4310 m) granite cores where the ambient temperature increases from 105°C to 313°C at the greatest depth. As a follow-up to those experiments we now report the results of differential helium retention in similar zircons extracted from the same granite core samples which were used in the lead analyses (Laney and Laughlin, 1981).

The procedure for separating the zircons from the six different granite cores (from depths of 960, 2170, 2900, 3502, 3930, and 4310 m) was the same as that used in the previous experiments. The high density fractions, obtained by passing the crushed core samples through different methylene iodide separating funnels, were thoroughly washed with acetone before being placed on a standard microscope slide. A fine-tipped needle was used to pick out the individual zircons with the aid of a polarizing microscope. Groups of these separated zircons, usually about 10 in number, were then loaded onto the platinum filament of the thermal inlet probe of the mass spectrometer for differential helium analysis.

The helium measurements were performed on a Leybold-Heraeus model F helium leak detector that had a Chemical Data Systems Pyrolysis unit interfaced to the test port. The leak detector has a detection limit of less than 10^{-10} cm^3/sec when operating in the dynamic mode. (The instrument could have been operated in a near-static mode with increased sensitivity down to $\sim 10^{-11}$ cm^3/sec of He, but our experiments did not necessitate this increased sensitivity.)

*Also: Research Assistant Professor, Physics Department, University of Tennessee, Knoxville, TN 37916.

Copyright 1982 by the American Geophysical Union.

Paper number 2L1385.
0094-8276/82/002L-1385$3.00

In our initial series of measurements our spectrometer was calibrated against a 5 (±0.5) × 10^{-8} cm^3/sec standard He leak. A subsequent recalibration with a more precise 5 (±0.5) × 10^{-10} cm^3/sec standard He leak revealed the total helium liberated during these initial measurements was slightly underestimated. The general procedure was to measure helium evolution from a group of zircons at progressively higher temperatures of 400°C, 600°C, and 1000°C for 20 sec intervals. (Previous studies of helium diffusion (Magomedov, 1970) from zircons indicated 1000°C was sufficient to liberate the helium with an activation energy of 15 kcal/mol.) We did not include the small amount of He observed at 400°C in the total He summation because of possible atmospheric contamination. Between six and eight groups of zircons were analyzed at each depth. Runs were repeated at a given temperature until background helium levels were observed. Data recordings and integration under the peaks were done with a Nicolet 1170 signal averager.

The third column in Table 1 shows, as a function of depth, the total amount of He liberated per µg of zircon for zircon groups comprised of approximately equal-size ($\sim$50-75 µm) zircons. The fourth column in Table 1 shows the ratio of the amount of He actually measured in zircons from any particular depth to the estimated amount of He which should have accumulated in those same zircons assuming negligible diffusion loss. For the zircons taken from a surface outcrop we assumed this ratio was one because the specimens we used were small fragments from the interior of larger zircon crystals.

For the other zircons from the granite and gneiss cores, we made the assumption that the radiogenic Pb concentration in zircons from all depths was, on the average, the same as that measured (Zartman, 1979) at 2900 m, i.e., $\sim$80 ppm with ^{206}Pb/^{207}Pb and ^{206}Pb/^{208}Pb ratios of ten (Gentry, et al., 1982; Zartman, 1979). Since every U and Th derived atom of ^{206}Pb, ^{207}Pb, and ^{208}Pb represents 8, 7, and 6 α-decays respectively, this means there should be $\sim$7.7 atoms of He generated for every Pb atom in these zircons.

Knowledge of the zircon mass and the appropriate compensation factor (to account for differences in initial He loss via near-surface α-emission) enabled us to calculate the theoretical amount of He which could have accumulated assuming negligible diffusion loss. This compensating factor is necessary because the larger (150-250 µm) zircons lost a smaller proportion of the total He generated within the crystal via near-surface α-emission than did the smaller (40-50 µm) zircons. For the smaller zircons we estimate as many as 30-40% of the α-particles (He) emitted within the crystal could have escaped initially whereas for

TABLE 1: The values listed below show first, as a function of depth and temperature, the amount of helium liberated from various groups of zircons in units of 10^{-8} cc per μg and second, the ratio of the amount of helium liberated to the theoretical amount which would have been retained assuming no diffusion loss. The near equality of the He concentrations in the surface and 960 m depth zircons is not particularly meaningful because the surface zircons were from an entirely different geological unit and doubtless have different U-Th-Pb concentrations than the zircons from the core samples.

Sample Depth(m)	Sample Temp.(°C)	He $(10^{-8}$ cc/μg)	$\frac{\text{He(measured)}}{\text{He(theoret.)}}$
Surface	20	8.2	1
960	105	8.6	0.58
2170	151	3.6	0.27
2900	197	2.8	0.17
3502	239	7.6×10^{-2}	1.2×10^{-2}
3930	277	$\sim 2 \times 10^{-2}$	$\sim 10^{-3}$
4310	313	$\sim 2 \times 10^{-2}$	$\sim 10^{-3}$

the larger zircons we studied only 5-10% of the total He would have been lost via this mechanism. The ratio of the measured to the theoretical amount of He is shown in the last column of Table 1. The uncertainties in our estimates of the zircon masses and compensation factors probably mean these last values are good only to ±30%.

In spite of these uncertainties, it is quite evident from Table 1 that the zircons from 960 m seem to have retained considerable amounts of He, and perhaps more significantly, differential He loss with increasing depth (and temperature) has occurred rather slowly down to 2900 m (197°C) before a precipitous drop is observed at 239°C (3502 m). In fact, at present we are not certain whether the minute amounts of He recorded from the deepest zircons (3930 and 4310 m) are actually residual He in the zircons or derived from some other source. That is, in the two deepest zircon groups (3930 and 4310 m), we observed only short bursts of He ($\sim$1-2 sec) in contrast to the prolonged 20 sec or more evolution of He which was typical of He liberation from zircon groups down to and including 3502 m. In fact, it was this prolonged He liberation profile seen in two 150-250 μm size zircon groups from 3502 m which convinces us that some residual He is still trapped in the zircons down to that depth (239°C).

Now it was recently noted that the high retention of Pb in even the deepest granite cores had favorable implications for nuclear waste containment in deep (1000 to 3000 m) granite holes (Gentry, et al., 1982). The rationale for these implications is straightforward: If zircons, which have been exposed to the same type of elevated temperature environment anticipated in deep granite burial, show no detectable Pb loss either from higher temperatures or from

aqueous solution corrosion effects, then nuclear wastes buried in that same granite should, if anything, experience even greater retention because of the comparative immobility of waste-type elements as compared to Pb.

The present results are important in that they provide clear evidence that the dominant factor in slow He loss down to 2900 m is attributable to greater diffusion loss at higher temperatures rather than any corrosion induced losses from the zircons. This is not at all surprising because microscopic examination shows first that zircons from all depths exhibit well-defined prismatic faces without any evidence of external corrosion, and secondly that the delicate internal inclusions within the zircons do not show any evidence of alteration from aqueous intrusion via any microstructural defects. Indeed, the relatively slow liberation of He over several 20 sec intervals observed in zircons from the surface all the way down to 2900 m is strong evidence that these zircons are virtually free of any microfractures which would have permitted a more rapid He escape. In fact, considering the Precambrian age of the granite cores (Zartman, 1979), our results show an almost phenomenal amount of He has been retained at higher temperatures, and the reason for this certainly needs further investigation for it may well turn out to have a critical bearing on the waste storage problem.

Thus the additional evidences reported herein considerably reinforce the view that deep-granite storage should be a very safe corrosion-resistant waste containment procedure. The certainty of these results stands in clear contrast with the uncertainties about how well alternative storage sites (e.g., salt domes) could withstand corrosion and/or dissolution from intruding aqueous solutions.

Acknowledgments. This research was sponsored by the U. S. Department of Energy, Division of Basic Energy Sciences, under contract W-7405-eng-26 with Union Carbide Corporation. We thank A. W. Laughlin of the Los Alamos National Laboratory for providing the core samples.

References

Gentry, R. V., T. J. Sworski, H. S. McKown, David H. Smith, R. E. Eby, and W. H. Cristie, Differential lead retention in zircons: Implications for nuclear waste containment, Science, 216, 296-298, 1982.

Laney, R., and A. W. Laughlin, Natural annealing of pleochroic haloes in biotite samples from deep drill holes, Fenton Hill, New Mexico, Geophys. Res. Lett., 8, 501-504, 1981.

Magomedov, S. A., Migration of radiogenic products in zircon, Geokhimiya, 2, 263-267, 1970.

Zartman, R. E., Uranium, thorium and lead isotopic composition of biotite granodiorite (Sample 9527-2b) from LASL Drill Hole GT-2, Los Alamos Sci. Lab. Rep. LA-7923-MS, 1979.

(Received August 6, 1982;
accepted September 3, 1982.)

STATE OF ARKANSAS
OFFICE OF THE ATTORNEY GENERAL
JUSTICE BUILDING, LITTLE ROCK 72201

STEVE CLARK
ATTORNEY GENERAL

(501) 371-2007

The Honorable Dale Bumpers
United States Senator
New Senate Office Building
Washington, D.C. 20515

Dear Senator Bumpers:

In my recent defense of Act 590 of 1981 (better known as the Creation-Science Law), I had the opportunity to become acquainted with several of the world's leading scientists who testified on behalf of both the State and the American Civil Liberties Union. Of all the scientists involved on both sides of the lawsuit, no one impressed me anymore than Robert Gentry, who for the past several years has been a guest scientist at the Oak Ridge National Laboratoies in Oak Ridge, Tennessee. This letter is written to bring to your attention Mr. Gentry's work and to enlist your aid on his behalf.

Mr. Gentry's testimony at trial concerned the presence of radioactive polonium halos in granite. The significance of these halos is that their presence in the granites is fundamentally inconsistent with the conventional wisdom that the granites underlying the earth's structure cooled over thousands of years. Mr. Gentry is acknowledged as the world's foremost authority on this particular subspecialty.

From every indication available to me, Gentry's work at the National Laboratory has been of a uniformly high quality and has added signifcantly to the progress made at that facility. Furthermore, as a guest scientist, Gentry has been paid only $1.00 per year by the government. (A college of which he is a faculty member has paid his salary.) Thus, the government has been able to avail itself of his services essentially free of charge.

However, Mr. Gentry has recently learned that his contract as a guest scientist will not be renewed for next year. As one admittedly viewing these events from afar, it appears to me that Gentry is being penalized for his generous offer of assistance to help the State of Arkansas and his own religious beliefs. Bob Gentry is very frank and forthright in stating his religious beliefs, of that there can be no doubt. His religious beliefs are, however, irrelevant to the work which he performs at Oak Ridge. His work in studying granites was recently quoted in the Congressional Record in connection with a discussion of possible sites for storage of low level radioactive wastes.

The Honorable Dale Bumpers

page 2

Obviously, this is an important issue and one on which Gentry has been on the cutting edge.

I want to ask for your assistance to assure that Robert Gentry will not be a victim of religious discrimination at the hands of his supervisors. The Oak Ridge National Laboratory, although operated by a private corporation under a contract, is, as I understand it, under the jurisdiction of the U.S. Department of Energy. I solicit your help in contacting the Energy Department through appropriate channels and requesting that the decision to not renew Gentry's contract be reviewed personally by the Secretary of Energy to assure that this decision was based solely upon the merits of his work, and not upon the subjective prejudices of his supervisors. It will be a sad day, indeed, if the First Amendment's guarantee of freedom of religion and the supposed freedom of scientific inquiry have both become hollow promises for men like Bob Gentry.

If I can supply you wiht any additional information regarding this matter, Please call upon me at your convenience.

Yours truly,

STEVE CLARK

SC/clr

PROCEEDINGS
of the
63rd ANNUAL MEETING
of the
PACIFIC DIVISION, AMERICAN ASSOCIATION
FOR THE ADVANCEMENT OF SCIENCE

Volume 1, Part 3   April 30, 1984

EVOLUTIONISTS CONFRONT CREATIONISTS

Edited by
Frank Awbrey
and
William M. Thwaites

Pacific Division
American Association for the Advancement of Science
San Francisco, California
1984

RADIOACTIVE HALOS IN A RADIOCHRONOLOGICAL
AND COSMOLOGICAL PERSPECTIVE

Robert V. Gentry*
Columbia Union College
Takoma Park, Maryland 20012

If the earth was created, it is axiomatic that created (primordial) rocks must now exist on the earth, and if there was a Flood there must now exist sedimentary rocks and other evidences of that event. But, if the general uniformitarian principle is correct, the universe evolved to its present state only by the unvarying action of known physical laws and all natural phenomena must fit into the evolutionary mosaic. If this fundamental principle is wrong, all the pieces in the evolutionary mosaic become unglued. Evidence that something is drastically wrong comes from the fact that this basic evolutionary premise has failed to provide a verifiable explanation for the widespread occurrence of Po halos in Precambrian granites, a phenomena which I suggest are in situ evidences that those rocks were created almost instantaneously in accord with Psalm 33:6,9: "By the word of the Lord were the heavens made; and all the host of them by the breath of his mouth. For he spake, and it was done; he commanded, and it stood fast." I have challenged my colleagues to synthesize a piece of granite with ^{218}Po halos as a means of falsifying this interpretation, but have not received a response. It is logical that this synthesis should be possible if the uniformitarian principle is true. Underdeveloped U halos in coalified wood having high U/Pb ratios are cited evidences for a Flood-related recent (within the past few thousand years) emplacement of geological formations thought to be more than 100,000,000 years old. Results of differential He analyses of zircons taken from deep granite cores are evidence for a recently created, several-thousand-year-age of the earth. A creation model with three singularities, involving events beyond explanation by known physical laws, is proposed to account for these evidences. The first singularity is the ex nihilo creation of our galaxy nearly 6000 years ago. Finally, a new model for the structure of the universe is proposed based on the idea that all galaxies, including the Milky Way, are revolving about the Center of the universe, which from Psalm 103:19 I equate with the fixed location of God's throne. This model requires an absolute reference frame in the universe whereas modern Big Bang cosmology mandates there is no Center (the Cosmological Principle) and no absolute reference frame (the theory of relativity). The motion of the solar system through the cosmic microwave radiation is cited as unequivocal evidence for the existence of an absolute reference frame.

* Present address: P. O. Box 12226, Knoxville, TN 37912.

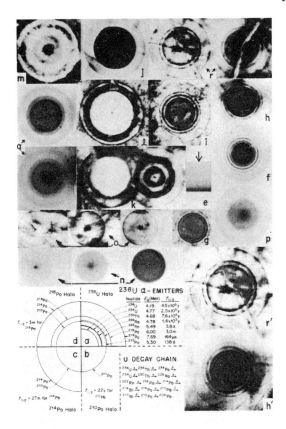

Figure 1. The scale for all photomicrographs is 1 cm = 25 μm, except for (h') and (r'), which are enlargements of (h) and (r). (a) Schematic drawing of ^{238}U halo with radii proportional to ranges of alpha particles in air. (b) Schematic of ^{210}Po halo. (e) Coloration band formed in mica by 7.7-MeV ^{4}He ions. Arrow shows direction of beam penetration. (f) A ^{238}U halo in biotite formed by sequential alpha decay of the ^{238}U decay series. (g) Embryonic ^{238}U halo in fluorite with only two rings developed. (h) Normally developed ^{238}U halo in fluorite with nearly all rings visible. (h') Same halo as in (h) but at higher magnification. (i) Well developed ^{238}U halo in fluorite with slightly blurred rings. (j) Overexposed ^{238}U halo in fluorite, showing inner ring obliteration. (k) Two overexposed ^{238}U halos in fluorite, showing outer ring reversal effects. (m) Second-stage reversal in a ^{238}U halo in fluorite. The ring sizes are unrelated to ^{238}U alpha particle ranges. (n) Three ^{210}Po halos of light, medium, and very dark coloration in biotite. Note the difference in radius. (o) Three ^{210}Po halos of varying degrees of coloration in fluorite. (p) A ^{214}Po halo in biotite. (q) Two ^{218}Po halos in biotite. (r) Two ^{218}Po halos in fluorite. (r') Same halos as in (r) but at higher magnification. (Reprinted from ref. (2) by permission of the AAAS.)

39

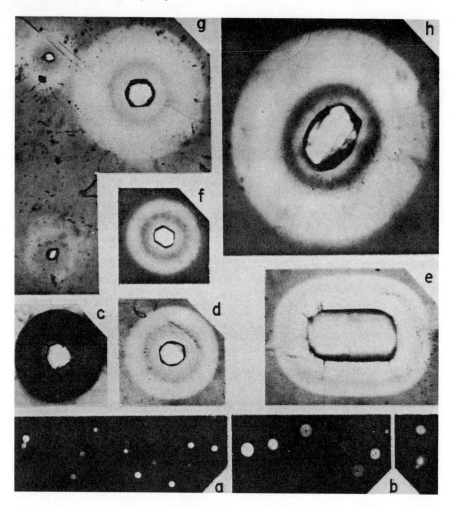

Figure 2. The scale for all photographs is 1 cm = 30 μm. (a) Dwarf halos
(≈2 μm radius) in Ytterby mica. (b) Dwarf halos (3 μm < r < 9 μm) in Ytterby
mica. (c) Overexposed Th halo in ordinary biotite. (d) Th halo in Madagascan
mica. (e) Th halo in Madagascan mica with a larger inclusion. (f) U halo in
Madagascan mica. (g) Giant halo of ≈ 65 μm radius, and two light Th halos
(Madagascan mica). (h) Giant halo of ≈ 90 μm radius Madagascan mica. (Reprinted
from ref. (1) by permission of the ARNS.)

Table 1. Comparison of sizes of induced bands (columns 1 to 5) with halo radii (columns 8 to 21). Column 6 gives the ^{4}He ion energies at which the induced bands were formed, or the α-particle energies corresponding to the nuclides in column 7. Thus, the nuclide or α-particle energy that produced any halo ring in columns 8 to 21 can be found from column 6 or 7. The letters K-L, H, S, M, and G represent halo measurements by Kerr-Lawson (5), Henderson (& 6, 7), Schilling (9), Mahadevan (10), and Gentry. Subscripts L, M, and D indicate light, medium (dose 10 to 20 times coloration threshold), and dark (dose about 50 times coloration threshold) induced bands; L→D and L→M indicate light to dark and light to medium; these were visually determined. Gentry's measurements were made with a filar micrometer readable to 0.07 μm. The estimated overall uncertainty was ±0.3 μm. Other abbreviations: N.M., not measured; N.R., not resolved; N.P., not present. Reprinted from Ref. (2) by permission of the AAAS.

| Coloration band size (μm) | | | | | | | U halo radius (μm) | | | | | | Po halo radius (μm) | | | | | | | |
| Biotite | | | Fluorite | Cordierite | E (MeV) | Nuclide | Biotite | | | Fluorite | | Cordierite | ^{210}Po | | ^{214}Po | | ^{218}Po | | Fluorite ^{210}Po | Fluorite ^{218}Po |
1. G_L	2. G_M	3. G_D	4. G	5. G	6.	7.	8. K-L	9. H	10. G	11. S	12. G	13. M	14. H	15. $G_{L\to D}$	16. H	17. $G_{L\to M}$	18. H	19. G_M	20. G	21. G
13.4	13.8	14.2	14.1	16.2	←4.2	^{238}U→	12.3	12.7	12.2→13.0	14.0	14.2	16	N.P.	N.P.	N.P.	N.P.	N.P.	N.P.	N.P.	N.P.
N.M.	16.7	N.M.	17.3	19.2	←4.77	^{226}Ra→	15.4	15.3	14.8→15.8	16.9	17.1	19	N.P.	N.P.	N.P.	N.P.	N.P.	N.P.	N.P.	N.P.
N.M.	N.M.	N.M.	17.3	19.2	←4.66	^{230}Th→	N.R.	N.R.	N.R.	15.8	17.1	19	N.P.	N.P.	N.P.	N.P.	N.P.	N.P.	N.P.	N.P.
N.M.	16.7	N.M.	17.3	22.5	←4.78	^{234}U→	15.4	15.3	14.8→15.8	16.9	17.1	19	N.P.	N.P.	N.P.	N.P.	N.P.	N.P.	N.P.	N.P.
N.M.	19.3	20.0	19.6	N.M.	←5.3	^{210}Po→	N.R.	N.R.	N.R.	19.1	19.5	N.R.	N.P.	18.3→19.9	20.0	18.1→19.1	19.9	19.3	19.8	19.8
N.M.	20.5	21.1	N.M.	N.M.	←5.49	^{222}Rn→	18.6	19.2	18.1→19.0	20.5	20.5	23.5	N.P.	N.P.	N.P.	N.P.	N.P.	N.P.	N.P.	N.P.
N.M.	23.0	23.9	23.6	26.7	←6.0	^{218}Po→	22.0	23.0	21.5→22.7	23.5	23.5	26.5	N.P.	N.P.	N.P.	N.P.	24.0	23.3	N.P.	23.7
33.1	33.9	34.4	34.6	38.7	←7.69	^{214}Po→	33.0	34.1	30.8→33.0	34.5	34.7	38.5	N.P.	N.P.	34.5	32.5→33.8	34.0	34.0	N.P.	34.9

URANIUM AND THORIUM RADIOHALOS IN MINERALS

A radioactive halo is generally defined as any type of discolored, radiation-damaged region within a mineral and usually results from either alpha or, more rarely, beta emission from a central radioactive inclusion. When the central inclusions, or radiocenters, are small (1 μm), the U and Th daughter alpha emitters produce a series of discolored concentric spheres, which in thin section microscopically appear as concentric rings whose radii correspond to the ranges of the various alpha emitters in the mineral.

Ordinary radiohalos are herein defined as those which initiate with ^{238}U and/or ^{232}Th alpha decay (1), irrespective of whether the actual U or Th halo closely matches the respective idealized alpha decay patterns. In a few instances the match is very good.

Compare, for example, the idealized U halo ring pattern in Fig. 1a with the well developed U halos in biotite (Fig. 1f) and fluorite (Fig. 1h,h'); these halos have ring sizes that agree very well (1,2) with the ^{4}He ion accelerator-induced coloration bands in these minerals (see Table 1). In general a halo ring can be assigned to a definite alpha emitter with confidence only when the halo radiocenter is about 1 μm in size.

In other cases, however, such as the halos in fluorite (1,2) shown in Fig. 1(g, i-m), much work was required before these halos could be reliably associated with U alpha decay (2). As explained elsewhere (2), reversal effects accompanying extreme radiation damage caused the appearance of rings that could not be associated with definite alpha emitters of the U decay chain. Thus some halos may exhibit a ring structure different from the idealized U and/or Th alpha decay patterns because of reversal effects. And even though most other halos exhibit blurred ring structures due to the large size of the inclusions, nevertheless the outer dimensions allow them to be classified as U and/or Th types.

Modern analytical techniques such as Scanning Electron Microscope X Ray Fluo-resence (SEMXRF) and Ion Microprobe Mass Spectrometry (IMMA) methods have been utilized to show that U and Th and their respective end-product isotopes of Pb are contained within the U and Th halo radiocenters. As is noted shortly, these modern analytical techniques have proved quite valuable in demonstrating that Po halo radiocenters in minerals contain little or no U or Th, which is in direct contrast to the abundance of these elements detected in the U and/or Th halo radiocenters (2,3).

RADIOACTIVE HALOS AND THE QUESTION OF INVARIANT DECAY RATES

A most important question pertaining to the evolution/creation issue is whether radioactive decay rates have remained invariant during the course of earth history. If they have, geochronologists are justified in interpreting various parent/daughter isotope ratios found in undisturbed rocks in terms of elapsed time. If on the other hand there have been periods in earth history where the decay rate was higher (i. e., during a singularity), then in general the isotope ratios in rocks would not reflect elapsed time except in the specific case where secondary rocks or substances containing only the parent radionuclide formed at the end of the most recent singularity. The practical significance of this last statement will be evident in the discussion of the secondary, U halos found in coalified wood specimens from the Colorado Plateau.

Even though most of Joly's (4) measurements of U and Th halos showed their radii were about the sizes expected from the alpha decay energies of the U and Th decay chains, nevertheless he claimed there were slight discrepancies which raised questions about whether the radioactive decay rate had been constant over geological time. His result was not confirmed however by later halo radii measurements (5-10), which agreed to within experimental error with the theoretical sizes. To eliminate any uncertainty about this correspondence I irradiated specimens of various minerals with He ion beams of varying energies to produce different size coloration bands whose widths corresponded to the various alpha energies of the U decay chain. The results of these experiments, presented in Table 1, show there is excellent agreement between the U and Th halo radii and equivalent He ion produced penetration depths (2).

The basis for thinking that standard size U and Th halos imply an invariant decay rate throughout geological time proceeds from the quantum mechanical treatment of alpha decay, which in general shows that the probability for alpha decay for a given nuclide is dependent on the energy with which the alpha particle is emitted from the nucleus. The argument is that if the decay rate had varied in the past, then the U and Th halo rings would be of different size now because the energies of the alpha particles would have been different during the period of change. This argument assumes that a change in the decay rate must necessarily be exlainable by quantum mechanics, which is of course an integral part of the uniformitarian framework. Thus, the usual proof of decay rate invariance based on standard size U and Th halos is nothing more than a circular argument which assumes the general uniformitarian principle is correct. In fact, the failure of the uniformitarian principle to explain the evidence for creation presented herein invalidates the basis for the above proof.

POLONIUM, DWARF, AND GIANT HALOS IN MINERALS

Of the three types of unusual halos that appear distinct from those formed by U and/or Th alpha decay, only the Po halos, Fig.1 (b-d, n-r, r'), can presently be identified with known alpha radioactivity (1-3,11-13). Po halos occupy a special niche in my creation model, and these halos will be discussed in more detail subsequently. Several lines of evidence which indicate the enigmatic dwarf halos (see Fig. 2) were produced by some presently unidentified radioactivity have been summarized (1,12,14,15). The rapid etch from HF and the K/Ca inversion are strongly characteristic of highly radiation-damaged regions.

The characteristics of the giant halos found in a certain Madagascan mica have also been summarized (1,14,16), and while no definitive evidence as yet exists for a radioactive origin, some halos with opaque inclusions in this same mica exhibit isotopic anomalies which raise questions about the uniformity of U and Th alpha decay. For example, the mass scans and x-ray fluorescence analyses shown in Fig. 3 clearly indicate that, whereas both the monazite and opaque inclusions exhibit ^{206}Pb and ^{207}Pb from U decay, the opaque inclusions exhibit a marked deficiency of ^{208}Pb from ^{232}Th decay (14).

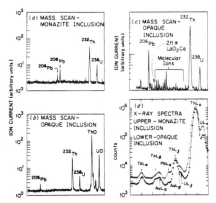

Figure 3. Mass scans and an x-ray fluorescence spectrum of a monazite and an opaque halo inclusion in Madagascan mica, showing Pb deficiency in the latter.

SECONDARY RADIOHALOS IN COALIFIED WOOD

All the various types of halos discussed thus far are termed primary halos because they developed from alpha radioactivity emanating from small accessory inclusions that were present when the mineral crystallized. But secondary halos also exist in pieces of coalified wood taken from highly uraniferous deposits in the Colorado Plateau. There is abundant evidence that U solutions infiltrated much of the sedimentary material in the geological formations of that region when the wood was still in a gel-like condition (17). When U-bearing solutions passed through pieces of wood, certain active sites within these specimens preferentially collected U, other sites collected rare earth type elements, and still others Se, Po, and Pb. It is quite significant that the U halos, which developed around the tiny U-rich sites, are all underdeveloped, which, on the

43

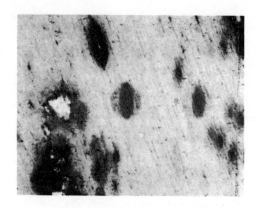

Figure 4. Elliptical (compressed) ^{210}Po halos in coalified wood from the Colorado Plateau. Reproduced from ref. (17) by permission of the AAAS. (x 250)

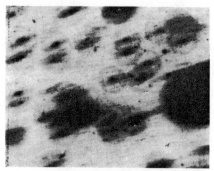

Figure 5.   Circular ^{210}Po halos in Colorado Plateau coalified wood. (x 250)

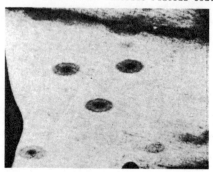

Figure 6. Circular and elliptical ^{210}Po halo in Colorado Plateau coalified wood. Reproduced from ref. (17) with AAAS permission. (x 250)

44

basis of a uniform decay rate (the rationale for using this assumption for these specimens will be explained subsequently), suggests only a relatively short time since U infiltration. Ion microprobe mass scans of these U halo centers have shown extremely high $^{238}U/^{206}Pb$ ratios, which, again on the assumption of a uniform decay rate, is consistent with a U infiltration within the last several thousand years (17).

Similar underdeveloped U halos have been found in the coalified wood from the Chattanooga Shale, and in fact recent ion microprobe analyses show, in agreement with earlier results (17), that the $^{238}U/^{206}Pb$ ratios of the U halos in the Colorado Plateau samples (Eocene, Triassic, and Jurassic) and the Chattanooga Shale (Devonian) are virtually indistinguishable. These results suggest that U-infiltration occurred concurrently in all these formations.

Another class of more sharply defined halos was also discovered in the Colorado Plateau coalified wood specimens (17). The centers of these halos exhibit a distinct metallic-like reflectance when viewed with reflected light. Three different varieties of this halo exist: one with a circular cross section, another with an elliptical cross section with variable major and minor axes, and a third most unusual one that is actually a dual halo, being a composite of a circular and an elliptical halo around exactly the same radio-center (see Figs. 4-6).

Although the elliptical halos differ radically from the circular halos in minerals, the circular type resembles the ^{210}Po halo in minerals and variations in the radii of circular halos approximate the calculated penetrated distances (26 to 31 μm) of the ^{210}Po alpha particle (energy E = 5.3 MeV) in this coalified wood (17). Henderson (18) theorized that Po halos might form in minerals when U-daughter Po isotopes or their alpha precursors were preferentially accumulated into small inclusions from some nearby U source. This hypothesis has not been confirmed for the origin of three distinct types of Po halos in U-poor minerals (1,2,11), but it does seem to provide a reasonable explanation for the origin of the ^{210}Po halos in U-rich coalified wood specimens.

Electron microscope x-ray fluorescence analyses showed these halo centers were mainly Pb and Se. This composition fits well into the secondary accumulation hypothesis for both of the U-daughters, ^{210}Po (half-life, $t_{1/2}$ = 138 days) and its beta precursor ^{210}Pb ($t_{1/2}$ = 22 y), possess the two characteristics that are vitally essential for the hypothesis: (i) chemical similarity with the elements in the inclusion and (ii) half-lives sufficiently long to permit accumulation prior to decay, a requirement related to the nuclide transport rate.

What is the meaning of the ^{210}Po halos in Figs. 4-6? Clearly, the variations in shape can be attributed to plastic deformation which occurred prior to coalification. Since the model for ^{210}Po formation thus envisions that both ^{210}Po and ^{210}Pb were accumulating simultaneously in the Pb-Se inclusion, a spherical ^{210}Po halo could develop in 0.5 to 1 year from the ^{210}Po atoms initially present and a second similar ^{210}Po halo could develop in 25 to 50 years as the ^{210}Pb atoms more slowly beta decayed to produce another crop of ^{210}Po atoms. If there was no deformation of the matrix between these periods, the two ^{210}Po halos would simply coincide. If, however, the matrix was deformed between the two periods of halo formation, then the first halo would have been compressed into an ellipsoid, and the second would be a normal sphere. The result would be a dual "halo" (Fig. 6). The widespread occurrence of these dual halos in both Triassic and Jurassic specimens can actually be considered corroborative

45

276   Creation's Tiny Mystery

evidence for a one-time introduction of U into these formations, because it is then possible to account for their structure on the basis of a single specifically timed tectonic event (17).

HALOS IN COALIFIED WOOD: A FLOOD-RELATED PHENOMENA

A worldwide Flood, which is postulated to have occurred about 1650 years after creation, is the third singularity in the creation model proposed herein. I have advanced the hypothesis that the underdeveloped U halos in both the Colorado Plateau and Chattanooga Shale coalified wood specimens exhibit very high U/Pb ratios because the uranium infiltration of the wood occurred only when those geological deposits were being emplaced at the time of the Flood several thousand years ago, instead of the 60 to 400 millions of years ago accepted by uniformitarian geology. I suggest at least part of the U-series disequilibria (19) found in the Colorado Plateau U deposits is because some U-daughter radionuclide separation occurred at the time of the Flood, and there has been insufficient time since then to reestablish equilibrium conditions.

The high U/Pb ratios and secondary ^{210}Po halos in the coalified wood samples from the Eocene epoch and the Triassic and Jurassic periods suggest to me that the wood in all these formations was in the same gel-like condition when infiltrated by the U-bearing solutions. To me these data represent evidence for a concurrent, single-stage invasion of U into all the different geological formations represented by the coalified wood samples. This is precisely what would be expected on the basis of a Flood-related phenomena.

The dual Po halos also fit well into the Flood scenario, i.e., the presence of a spherical and elliptical Po halo arround the same radiocenter suggests a tectonic event occurred within 50 years after the initial infiltration of uranium into the wood samples. A readjustment of the earth's crust after such a massive event is not unexpected. Another implication of the existence of ^{210}Po halos in these specimens is that the transformation of the wood to a semi-coal-like condition must have occurred within a period of about one year. This evidence for a rapid coalification process is in contrast to the generally accepted view that coalification is a long-term geological process.

THREE TYPES OF POLONIUM HALOS IN MINERALS

Now there are two other Po isotopes (^{214}Po and ^{218}Po) in the U decay chain besides ^{210}Po, but no halos representative of these other Po isotopes have been found in coalified wood. This is not surprising, because the half-lives of the other Po isotopes are rather short, i.e., $t_{\frac{1}{2}}$ = 3m for ^{218}Po and $t_{\frac{1}{2}}$ = 164 μs for ^{214}Po, as are the half-lives of the beta precursors of ^{214}Po, i.e., $t_{\frac{1}{2}}$ = 26.4 m for ^{214}Pb and $t_{\frac{1}{2}}$ = 19.8 m for ^{214}Bi (the precursor of ^{218}Po is the inert gas ^{222}Rn). What is surprising is that all the three types of Po halos occur in certain minerals which typically contain orders of magnitude less uranium than the U-rich coalified wood. Further, the minerals such as biotite and fluorite must have diffusion rates considerably lower than those expected for a U-solution-infiltrated specimens of gel-like wood. Figure 7 shows the idealized structure of the different Po halos in comparison with the U halo.

Photographic evidence relating to the existence of different types of Po halos in minerals is shown in Fig. 1. Figure 1(n) shows three ^{210}Po halos of

46

light, medium, and very dark coloration. The slightly higher radii for the darker halos is attributable to the higher dose. Figure 1(o) shows three different ^{210}Po halos in fluorite. Figure 1(p) shows a ^{214}Po halo in biotite, and Fig. 1(q) shows two ^{218}Po halos in biotite. Comparison of these halos with the idealized ring structure in Fig. 7 shows that Po halos in minerals can be clearly identified by ring structure studies alone. The data in Table 1 shows there is an excellent agreement between the experimentally produced He ion produced coloration bands and the Po halo ring radii.

An important observation from Fig. 7 is that in the idealized ^{238}U and ^{218}Po halo patterns, it is evident that the ^{222}Rn ring should be missing from the ^{218}Po halo and present in the ^{238}U halo. Figures 8 and 9 show the presence of the ^{222}Rn ring in the U halo in contrast to its absence in the ^{218}Po halo. This is unequivocal evidence that the ^{218}Po halo initiated with ^{218}Po rather than with any earlier alpha emitter in the U decay chain. Figures 10 and 11 show ^{214}Po halos and ^{218}Po halos in different types of biotite.

Henderson's (18) original idea that Po halos in minerals may have originated from a secondary source of radioactivity encounters formidable obstacles when closely examined. In most cases the minerals contain only ppm abundances of uranium, which means only a negligible supply of Po daughter atoms is available for capture at any given time. To form a halo these daughter atoms must migrate or diffuse so they can be captured at a collecting site, a problem which is compounded by the low diffusion rates in minerals (11,20,21). Despite these objections, in 1979 several investigators suggested their results (22) might provide support for secondary Po halo formation in minerals after all. They were apparently unaware that three years earlier I had reported the experimental observation of secondary ^{210}Po halos in coalified wood (17). In that report I discussed how even under the most favorable conditions (i. e., an abundant supply of U-daughters in a highly mobile environment) for the formation of secondary Po halos, only the longer half-life ^{210}Po halos actually formed, the reason being that the shorter half-life Po isotopes generally decayed away before they could be captured at the tiny Pb-Se sites. If these other two Po halo types didn't form under the best conditions in the gel-like wood, how could it be expected they would form naturally in the granites where diffusion rates are vastly lower and the supply of Po atoms is negligible?

The identity of U, Th and Po halos in minerals has been confirmed by analyzing the various types of halo radiocenters using scanning electron microscope x-ray fluorescence (SEMXRF) and ion microprobe mass spectrometric (IMMA) techniques (2,3). Studies of various Po halo radiocenters in biotite and fluorite have generally shown little or no U in conjunction with anomalously high $^{206}Pb/^{207}Pb$ and/or Pb/U ratios which would be expected from the decay of Po without the U precursor which normally occurs in U radiohalo centers (2,3). These results were obtained clearly in the analysis (3) of the most unusual array of Po halos which I ever found. That array, shown in Figure 12, has the appearance of a pair of spectacles, hence the designation 'Spectacle Halo.' The Spectacle Halo appearance compounds the problem of explaining its existence on the basis of known physical laws. In conclusion, in spite of attempts to define them out of existence (23), there is demonstrable evidence that Po halos do exist as separate entities (1-3).

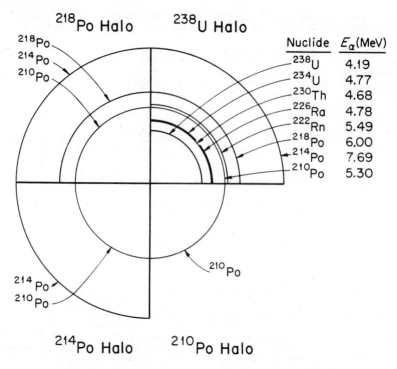

Nuclide	E_α(MeV)
^{238}U	4.19
^{234}U	4.77
^{230}Th	4.68
^{226}Ra	4.78
^{222}Rn	5.49
^{218}Po	6.00
^{214}Po	7.69
^{210}Po	5.30

Figure 7.   Idealized schematic of ^{238}U, ^{218}Po, ^{214}Po, and ^{210}Po halos.

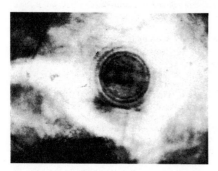

Figure 8. ^{238}U halo in fluorite.
(x 535)

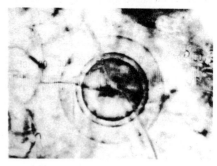

Figure 9. ^{218}Po halo in fluorite.
(x 535)

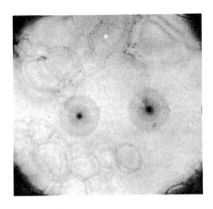

Figure 10. ^{214}Po halos in mica.
(x 250)

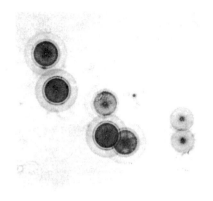

Figure 11. ^{218}Po halos in mica.
(x 250)

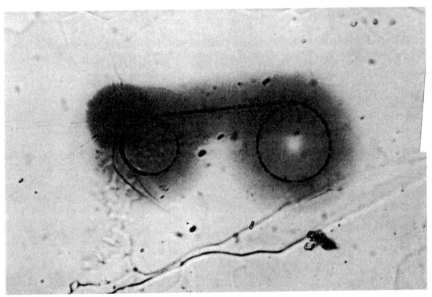

Figure 12. The Spectacle Halo, an overlapping series of ^{210}Po halos discovered in a piece of biotite from the Silver Crater mine, Faraday Township, Ontario. Reproduced from ref. (3) by permission of <u>Nature.</u> (x 560)

POLONIUM HALOS IN MINERALS: AN INDEPENDENT EVALUATION

Because of the implications which will be attributed to the presence of Po halos in minerals, it is important that my colleagues be apprised of the independent investigation of these phenomena by Professor Norman Feather. In an exhaustive theoretical treatment (24) of the problem concerning their origin in minerals, Feather concludes it is difficult to account for the existence of Po halos in certain minerals on the basis of known physical principles. His exact words, as given in the synopsis of his paper, are as follows:

> Ever since the discovery of Po-haloes in old mica (Henderson and Sparks 1939) the problem of their origin has remained essentially unsolved. Two suggestions have been made (Henderson 1939; Gentry et al. 1973), but neither carries immediate conviction. These suggestions are examined critically and in detail, and the difficulties attaching to the acceptance of either are identified. Because these two suggestions appear to exhaust the logical possibilities of explanation, it is tempting to admit that one of them must be basically correct, but whoever would make this admission must be fortified by credulity of a high order.

POLONIUM HALOS AND PRIMORDIAL ROCKS: A TEST OF THE HYPOTHESIS

I have advanced the hypothesis (25,26) that the three different types of Po halos in minerals represent the decay of primordial Po, in which case the rocks that host these halos, i.e., the Precambrian granites, must be primordial rocks (25,26). By this reasoning the Precambrian granites are identified as rocks that were created almost instantly as a part of the creation event recorded in Genesis 1:1 rather than rocks that are a product of the evolution of the earth. This rationale would be without scientific content if I had not also stated (25) that the laboratory synthesis of a hand-sized piece of granite or biotite would be accepted as falsifying my view that the Precambrian granites are created rocks and, likewise, that the subsequent production of 218Po halos in that synthesized specimen of granite or biotite would be accepted as falsifying my view that Po halos in Precambrian granites originated with primordial polonium. The only response to my repeated (25,26) challenges to perform these laboratory syntheses and falsify the aforementioned evidences for creation has thus far been silence. It is inescapable that these experiments should be successful if the uniformitarian principle is true. Thus, with so much at stake for evolution, I suspect the reason why my evolutionary colleagues have failed to achieve success is because the Precambrian granites never formed by the uniformitarian principle to begin with; hence, to attempt to utilize it now to produce a synthesized piece of granite is just a futile effort. The end result is that the uniformitarian principle is essentially falsified because of its failure to live up to its own predictions. But since all the pieces in the evolutionary puzzle are glued together by this principle, we must now come to the same conclusion about evolution itself.

A PROPOSED CREATION MODEL AND THE AGE OF THE EARTH

The evidence for creation cited above suggests there may have been special periods in earth history when physical laws as presently understood were insufficient to explain all the events transpiring within those periods. This evidence also undergirds the formulation of a creation model based on the Judeo-Christian ethic. The creation model proposed herein postulates that on at least three occasions (singularities) during the past 6000 years there were significant exceptions to the uniformitarian principle within our local cosmos (the Milky Way), viz., the ex nihilo creation of our galaxy about 6000 years ago, the Fall of man shortly thereafter, and the occurrence of a worldwide Flood about 4350 years ago. These ages are derived from Scriptural chronology. It is assumed that the creative act which brought the Milky Way into existence also caused the immediate propagation of light throughout the galaxy. No constraints are placed on the age of the universe.

Singularities and Uniformities: A Complementary Approach

It is essential to understand that uniform action of physical laws <u>between</u> singularities is an integral part of this creation model. Moreover, the occurrence of a singularity does not mean a completely chaotic condition without any laws to govern the operations of nature during that period. During the Flood singularity some physical processes may not have changed at all whereas there is evidence others varied considerably. An enhanced radioactive decay rate during the Flood singularity would have generated a considerable amount of heat, thus initiating volcanic and tectonic activity during and after that period. This three-singularity model appears to be the minimum framework that includes the essential features of the Genesis narrative. Possibly the continent-separating episode recorded in Genesis 10:25, when the earth was divided in the days of Peleg a few hundred years after the Flood, should also be included as a singularity; certainly it must figure prominently in any creation-based reconstruction of earth history that deals with continental drift. However, to simplify matters, the following comments exclude consideration of this event.

Singularities and the Interpretation of Radioactive Decay as Elapsed Time

In summary, the creation model envisions an initial creation singularity followed by a short period of uniformity until the the second singularity, an event which involved degenerative changes in the biological world and quite possibly modification of some of the original physical laws which governed the earth and our near celestial environment. Another period of uniformity follows, with the modified physical laws now in effect, for about 1600 years down to the longer-duration Flood singularity. The last period of uniformity extends down to the present. In this scenario U/Pb ratios are presently utilized as indicators of elapsed time <u>since</u> the last singularity. ^{238}U /^{206}Pb ratios are not used as time measures prior to this last singularity because of conflicting evidence of very high Pb and He retention in natural zircons subjected to a prolonged high temperature environment in deep granite. Those results, discussed below, are consistent with a very young age of the earth, and suggest that the radioactive decay rate may have been enhanced (indeed, had to be if this creation model is correct) during any one of the three singularities. (The Peleg episode potentially adds one more possibility.) The assumption of uniform decay since the Flood is the basis for interpreting the very high U/Pb ratios in coalified wood samples as evidence for a several-thousand-year age of specimens which conventional geology holds to bè about 60 to 400 million years old.

51

Possible Evidence of Enhanced Radioactive Decay from 'Blasting' Halos

Additional evidence for an enhanced radioactive decay rate comes from Ramdohr's observations on fractured radioactive halos in polished ore sections. He reports (27) that certain radioactive inclusions, which exhibit a considerable volume increase due to isotropization from radioactive decay, have in numerous cases been observed to fracture the surrounding mineral in a random pattern. Ramdohr points out that the surrounding mineral should expand slowly over geological time due to radioactive isotropization, and individual cracks should appear as soon as the elastic limit is reached. He further points out that, while these expansion cracks should occur first along cohesion minimums and grain boundaries, nothing like this happens. Individual cracks surrounding the radioactive inclusion are randomly distributed and evidently occur quite suddenly in the form of an explosive fracture and not a slow expansion. Ramdohr shows many photographs of instances wherein the central inclusion fractures the non-isotropic outer zone. The occurrence of this phenomenon is worldwide.

While there might be other alternatives, one possible explanation of these "fractures" or "blasting" halos is that the rate of radioactive decay was at one time far greater than that observed today. The isotropization of the host minerals would have occurred very rapidly due to an anomalous decay rate, and hence fracturing of the outer mineral would be expected.

The Age of the Earth and Pb Retention in Deep Granite Cores

Results pertaining more specifically to a recent creation of the earth come from studies of Pb retention in zircons taken from deep Precambrian granite cores (28). To understand the rationale for this last statement, it must first be understood that the Pb in these zircons is primarily a secondary trace component derived from the decay of small amounts of U and Th. Secondly, this radiogenic Pb has a tendency to migrate or diffuse out of the zircon crystals far more rapidly than the parent U and Th because these elements are relatively tightly bound in lattice sites, whereas the Pb atoms really do not fit into the zircon lattice. Further, since all elements show an exponential increase in the bulk diffusion rate with increasing temperature, and since the temperature in the granite cores increases significantly from near the top (105°C) to the bottom (313°C) of the granite portion of the drill hole, calculations show that 50 μm-size zircons taken from the bottom of the drill hole (313°C) should have lost 1% of their Pb content in about 300,000 years. Since the zircons were in cores taken from a Precambrian granite that is estimated to be 1.5 billion years old by conventional geochronology (29), the prediction based on uniformitarian geochronology would be that most of the Pb would have long ago diffused out of the zircons extracted from the deepest cores at 313°C. But the results of the experiments did not agree with this prediction; rather they showed equally high retention of Pb in zircons taken from all depths. In fact no Pb loss from zircons at 313°C would appear to place an upper limit to the age of this Precambrian granite, which, on the presumption that these granites are primordial rocks, in essence places the same limit on the age of the earth.

The Age of the Earth: Limited by Helium Retention in Deep Granite Cores

Another approach which seemed to hold greater prospects for more closely defining an upper limit for the age of these Precambrian granites (and hence of the earth) was the differential analysis of similar size zircons from these same cores for helium, the second most volatile chemical element known. The helium accumulates in these zircons in a manner similar to the radiogenic Pb, viz., from the alpha particles emitted from trace amounts of U and Th. However, the extreme volatility of this gas means that it diffuses out of the zircons at a far greater rate than Pb. On a purely uniformitarian basis the search for helium in these zircons would quite possibly never have been done because conventional geological wisdom suggests negligible helium retention in zircons subjected to even 100°C for the presumed 1.5 billion year age (29) of those granites. But having already discovered that the Pb retention in these zircons contradicted the age estimates determined by radiometric dating techniques, I decided that, from a creationist perspective, the search might just reveal something of exceptional interest. Groups of zircons from six different depths were repeatedly analyzed for helium using an extremely sensitive gas mass spectrometric system. The results (30) showed a helium retention of about 58% in the tiny 50 μm zircons from 960 meters depth (105°C), about 27% in zircons from 2170 meters (151°C) and a phenomenal 17% retention of helium even at 2900 meters where the temperature is 197°C. These results show a creation-based perspective of science does possess predictive capabilities which can be scientifically tested.

It is difficult to understand how such high retention (30) of helium can be accounted for except by restricting the age of these granites (and hence the earth) to something of the order of several thousand years. These results are consistent with an approximate 6000-year age of the earth and moreover are in direct conflict with the presumed 4.5-billion-year age of the earth determined by radioactive dating techniques. Evolutionary colleagues can prove this deduction for a young age of the earth is wrong if they can show just how this unusually high retention of helium can be deduced from the accepted 1.5-billion-year age (29) of those zircons by using only uniformitarian principles.

A CREATION MODEL OF THE STRUCTURE OF THE UNIVERSE

Decades of research in astronomy and cosmology have led to the general belief that the present state of the universe can ultimately be traced to an initial event popularly known as the Big Bang. Despite this popularity it should be remembered that the Big Bang cosmological model is only as valid as the fundamental premises which support it. Thus the discussion of the proposed creation model of the universe must necessarily also focus on the validity of the Big Bang theory, whose basic framework consists of the cosmological and uniformitarian principles together with the general theory of relativity. The previous sections of this article have documented the failure of the uniformitarian principle to provide confirmation for the geological evolution of the Precambrian granites. If this principle cannot account for the evolution of the earth, is it difficult to understand how it can provide a rational basis for constructing an evolutionary model of the universe. It may be argued, however, that the edifice of modern cosmology fits together too well for there to be something wrong with basic assumptions. This point will receive close examination in the following discussion of the hot Big Bang Model (31,32).

The Big Bang Model and the Hubble Relation

About 50 years ago Hubble proposed that the astronomical data then available seemed to linearly relate the redshift z of a galaxy with the distance R to the galaxy, and this has become known as the Hubble relation. Since then galactic redshifts have been mainly interpreted as Doppler shifts resulting from high recessional velocities of the distant galaxies and, moreover, have been generally thought to provide some of the strongest evidence for the hot Big Bang model of an expanding universe. (See, however, Hetherington's evaluation (33) of the Hubble relation.) The reason for confidence in this interpretation is that by using the general theory of relativity as the mathematical basis for calculating the space-time development of the primeval fireball, it is possible to derive the $z \propto R$ Hubble relation (31,32) provided certain assumptions are made.

Notwithstanding the general belief that the accumulated astronomical data do support a $z \propto R$ relation, the fact is that over the past two decades several detailed studies of redshift distributions have been published which call the Hubble relation into question. As early as 1962 Hawkins (34) claimed that the redshift data indicated an approximate quadratic-distance redshift relation, in particular $z \propto R^{2.22}$. More recently the case for a $z \propto R^2$ relation (for low z) was considerably reinforced by the extensive statistical analyses of Segal (35) and of Nicoll and Segal (36). Even though these latter results have been disputed by Sandage et al. (37), it appears that Nicoll and Segal (38) have responded with stronger evidence for a $z \propto R^2$ relation. In fact, Nicoll et al. (39) have gone so far as to claim statistical invalidation of the Hubble relation for low values of z. At a minimum the foregoing results make it very difficult to believe that the redshift data as presently interpreted actually support the Hubble relation, which is the cornerstone of Big Bang cosmology.

As noted above, the latest analyses of Nicoll and Segal (38) show the redshift data more closely fit what is thought to be the equivalent of a quadratic rather than a linear distance relation. The reason for qualifying the last statement is because astronomers measure not distances but apparent magnitudes, which are first corrected for various factors before being used as a basis for establishing the magnitude-redshift relation. One important correction involves the assumption that the galactic light intensity (for any given frequency interval) as observed on earth is reduced by two factors of 1+z, one for the redshift itself, and the other for the presumed galactic recession. Of course if the galaxies are not receding, then an unwarranted factor has been introduced into the magnitude correction procedures, and this would affect the perceived redshift distributions.

The Big Bang Model and the Cosmic Microwave Radiation (CMR)

In 1978 Penzias and Wilson received the Nobel prize in physics for their discovery of the CMR in 1965. Since then it has been widely claimed that this pervasive radiation field is a relic of the time eons ago when radiation quanta decoupled from matter in the primeval fireball (31). According to this theory, the decoupling presumably occurred about 300,000 years after the Big Bang when the primeval fireball had expanded and its temperature had dropped to the point where matter and radiation ceased to interact as it had before. After this time, supposedly about 15 billion years ago, it is believed that this radiation propagated throughout space in an unobstructed fashion to eventually become the CMR. It is essential to note that the radiation leaving the primeval fireball at the time of decoupling was presumably still quite hot (about 3000°K). The experimental measurements of the CMR temperature at present reveal that it is very cold (3°K). But if the radiation from the primeval fireball is assumed not to interact with matter after the time of decoupling, then how did this initially hot radiation lose its energy, or temperature, to later become the 3°K CMR. The standard explanation is that the general relativistic analysis of the space-time expansion of the primeval fireball predicts that the decoupled radiation quanta will lose energy just as a result of the expansion of the universe. There is, however, nothing in modern experimental physics which suggests that radiation quanta change energy by moving through free space. Thus, the standard explanation for this remarkable thousand-fold energy loss in the decoupled radiation quanta depends upon an aspect of general relativity that is unsupported by scientific evidence.

To avoid possible misunderstandings, some recent experimental results of gravitational effects on photons will be discussed. Einstein's principle of equivalence, which is independent of general relativity, does not distinguish whether a photon traversing a gravitational potential gradient undergoes a change in energy in transit, or whether its energy is uniquely determined by the gravitational potential at the point of emission. The earliest Mossbauer experiments (40) on the gravitational redshift could not distinguish between these two alternatives, and it was widely believed that the photon energy could change when passing through a difference in gravitational potential. But recent experimental results (41) suggest the photon energy is characterized by the gravitational potential at the point of emission rather than varying as the photon moves to a different potential. In the light of these results it is quite difficult for me to believe that radiation quanta can undergo energy loss in free space as predicted in the general relativistic Big Bang model. At this point my views on the theory of relativity need to be clarified.

I recognize there are some notable experimental results in physics such as apparent time dilation, the transverse Doppler effect, the increase in mass with velocity, and the gravitational bending of light, which are in accord with the predictions of the theory of relativity. However, these experimental results cannot be used as confirmations of the special or general theory of relativity because there are other (albeit far lesser known) theories which predict similar results. (See for instance North's (42) review of various alternative theories of gravitation and their predictions.) Further, recently Rastall (43) and especially Marinov (44) have shown independently that it is not necessary to assume the general relativistic framework to obtain many of the same mathematical results. On the other hand, the question of whether the Big Bang model is a correct description of the origin and evolutionary development of the universe is entirely hinged on the ultimate validity of general relativity's fundamental postulate, which in principle denies that privileged reference frames exist. Very germane to this discussion is the recent admission (45) of an eminent physicist to the effect that the CMR presents undeniable experimental evidence for the existence of an absolute reference frame in the universe, a result which is consistent with Marinov's (44) evidence for absolute space-time and also with at least one of the earlier gravitational theories reviewed by North (42). This point is treated in more detail subsequently and it is shown that the existence of the CMR as an absolute reference frame is perhaps the most important evidence that can be adduced for the creation model of the universe as proposed herein. Before engaging in this discussion further, it is necessary to complete the present discussion of the CMR and the Cosmological Principle.

Measurements have shown the spatial distribution of the CMR is so uniform that it is questionable whether it could have been produced by the Big Bang scenario as it was originally conceived. Weisskopf (45) has recently reviewed the nature of this and other problems with the Big Bang model, and has discussed the provisional solutions offered by postulating an explosive expansion in the very early stages of the Big Bang. Questions still remain, however, not the least being that the entire scenario assumes some type of grand unification theory which has yet to be verified. But is it consistent for cosmologists on one hand to claim that the universe evolved only through the action of known physical laws and on the other hand to devise solutions to cosmological problems by using unverified hypotheses as a basis for those solutions? We have already noted the failure of the uniformitarian principle to successfully account for the origin of Po halos in Precambrian granites, or to provide a basis for synthesis of a piece of granite. In a similar manner it seems the introduction of unverified physical concepts as the basis for possible solutions to difficult evolutionary cosmological problems is just the inevitable result of the failure to explain the creation of the universe on the basis of the uniformitarian principle. In any event, the newly proposed expansionary modification to the Big Bang only deals with the earliest instants of the Big Bang, after which it is supposed the expansion of the primeval fireball continues as envisioned in the original Big Bang model. As we shall soon see, it appears there may be a contradiction involved in the theoretical development of expansion of the fireball.

The Big Bang Model and the Cosmological Principle

In spite of the foregoing difficulties it might still be argued that Big Bang model must be correct because it predicts a universe in accord with the Cosmological Principle, viz., that the universe appears the same irrespective of the location of the observer in the universe. The problem with this argument is that we really do not know the Cosmological Principle is true. In fact, all that we know is that the large scale structure of the universe appears to be approximately isotropic (i. e., the same in all directions) from our present point of observation. Modern cosmology justifies the Cosmological Principle by coupling the observation of isotropy about our position with the assumption that our galaxy does not occupy a special position in the universe. That is, if our galaxy occupies a non-specific or arbitrary position in the universe, then it follows the universe must be isotropic everywhere and hence homogeneous as well.

But what if our galaxy does occupy a privileged position in the universe? First, it would no longer be logical to extrapolate the isotropy which we observe to the other parts of the universe, which means it would no longer be possible to justify either the condition of homogeneity or the cosmological principle. Second, the simplest deduction of the observed isotropy of the universe from our location is that the universe must be spherically symmetric about either the Milky Way or some point which is astronomically nearby. But spherical symmetry about any point in the universe implies that point is the Center, and this brings us to the discussion of the creation model.

A Creation Model of the Universe: The Fundamental Postulate

The fundamental premise of the Judeo-Christian creation model of the universe is determined by the scripture, "The Lord has established His throne in the heavens, and His kingdom ruleth over all." Psalm 103:19 (RSV). On the basis of this statement it is evident that the Creator has established, or fixed, His throne at some point in the universe, which in my view is none other than the Center of the universe. It is axiomatic that a fixed point in the universe requires the existence of a fixed or absolute reference frame. Pre-

viously it was noted that the CMR has been recognized as establishing an absolute reference frame (45); so it is quite clear that the fundamental postulate
of this creation model of the universe is based on tangible scientific evidence.

The Revolving Steady State Model of the Universe: A Brief Description

Assuming there is a Center (C) to the universe, I propose that the galaxies
are not receding from each other as presently supposed, but instead are revolving at different distances and at different tangential speeds around C. On
this basis all galaxies must have a tangential velocity around C. Measurements
have shown that our solar system, and hence the Milky Way, has a cosmic velocity through the CMR (46), and it is this velocity which is identified with the
tangential velocity of the Milky Way around C. In this view C must lie somewhere in that plane which passes through the MW which is also perpendicular to
the cosmic velocity vector of the MW. It is evident that the RSS model pictures
the galaxies orbiting C in any one of many different-sized concentric shells,
which suggests the alternate designation 'Shell Model of the Universe.'

As originally conceived this Revolving Steady State (RSS) model envisions a
universe with galaxies which move in circular orbits under the gravitational
field produced by all of them. The field is assumed to be stationary and spherically symmetric. Decades ago Einstein made a general relativity study (47) of
circulating particles constrained by this type of gravitational field, but his
analysis did not mention redshifts, nor was there any hint that he considered
his analysis had any reference to the structure of the universe.

The RSS Model and Galactic Redshifts

Assuming the galaxies are revolving in different orbital planes and with
different tangential velocities v around some universal center C, initially I
thought that if the Milky Way was one of the innermost galaxies, then most of
the galactic redshifts as observed on earth might be due to a combination of
gravitational and transverse Doppler effects. (A literature search showed that
Burcev (48) had proposed over a decade ago that quasars were possibly stellar
objects whose redshifts might be attributable to the transverse Doppler effect.)

Although questions have arisen about this explanation for the galactic
redshifts in the RSS model, it seems worthwhile to explain my original rationale
and the objections which now appear to present themselves. In particular, in
the Newtonian-based RSS model the galaxies of mass m and tangential velocity v
remain in circular orbits by gravitational attraction of the total mass M within
the sphere of orbital radius R. In this scenario, $mv^2/R = mMG/R^2$, or $v^2 = GM/R$,
where G is the gravitational constant. Thus an observer on an innermost galaxy
located at a distance R_1 from C would in theory see light from a more distant
galaxy (at R_2 from C) shifted in frequency because of the transverse Doppler
effect and the change in gravitational potential $V(R) = - GM/R$. The presumed
limiting distance R' at which galaxies could remain in stable orbits would be
when the tangential velocity v = c, the velocity of light. Beyond this presumed
galactic cutoff distance the RSS model tentatively assumes a rapidly diminishing mass/energy density so that we do not encounter an infinite gravitational
potential (see discussion of equations (2) and (3) for more details).

The frequency shifts expected in the RSS model can be compared to an earth-

57

bound observer comparing the frequency of a light signal emitted from his position on the rotating earth's surface, where the tangential velocity is v_1 and
the gravitational is V_1, with the frequency of the same signal emitted from an
overhead satellite which is orbiting with velocity v_2 in a gravitational potential V_2. The experimentally confirmed (41) equation for the redshift, as derived
from the principle of equivalence, is:

(1) $$z = (V_1 - V_2)/c^2 - (v_1^2 - v_2^2)/2c^2.$$

The same equation applies in the RSS model except that v_1 and V_1 are the cosmic
velocity and gravitational potential of the Milky Way at R_1 from C whereas v_2
and V_2 represent the same quantities for a more distant galaxy at R_2 from C.

Another source of frequency shifts arises because the Milky Way (MW) is not
exactly at C. In this case the more distant galaxies, which are rotating away
from or toward the MW, produce first order Doppler redshifts or blueshifts. The
blueshifts, which would be most pronounced for nearby galaxies, can be eliminated for all practical purposes if it is assumed that the more distant galaxies
are rotating away from the MW. This scenario would result in a recessional
redshift which, because it depends on the cosine of the angle between the
velocity vector of the outer galaxy and the line of sight from the MW to that
galaxy, would diminish with distance. Thus, of itself this redshift could at
most be only a part of the total galactic redshift observed on the earth. Of
course, a significant distance-related redshift, irrespective of its origin,
could overshadow most blueshifts expected from galaxies rotating toward the MW
and eliminate the need for assuming rotation away from the MW.

We now return to the discussion of the redshifts expected on the basis of
eq.(1). If the ρ, the mass/energy density of the universe, is assumed to be
constant then $M = 4 \pi \rho R^3/3$, and substitution of the appropriate quantities into
eq. (1) leads to the formal result that z is proportional to R^2, which is of
the same form of the redshift relation proposed in references (33,34,37-39). On
a similar basis, if the density is assumed to vary inversely as R, then one can
obtain an expression for z which is proportional to R, which is of the same form
as the Hubble relation (49).

Of course, astronomers measure apparent magnitudes, not distances, and, for
there to be a quantitative comparison between the above results and the redshift
distribution, the light flux relation for the RSS model must be formulated so as
to include the combined effect of the redshift and gravitational focusing. This
formulation has yet to be done; thus on this basis alone it would be premature
to claim the forgoing results are consistent with the galactic redshift relation
proposed by Nicoll and Segal (38). Moreover it should be remembered that if the
universe is revolving, then an extraneous factor has been included into the data
which comprise the redshift distribution, and this would preclude any immediate
comparison. But regardless of the outcome of the above calculations, there
seems to be a more fundamental objection to the preceding formulation.

In particular, we must carefully investigate whether the gravitational
potential $V = - GM/R$ used in the above calculations is the correct expression
for the potential function. It is of crucial importance to know whether it is
correct for it is used as the basis for the derivation of the Hubble relation
(31,32) in Big Bang cosmology. According to Silk (31) and Weinberg (32), its
use in computing the potential at the surface of an arbitrarily large, but
finite sphere, of radius R within an infinite universe is justified by a theorem

due to Birkhoff. Part of the proof of this theorem implicitly assumes that the universe is structured according to the Cosmological Principle. Now the creation model of the universe proposed herein is also of infinite extent, but the Cosmological Principle does not hold, so that there is no basic reason why this theorem should yield the correct gravitational potential in the RSS model. But should it hold for the Big Bang model?

To answer this question we first note that the negative gradient of the potential $V = - GM/R$ yields a repulsive force per unit mass $F/m = GM/R^2$, whereas there is an experimentally confirmed theorem in classical mechanics which definitely requires an attractive force per unit mass $F/m = - GM/R^2$ to exist at any point R within a sphere enclosing a uniform mass distribution. This latter result is an integral part of both the RSS and the Big Bang models. Thus the potential $V = - GM/R$ is just as wrong for the Big Bang model as it would be for the RSS model because it yields an incorrect sign for the force. Even Silk's (31) elementary treatment (see page 332) makes it clear that the derivation of the Friedmann equation for the Big Bang expanding universe is based on the potential $V = - GM/R$. Here we have a logical contradiction in the theoretical development of the primeval fireball, which is of course the basis for predicting the Hubble relation in the Big Bang.

An expression for the potential (50,51) which does yield the correct attractive force is given by

$$(2) \qquad V(R) = - GM/R \quad - G \int_R^\infty 4 \pi \rho \, r \, dr \quad \text{where} \quad M = 4 \pi \int_0^R \rho r^2 dr \, .$$

The problem here is that for a finite, uniform density we encounter an infinite potential due to the presumed infinite size of the universe. This result is the same for both the Big Bang model and the RSS model.

Alternatively, a finite potential can be obtained from eq. (2) by assuming the density diminishes more rapidly than $1/R^3$ after R', where v = c. As a first approximation this assumption truncates the potential at R'. In this case the upper integration limits in eq. (2) must be changed from infinity to R', and we have the following potential:

$$(3) \qquad V(R) = - GM/R \quad - G \int_R^{R'} 4 \pi \rho \, r \, dr \quad \text{where M is defined in eq. (2).}$$

If this potential is used in eq. (1) to compute z for the RSS model, then for a uniform density for all R less than R', we find the redshift is zero. If, however, the density increases as $R^{0.22}$, then we can formally obtain a relation (51) similar to that deduced by Hawkins (34). Again, however, it is premature to make any claims about this result until more work is done.

Another possibility for obtaining redshifts in the RSS model is to assume the mass/energy density diminishes as $1/R^4$. In this case the galactic orbits are no longer circular but spirals, and there is a recessional component to the velocity which leads to a first order Doppler shift and a Hubble type $z \propto R$ relation. For this view to have any credibility most of the mass/energy of the universe must be in a form other than the matter and radiation energy presently observed and/or inferred in stellar systems and intergalactic dust. In this context it is perhaps worth mentioning that Ellis (52) has proposed that there may be a large amount of undetected mass/energy in other forms (e. g., neutrinos) which could raise the cosmic mass/energy density to more than a million times the present density estimates of 10^{-31} to 10^{-29} g/cm^3.

Of course the RSS model does not require that the redshifts are velocity dependent. In this respect it is well known that years ago proponents of a static or steady state universe proposed a variety of distance-dependent interpretations of the redshift which were non-recessional in nature (see North's (42) review for details and references). The investigation of the origin of the redshifts in the RSS model should include a reexamination of these alternatives.

Estimates of the Distance from the Milky Way to the Center

Earlier it was implied that the Milky Way could be one of the innermost galaxies in the RSS model. This view is based on the assumption that the Milky Way's cosmic galactic velocity of 550 km/s through the CMR (46) is just the tangential velocity of the Milky Way (MW) around C. Galactic peculiar motions may also be of the same nature. On this basis we can compute the angular velocity ω of the MW around C from $v^2 = \omega^2 R^2 = GM/R$, which leads to the result that $\omega = 2(\pi \rho G/3)^{1/2}$.

For a constant $\rho = 10^{-29}$ g/cm^3, then $\omega = 5 \times 10^{-11}$ rad/y, and the distance from C to our galaxy would be about 3.7×10^7 light-years. (C of course would be located somewhere in the plane perpendicular to the direction of the motion of the MW through the CMR.) If $\rho = 10^{-27}$ g/cm^3 then $\omega = 5 \times 10^{-10}$ rad/y (or 5×10^{-5} arc-s/y), which means that differential angular motions of the more distant galaxies (as observed at the MW) would still be below the present detection limit of light telescopes ($\approx 10^{-3}$ arc-s/y). In the latter case the distance from the MW to C is about 3.7×10^6 light-years and is considered the preferred value so as reduce potential blueshift effects. This distance places C outside our galaxy but still in the plane which is perpendicular to the MW's cosmic velocity vector. No observational data as yet seems to locate the direction of C in that plane. On the other hand Orion is in that plane, and is prominently mentioned in Scripture (Job 9:9;38:31, Amos 5:8). As a working hypothesis I suggest that C may lie a few million light years beyond Orion. One density used in the preceding calculations is higher than current estimates but, as previously noted, Ellis (52) has suggested there may be a large amount of undetected mass/energy which may raise the value to more than 10^{-24} g/cm^3. On this basis the higher density estimate is not unreasonable. In the RSS model the value of the density cannot much exceed 10^{-26} g/cm^3 or else the angular velocity will increase to the point where differential motions of distant galaxies would be observed.

The RSS Model and Olber's Paradox

We briefly digress to note that Olber's Paradox is resolved if the universe is structured according to the RSS model because the finite number of galaxies within a sphere of radius R' will only produce a finite light flux at the Milky Way. Even if there is luminous matter beyond R', the density is assumed to diminish so rapidly that the light flux received at the Milky Way from beyond R' will also be finite.

The RSS Model and Varshni's Analysis of Quasar Redshifts

In the context of the present proposal for the structure of the universe, it is most appropriate to refer to Varshni's (53) investigation of the redshift distribution of 384 quasars. From a probability analysis of those 384 qusars he found an astounding 57 sets of redshift coincidences within small redshift intervals. Varshni calculates the probability of chance coincidence of these groups to be about 10^{-85}. He concludes that if quasar redshifts are real (he thinks they are not) and are of cosmological origin (i. e., distance related), then the only logical deduction from the data is, in his own words, as follows:

> The Earth is indeed the center of the Universe. The arrangement of quasars on certain spherical shells is only with respect to the Earth. These shells would disappear if viewed from another galaxy or a quasar. This means that the cosmological principle will have to go. Also, it implies that a coordinate system fixed to the Earth will be a preferred frame of reference in the Universe. Consequently, both the Special and the General Theory of Relativity must be abandoned for cosmological purposes.

These deductions are amazingly similar to the deductions of the RSS model except that, first, the earth, or MW, is only astronomically close to rather than being exactly at the Center, and, second, the absolute reference frame is defined by the CMR and not the position of the earth. And from earlier discussions in this article, it should now be clear that the special and the general theory of relativity are not credible theories in the RSS model. In fact, as shown below, if anything it now appears that the results of one of the most celebrated experiments in the history of physics contradict the basic premises of both special and general relativity so directly that, to me at least, it seems these theories are no longer tenable. As noted earlier, however, just because special and general relativity are shown to be untenable does not invalidate all the mathematical results obtained by these theories. It suggests rather that there must exist an absolute space-time framework which would encompass all the results of relativity which do accord with experiment, but different results where relativity theory makes incorrect predictions. Several investigations pertaining to this alternative framework have already been cited (42-44). In addition we should also mention Clube's (54) work and his exchanges with others (55) on neo-Lorentzian relativity.

The RSS Model, the CMR, and the Theory of Relativity

Clube's (54) explanation for the CMR is undergirded by the assumption of a non-relativistic Lorentz invariant material vacuum. It is intriguing to consider that the CMR may be the result of emissions from a cold material vacuum. On a related matter, Clube cites other work (56) as evidence that observations are not at all inconsistent with an essentially Euclidean infinite cosmos. Certainly these ideas appear easily reconcilable with the RSS model since they assume the existence of an absolute reference frame. However, the details of Clube's theory have yet to be worked out so it is premature to make any claims until further work is done. Of course there is also the possibility that the CMR may be a part of the 'light' that was created in Gen. 1:3. Interestingly, Weisskopf (45) alludes to that very possibility in the closing paragraph of his recent article:

> Indeed, the Judeo-Christian tradition describes the beginning of the world in a way that is surprisingly similar to the scientific model. Previously, it seemed scientifically unsound to have light created before the sun. The present scientific view does indeed assume the early universe to be filled with various kinds of radiation long before

the sun was created.   The Bible says about the beginning: "And God
said, 'Let there be light'; and there was light. And God saw the light,
that it was good."

Irrespective of how it originated, the most important fact about the CMR
is that it represents unequivocal evidence of an absolute reference frame in the
universe, a very necessary condition in the RSS model, but an inconsistent
condition for the relativistic foundations of the Big Bang model. To explicitly
show exactly how this inconsistency arises, it is most helpful to include another quote from Weisskopf's recent article:

> It is remarkable that we now are justified in talking about an absolute
> motion, and that we can measure it.   The great dream of Michelson and
> Morley is realized.   They wanted to measure the absolute motion of the
> earth by measuring the velocity of light in different directions.
> According to Einstein, however, this velocity is always the same. But
> the 3 K radiation represents a fixed system of coordinates.   It makes
> sense to say that an observer is at rest in an absolute sense when the
> 3 K radiation appears to have the same frequencies in all directions.
> Nature has provided an absolute frame of reference.   The deeper signi-
> ficance of this concept is not yet clear.

With all due respect to my eminent colleague I suggest the meaning of this
fact is not obscure at all. I suggest the evidence (the CMR) which has received
worldwide acclaim as confirmation of the Big Bang is in reality its death knell
for, ironically, it is now clear that the existence of the CMR essentially
falsifies the fundamental postulates of the theory of relativity.   The logic is
quite straightforward. Referring to the last quotation by Weisskopf, we note he
mentions the famed Michelson-Morley experiment, which achieved only a null
result.   Lorentz's efforts to explain this null result on the basis of an
absolute reference frame were supposedly untenable. The real explanation, according to almost every physics textbook written in the past 60 years, was given
by the theory of relativity, namely that: <u>Given the null result of the Michelson-Morley experiment, if the fundamental principles of relativity are true,
then there is no absolute reference frame.</u> But the CMR is an absolute reference
frame, so the original relativistic deductions about the Michelson-Morley experiment are in error. More precisely, since logic requires the contrapositive of a
statement to be equivalent to the statement itself, the preceding "if relativity
is true, then no absolute reference frame" statement must be equivalent to "if
an absolute reference frame exists, then the fundamental principles of relativity are untrue." <u>In simpler terms the theory of relativity has been falsified
because a major prediction of the theory is now known to be contradicted by an
unambiguous experimental result.</u> Without relativity theory there is no Big
Bang, no Hubble relation for the redshift, and no explanation for the CMR in an
evolutionary cosmological model.

ACKNOWLEDGMENTS, REFLECTIONS, AND CONCLUSIONS

Special thanks goes to Drs. Frank Awbrey and Bill Thwaites, Biology Department, San Diego State University, for extending to me the opportunity of
participating in this symposium, and for their understanding and patience during
the revision of this contribution. Special thanks also to Dr. Alan Leviton,
Director, Pacific Division of the AAAS, who very kindly undertook the task of
translating my computer disks into a finished manuscript.

Several years ago the American Physical Society sent its members a copy of the National Academy of Sciences resolution of April 1976, "An Affirmation of Freedom of Inquiry and Expression," which reads in part ". . .That the search for knowledge and understanding of the physical universe and of the living things that inhabit it should be conducted under conditions of intellectual freedom, without religious, political or idealogical restrictions. . . . That freedom of inquiry and dissemination of ideas require that those so engaged be free to search where their inquiry leads. . .without political censorship and without fear of retribution in consequence of unpopularity of their conclusions. Those who challenge existing theory must be protected from retaliatory reactions."

In recent years the lofty aim of that resolution has not been realized as I have tried to pursue my research. In my opinion some of my more influential colleagues have found it easier to support this NAS resolution for foreign dissident scientists than for an American scientist who dissents from evolution. In fact I read in a recent issue of Science (57) that the NAS itself has recently stepped up its anti-creation campaign by the widespread distribution of a publication which claims that creationism is not science. I will present the opposite viewpoint in my forthcoming book (58) while also relating some details concerning my difficulties in pursuing research in this somewhat controversial field. The impact of aforementioned NAS resolution on my research efforts receives special attention.

In closing I wish to express my gratitude to those of my evolutionary colleagues who on so many occasions have assisted me, and on other occasions have collaborated with me in my research. Of one thing I am certain: Only in America could my research over the past two decades have been accomplished. I close by expressing gratitude to my Creator for allowing me the privilege of being an American. I submit this article to the scientific community not as an antagonist who purports to have the last word on the subject, but as a colleague who, in the spirit of free scientific inquiry, genuinely seeks a vigorous, critical response to the evidence presented herein. Perhaps a future "Evolutionists Confront Creationists" AAAS symposium would be the ideal forum for this exchange to occur.

REFERENCES

1. Gentry, Robert V. 1973. Annual Rev. Nucl. Sci. 23: 347.
2. Gentry, Robert V. 1974. Science 184: 62.
3. Gentry, Robert V. et al. 1974. Nature 252: 564.
4. Joly, J. 1917. Phil. Trans. Roy. Soc. London Ser. A. 217: 51-79; Idem. 1917. Nature 99: 457-58, 476-78; Idem. 1923. Proc. Roy. Soc. London Ser. A 102: 682-705; Idem. 1924. Nature 114: 160-64.
5. Kerr-Lawson, D. E. 1928. Univ. Toronto Stud. Geol. Ser. No. 27:15
6. Henderson, G. H., C. M. Mushkat, D. P. Crawford. 1934. Proc. R. Soc. Lond. Ser. A. Math. Phys. Sci. 158: 199.
7. Henderson, G. H., and L. G. Turnbull. 1934. Proc. R. Soc. Lond. Ser. A. Math. Phys. Sci. 145: 582.
8. Henderson, G. H., and S. Bateson. Ibid., Proc. R. Soc. Lond. Ser. A. Math. Phys. Sci. 145: 573.

9.   Schilling, A., 1926. Neues Jahrb. Mineral. Abh. 53A: 241. See also
     Oak Ridge Nat. Lab. Rep. ORNL-tr-697.
10. Mahadevan, C., 1927. Indian J. Phys. 1: 445.
11. Gentry, Robert V. 1968. Science 160: 1228.
12. Gentry, Robert V. 1971. Science 173: 727.
13. Gentry, Robert V. et al. 1973. Nature 244: 282.
14. Gentry, Robert V. 1978. Are Any Unusual Radiohalos Evidence for SHE? In
     Proc. International Symposium on Superheavy Elements. Lubbock, March 1978.
     Pergamon Press, New York, Oxford.
15. Gentry, R. V., W. H. Christie, D. H. Smith, J. W. Boyle, S. S. Cristy, &
     J. F. McLaughlin. 1978. Nature 274: 457.
16. Gentry, R. V. 1970. Science 169: 670.
17. Gentry, R. V., W. H. Christie, D. H. Smith, J. E. Emery, S. A. Reynolds, R.
     W. Walker, S. S. Cristy, & P. A. Gentry. 1976. Science 194: 315.
18. Henderson, G. H. and F. W. Sparks 1939. Proc. R. Soc. Lond. Ser. A Math.
     Phys. Sci. 173: 238. Henderson, G. H., ibid., p. 250.
19. Lind, S. C., and C. F. Whittemore. 1915. U. S. Bur. Mines Tech. Pap. 88:1.
     Stern, T. W., and L. R. Stieff, 1959. U.S. Geol. Surv. Prof. Pap. 320:151.
     Rosholt, J. N., 1958. In Proceedings of the Second U. N. International
     Conference on the Peaceful Uses of Atomic Energy, Geneva. United Nations,
     New York, vol. 2, p. 321.
20. Gentry, Robert V. 1975. Nature 258: 269.
21. Fremlin, J. H. 1975. Nature 258: 269.
22. Hashemi-Nezhad, S. R., et al., 1979. Nature: 178: 333-335.
23. Moazed, C., R. M. Spector, and R. F. Ward. 1973. Science 180: 1272.
24. Feather, N. 1978. Roy. Soc. Edinburgh Commun. 11: 147.
25. Gentry, Robert V. 1979. EOS 60: 474. Idem. 1980. EOS 61: 514.
26. Gentry, Robert V. 1982. Physics Today 35: No. 10, 13. Ibid., 36: No. 4,
     13.
27. Ramdohr. P. 1957. Abb. der Deutsch. Adad. d. Wiss., Berlin, Kl. f. Chem.,
     Geol. u. Biologie, no. 2: 1. See also Oak Ridge National Laboratory
     Translation (ORNL-tr-755).
28. Gentry, Robert V., T. J. Sworski, H. S. McKown, D. H. Smith, R. E. Eby, W.
     H. Cristie. 1982. Science 216: 296. See also R. V. Gentry. 1984. Sci-
     ence 223:835.
29. Zartman, R. E. 1979. Los Alamos Sci. Lab. Rep. LA-7923-MS.
30. Gentry, R. V., G. Glish, & E. R. McBay. 1982. Geophys. Res. Lett. 9:1129.
31. Silk, J. 1979. The Big Bang. W. H. Freeman & Co., San Francisco.
32. Weinberg, S. 1972. Gravitation and Cosmology. Wiley, New York.
33. Hetherington, Norriss. 1971. Astron. Soc. of the Pacific, Leaflet No.
     509, November.
34. Hawkins, G. S. 1962. Nature 194: 563.
35. Segal, I. E. 1976. Mathematical Cosmology and Extragalactic Astronomy.
     Academic Press. Idem. 1975. Proc. Nat. Acad. Sci. 72: 2473.
36. Nicoll, J. F. and I. E. Segal. 1975. Proc. Nat. Acad. Sci. 72: 4691.
37. Sandage, A., G. A. Tammann, and A. Yahil. 1979. Ap. J. 232: 352.
38. Nicoll, J. F. and I. E. Segal. 1982. Proc. Natl. Acad. Sci. 79: 3913.
     Idem. 1982. Ap. J. 258: 457. Idem. 1982. Astron. & Astrophys. 115: 398.
     See also Segal, I.E. 1982. Ap. J. 252: 37.
39. Nicoll, J. F. et al. 1980. Proc. Natl. Acad. Sci. 77: 6275.
40. Pound, R. V. and J. L. Snider. 1964. Phys. Rev. Lett. 13: 539. Idem.
     1965. Phys. Rev. 140: B788.
41. Alley, C. O. 1982. Proper Time Experiments in Gravitational Fields with
     Atomic Clocks, Aircraft, and Laser Light Pulses. In Quantum Optics,

Experimental Gravitation, and Measurement Theory. Edited by P. Meystre and M. O. Scully, Plenum Pub. Corp., New York.
42. North, J. D. 1965. The Measure of the Universe, Clarendon Press, Oxford.
43. Rastall, P. 1978. Astrophys. J. 22: 745. Idem. 1979. Can. J. Phys. 57: 944.
44. Marinov, S. 1981. Eppur Si Muove. East West Publishers, Graz, Austria.
45. Weisskopf, V. F. 1983. Am. Sci. 71, No. 5: 473.
46. Smoot, G. F. et al. 1977. Phys. Rev. Lett. 39: 898. Idem. 1979. Ap. J. 234: L83.
47. Einstein, A. 1939. Ann. Math. 40: 922.
48. Burcev, P. 1968. Phys. Lett. 27A: 623.
49. Gentry, R. V. 1983. Bull. of the Am. Phys. Soc. 28: 30.
50. Landsberg, P. T. and D. A. Evans. 1979. Mathematical Cosmology. Clarendon Press, Oxford.
51. Gentry, R. V. 1983. Phys. Today 36, No. 11: 124.
52. Ellis, G.F.R. et al. 1978. Mon. Not. R. Astr.Soc. 184: 439.
53. Varshni, Y. P. 1976. Astrophys. Space Sci. 43: 3. 1977. Ibid. 51: 121.
54. Clube, S.V.M. 1980. Mon. Not. R. Astr. Soc. 193: 385. Idem. 1982. Proceedings of an International Colloquium on the Scientific Aspects of the Hipparcos Mission, Strasbourg, France (ESA SP-177).
55. Clube, S.V.M. et al. 1980. Comm. R. Observatory, Edinburgh, No. 383: 467.
56. Jaakkola, T., M. Moles & J. P. Vigier. 1979. Astr. Na. 300: 229.
57. Holden, C. 1984. Science 223: 1274.
58. Gentry, Robert V. 1984. Creation's Tiny Mystery (to be published).

UNITED STATES DEPARTMENT OF THE INTERIOR
GEOLOGICAL SURVEY
Branch of Isotope Geology (ms 937)
345 Middlefield Road
Menlo Park, CA 94025

26 March 1985

Mr. Kevin H. Wirth
Director of Research
Students for Origins Research
P. O. Box 203
Santa Barbara, CA 93116

Dear Mr. Wirth:

   Tom Jukes sent me copies of your exchange of correspondence
and your Document 32 to PACIFIC DISCOVERY. I note that in the
latter you have incorrectly characterized my testimony at the
Arkansas trial by omitting nearly the entire substance of my
comments on Gentry's pleochroic halos. About my testimony you
say, "Although it is conceivable to test Gentry's findings, his
challenge was dismissed at the trial by USGS expert witness for
the ACLU Brent Dalrymple as a 'minor mystery' with prohibitively
high costs for building a test mechanism and that was that. No
more discussion." Your statement is untrue and since my
testimony is a matter of public record I am mystified as to the
source of your information. Please allow me to summarize what I
have said about Gentry's conclusions, both at the trial and at
the Santa Barbara AAAS meeting.
   First and foremost, Gentry's "hypothesis" is unscientific
because he proposes fiat creation of the Earth, and a
fiat-induced flood. In science, miracles are a no-no. Miracles
may occur (although I am unconvinced that they do), but they are
the subject of philosophy and religion, not science.
   Second, there have been several credible alternate
hypotheses advanced in the scientific literature for the origin
of anomalous Po-halos, including erasure or modification of the
inner halos by the Alpha radiation from another isotope, such as
Po-210, migration of uranium-series elements through the rocks by
either fluid migration or diffusion, and modification of halos by
heat, pressure, and chemical change during metamorphism. Gentry
has disputed all of these explanations but has disproved none of
them and I find his arguments unconvincing.
   Third, there are numerous problems with halo interpretation
and there is a distinct possibility that the Po halos are not
what they appear to be. Known difficulties include coloration

reverals due to saturation effects, attenuation of alpha particle
ranges by the radioactive inclusion, dose dependence of halo
radii, the lack of adequate data on the relation between energy
and distance in the various mineral types in which halos are
found, and the probable but unknown effects of crystal
imperfections and chemical impurities.

Thus, the identification and interpretation of the so-called
Po-halos is a very uncertain business. We don't know with
certainty, a) that they are Po-halos, or b) how they are formed.
This is the background of my description to the court of Po-halos
as one of sciences many "tiny mysteries"---mysterious because
their explanation is uncertain and tiny because I think that they
are a problem of minor importance (which explains why few
scientists bother with them). The fact that the origin and
interpretation of Po-halos is uncertain lends no credibility to
Gentry's unscientific "hypothesis". Science is full of
mysteries, that is why there are still employment opportunities
for scientists.

As for Gentry's "challenge", it is nonsense for several
reasons. First, the synthesis of a hand-sized piece of granite
in the laboratory would neither prove nor disprove Gentry's
"hypothesis", only demonstrate that someone had figured out the
technology and spent the money to make large pieces of granite.
Thus, for its stated purpose it would be a worthless experiment
and that is one reason why no serious scientist will take
Gentry's challenge seriously. The problems in completely
crystallizing igneous rocks in the laboratory are well known and
are due to scale, i.e., it is not always easy to reproduce in the
laboratory what nature requires hundreds of thousands or millions
of years to do. The principal difficulties are nucleation,
kinetics, time, and volume.

Gentry's insistence on a hand-sized piece also is somewhat
of a problem. Experimental petrologists find it most convenient
and sufficient to work with equipment of reasonable size and
cost. As a result, most high-pressure and temperature bombs use
charges of less than a gram in weight. An apparatus to
synthesize a hand-sized piece of an igneous rock would be immense
and costly, and as far as I know, no experimental petrologist has
found it necessary to build and utilize such an apparatus when
the smaller, less costly equipment serves the purposes of
science.

Second, Gentry's "challenge" is absurd because it is not
necessary to perform Gentry's experiment to prove his hypothesis
incorrect because it is already proven false . Gentry's main
point seems to be that granites (and here we don't know whether
he is using the term loosely to include all granitic rocks or
specifically to include only granite in the strict compositional
sense) do not cool from a liquid rock melt, but there is ample
proof that he is wrong. Igneous textures are distinct and can be
duplicated in the laboratory using a variety of materials,
including rocks. Igneous rocks with igneous textures can also be
observed forming in nature. One example is Kilauea Iki lava

lake, where drilling over a period of several years has recovered
a continuum of samples in progressive stages of crystallization
by cooling of a rock melt. Other examples include lava flows,
many of which crystallize completely within a year or so. The
textures of these lavas are virtually identical to granites,
which is not too surprising. Recall, if you will, that the
primary difference between volcanic and batholithic intrusive
rocks (granites) is that the former reaches the surface whereas
the latter does not. Furthermore, the sequence of
crystallization of minerals in granite, which can be determined
by any experienced petrologist for any given rock, invariably
agrees with the order predicted by thermodynamic calculations and
laboratory phase equilibria studies for minerals crystallizing
from a rock melt of granitic composition.

Third, Gentry seems to think that the Precambrian consists
entirely of "primordial" granites and that this "primordial
basement" is overlain by the stratified rocks of the world
(deposited during the Flood, of course). At best this view is
naive. The Precambrian consists of rocks of virtually every
type, including lava flows, glacial deposits, and continental
and oceanic sedimentary rocks. In fact, the oldest rocks on
Earth (3.5 to 3.8 billion years---Western Australia, Greenland,
southern Africa) are lava flows and shallow marine sedimentary
rocks. These are intruded by younger granitic rocks. There are
no "primordial granites" as per Gentry, or at least none have
been found.

As far as I am concerned, Gentry's challenge is silly. He
has proposed an absurd and inconclusive experiment to test a
perfectly ridiculous and unscientific hypothesis that ignores
virtually the entire body of geological knowledge. Science is
not required to respond to such a challenge and the fact that
Gentry's proposal has been ignored does not entitle him to any
claim to victory.

As you can see from the above (and as you should have
recalled from my remarks at the AAAS symposium), my objections to
Gentry's interpretation of Po-halos are far more numerous and
substantive than, as you say in your document, the "high costs
for building a test mechanism". You also imply that I have no
substantive objections to Gentry's proposal. In so doing you
have falsely represented my position and incorrectly reported my
testimony at the Arkansas trial. I trust that you will take
immediate steps to correct the error.

                              Yours truly,

                              G. Brent Dalrymple

cc: S. Warrick
     W. Bennetta
     T. Jukes
     W. Meikle

GENTRY RESPONDS TO DALRYMPLE'S LETTER
TO KEVIN WIRTH
Dalrymple's Letter Outlined Along With
Gentry's Comments to Each Point

I.   [Gentry's] hypothesis is unscientific because it assumes fiat creation of the earth and a worldwide flood, according to Genesis.

     Gentry: In his talk at the AAAS symposium at Santa Barbara Dalrymple referred to science as "...that magnificent field of objective inquiry whose only purpose is to decipher the history and laws of the physical universe..." A "field of objective inquiry" implies that scientists are searching for the truth. If Dalrymple is really looking for the truth, he has no logical basis for a priori excluding the possibility of creation. Scientists should keep an open mind to all possibilities and make decisions on the weight of the evidence.

II.   Several credible alternate hypotheses advanced to explain Po halos:

   A. Erasure of inner halos by alpha-radiation from another isotope, such as Po-210

     Gentry: I have shown (Gentry 1978a and 1978b) that erasure of inner halo rings occurs at extremely high doses when reversal effects are apparent. This is an extremely rare occurrence that is easily discernible by microscopic observation and can be confirmed by scanning electron microscope x-ray fluorescence methods which show the reversed region is characterized by a calcium and potassium inversion anomaly. There is no basis for claiming erasure of inner halo rings has caused misidentification of Po halos.

   B. Migration of U-series elements through rocks by diffusion (secondary hypothesis)

     Gentry: I have investigated the hypothesis of the secondary origin of polonium halos in granites from uranium daughter activity and have shown (Gentry 1968; Gentry 1976a; Gentry 1984a) that this hypothesis has no basis in fact. Dalrymple presents no new data to support his comment; so it must be assumed that this comment is similar to the one he made at the Arkansas trial. When Attorney Williams asked whether he had done any investigations to support such comments, he responded negatively. It appears he is doing the same in his correspondence to Kevin Wirth.

   C. Modification of halos by heat, pressure, and chemical change during metamorphism

     Gentry: Halos occur in many mica samples which have not undergone metamorphism of any kind; so it is useless to object that halos have been modified for that reason. Proof of this is demonstrated by the fact that Po halos occur in the same mineral specimens with well-defined uranium and thorium halos, which show no modification of their ring structure. (See, for example, Gentry 1968; Gentry 1971b; Gentry 1973a; Gentry 1974a; Gentry 1978; Gentry 1984a.)

III. Problems with Interpretation of Po Halos

   A. Coloration reversals due to saturation effects

     Gentry: Years ago I showed (Gentry 1973a and Gentry 1974a) that Po halos exist in all stages of coloration, ranging from those which are barely discernible to those which are densely colored. Saturation effects, if they were to exist at all, would only occur with those halos that were densely colored. There is no basis to the claim that saturation effects are a cause for misinterpreting Po halos.

   B. Attenuation of alpha-particle ranges by radioactive inclusion

Gentry: At the Arkansas trial, my colleague admitted he had read virtually none of my technical reports on radioactive halos. His suggestion that alpha-particles may be attenuated by the finite size of the Po halo radiocenter suggests that he still hasn't read my reports, or if he has, he hasn't read them very carefully. Had he done so, he would have learned that Po halo radiocenters in micas are typically extremely small, about just one or two micrometers in size. Uranium halos with radiocenters this small show excellent definition of all the uranium halo rings because there is virtually no attenuation of the alpha particles. Since the energies of a number of the uranium-series daughters are the same as for the Po halos, it likewise follows there is virtually no attenuation of the alpha particles by Po-halo inclusions.

C.　Dose dependence of halo radii

Gentry: I have reported on a long series of helium-ion irradiations of several minerals and documented in detail the dependence of coloration on the alpha dose (Gentry 1973a and Gentry 1974a). The coloration bands measured at various doses and energies were then shown to correspond almost exactly to the measured values of the corresponding halo radii. Thus there is no uncertainty about Po-halo identification relating to the alpha dose.

D.　Lack of data on relation of energy and distance in the various mineral types in which halos are found

Gentry: The same comments apply here as in (C) above with the additional statement that the reports mentioned contained exactly the information on the relationship of energy and distance that Dalrymple seems to feel is in question. Again I ask: Has he even read my reports?

E.　Unknown effects of crystal imperfections and chemical impurities

Gentry: As Dalrymple well knows, there are crystals of various minerals, which are well-nigh perfect and others which have many crystallographic imperfections and chemical impurities. I have made it a practice to perform my halo studies using good mineral specimens. It is a simple matter to avoid the poor specimens. Moreover, I should again point out that Po halos are found in the same mineral specimens with well-defined uranium and thorium halos. Crystal imperfections did not affect the structure of the uranium and thorium halos; neither did they affect the structure of the Po halos.

F.　Conclusion of Interpretation - Tiny Mystery

Tiny - because halos are problem of minor importance

Mysterious - explanation is uncertain

This explains why few scientists bother with them.

Gentry: The net result of Dalrymple's evaluation is that Po halos in granites are only a tiny mystery. To him and many others, they may be only this, but the fact remains they cannot be explained on the basis of uniformitarian evolutionary principles. Something so tiny should already have found a rational explanation within the realm of conventional science, if indeed one was to ever have been found. No, more than that, since the secondary origin of Po halos from uranium is the favorite candidate for explaining Po halos in granites, we must ask why no one has artificially produced a Po-218 halo in granite. The radioactivity necessary to do the experiment is available as is the rock itself. So what is the barrier in reproducing a tiny mystery such as a Po-218 halo if indeed it can be done by man? I suggest the Po halos are mysterious only to those who wish to exclude the activity of the Creator of the universe to His own creation. Perhaps scientists should awaken to the possibility that the Creator is attempting to attract their attention by this paradoxical, tiny mystery that continues to confound giant intellects in science.

IV. Discussion of Challenge
   A. Nonsense for several reasons
      1. Synthesis of hand-sized piece of granite would neither prove nor disprove hypothesis.

       Gentry: As has been pointed out a number of times in this book, confirmed evolutionists have essentially dug their own graves by insisting on the universal application of the uniformitarian principle. If evolutionary theory is right, the Precambrian granites formed numerous times over the vast expanse of time during which the earth was evolving, and this was presumably being done solely by the action of the same physical laws that are operating today. It is inescapable, therefore, that it should be possible to reproduce today by design what nature presumably did just by chance.

      2. Problems in crystallizing igneous rocks in laboratory are a) due to scale; i.e. - nucleation, kinetics, time, and volume; and b) hand-sized piece is a problem because it would involve immense and costly apparatus.

       Gentry: As we showed in Chapter 9, Dalrymple's contention that he knows why it has thus far been impossible to synthesize a granite is based on his own view of Earth's history, namely, that the granites crystallized slowly over geological time. There we also noted that if nature was supposedly successful in overcoming the obstacles of nucleation and kinetics numerous instances during the course of geological time, there is no reason why these obstacles should not be surmounted in the modern scientific laboratory. He refuses to admit that the impossibility lies, not in technological factors, such as those he mentioned, but in the fact that the Precambrian granites are the Genesis rocks of the earth, made by the Creator in such a way it is impossible to reproduce them without His intervention. Finally, at the Arkansas trial Dalrymple admitted that geologists had failed to synthesize even a tiny piece of granite. So why does he now claim that the problem in granite synthesis is related to its size?

   B. Absurd because:
      1. Unnecessary to falsify hypothesis because it is already proven false

       Gentry: My hypothesis is that the Precambrian are the Genesis rocks of the earth, created by God in such a way that they cannot be duplicated without His intervention. Dalrymple apparently is claiming my view of these rocks has already been proven false. Where is the proof? There is no proof! What Dalrymple calls a disproof of my views relates to his flawed comparison of the Kilauea-Iki lava specimens with granites, as was discussed in Chapter 10.

      2. Wrong in saying granites do not cool from a liquid melt
        a. Igneous textures are distinct and can be duplicated in laboratory using rocks. Igneous rocks with igneous textures observed forming in nature; e.g., (1) Kilauea Iki lava and (2) lava flows - texture virtually identical to granites.

       Gentry: This is the so-called "proof" that my hypothesis is wrong. The inference of these comments is that there is a lot of similarity between the Kilauea-Iki samples and granites. True, Dalrymple claims that only the texture is the same, but in Chapter 10 we showed that only one aspect of the texture is similar—the intergranular structure—whereas the grain size is considerably different between the lava lake samples and the granites. Moreover, we also showed in Chapter 10 that the samples are grossly different in bulk composition and mineralogy, meaning there is little similarity between the Kilauea-Iki lava lake samples and the granites.

    b.   Sequence of crystallization of minerals in granite agrees with the order predicted by thermodynamic calculations and laboratory phase equilibria studies for minerals crystallizing from rock melt of granitic composition.

> *Gentry: In Chapter 10, I pointed out that my creation model envisions a primordial liquid as a precursor of the Precambrian granites. But there is nothing in my model which prohibits the Precambrian granites from having a sequence of crystallization that agrees with thermodynamic calculations. So Dalrymple's argument that granites came from a melt is no argument at all against the Precambrian granites being among the primordial Genesis rocks of our planet.*

  C. Naive because Gentry claims Precambrian consists entirely of "primordial" granites, overlaid by stratified rocks of the world deposited by flood. Actually, Precambrian consists of every type of rock including lava flows, glacial deposits, and sedimentary rocks. Oldest rocks in world (3.5 - 3.8 b.y.) are shallow marine sedimentary rocks. These are intruded by younger granitic rocks.

> *Gentry: Here Dalrymple argues against a "straw man" creation model. In Chapter 10 I explained in detail that my creation model is much broader and envisions many more possibilities for the formation of various rock types than Dalrymple considers to be the case. In particular, I explained that the Genesis record of creation week and the subsequent events of the world-wide flood encompass, in addition to the primordial created rocks such as the Precambrian granites, the formation of pristine sedimentary rocks, lava-like rocks, the intrusion of granite-like rocks into pristine sedimentary rocks, and almost unlimited possibilities of mixing these various rock types with secondary rocks that were formed at the time of the flood.*
>
> *Dalrymple also refers to Precambrian glacial material, apparently for the purpose of attempting to cast doubt on my creation model. The reader should understand that just because geologists designate something as Precambrian doesn't automatically mean it has any connection with the primordial events of Day 1, or for that matter, of creation week. In the case of the Precambrian granites it does have a connection; in other cases it may not. Investigation on a case-by-case basis is needed before it can be decided whether something called "Precambrian" can be connected to the events of creation week.*
>
> *So the mere existence of what Dalrymple refers to as Precambrian glacial deposits does nothing to detract from the solid identification of the Precambrian granites as the primordial rocks of our planet. I should also remark that whatever it is that Dalrymple is classifying as glacial material may or may not ultimately prove to be glacial material at all. Additional information about my creation model is given in Chapter 14. That model includes the possibility that some granites may have been created on Day 1 adjacent to and immediately after some primordial or pristine "sedimentary" rocks were created. Perhaps this is what Dalrymple refers to as granites intruding ancient sedimentary rocks.*

V. Conclusion

  A. Gentry's challenge is silly; synthesis test is absurd and inconclusive; hypothesis is perfectly ridiculous and unscientific, ignoring virtually the entire body of geological knowledge.

> *Gentry: I agree that my discoveries upset virtually the entire body of geological knowledge. My colleague is obviously concerned, as many other scientists have been over the past 20 years, because of the implications of my research. The falsification test puts evolutionists on the defensive, and naturally a human reaction is to recoil with negative rhetoric. The important point to be emphasized is that*

instead of relegating the phenomenon of polonium radiohalos to the realm of anomalies, scientists should admit that the evidence exists and deal with it objectively.

B.   Science is not required to respond to such a challenge, and the fact that Gentry's proposal has been ignored does not entitle him to any claim to victory.

Gentry: Science deals with reality. Polonium halos in granites are real—they will not disappear because evolutionists ignore them. I have not claimed victory— only the discovery of irrefutable evidence for creation.

SCIENCE

PUBLISHED BY THE AMERICAN ASSOCIATION
FOR THE ADVANCEMENT OF SCIENCE
1333 H STREET, N.W., WASHINGTON, D.C. 20005
(202) 326-6500

CABLE ADDRESS ADVANCESCI

30 August 1985

Dr. D. Russell Humphreys
Division 1252
Sandia National Laboratories
Albuquerque, NM 87185

Dear Dr. Humphreys:

Thank you for your letter of 30 July. It is true that we are not likely to publish letters supporting creationism. This is because we decide what to publish on the basis of scientific content.

The letters we received objecting to the study reported by Roger Lewin contained arguments that were largely conjectural or anecdotal. They were therefore not considered acceptable material for Science.

Yours sincerely,

Christine Gilbert
Letters Editor

CG:jp

TRANSCRIPTION OF ROBERT V. GENTRY'S
CROSS-EXAMINATION FROM AUDIO TAPE
McLean vs. Arkansas State Board of Education
Little Rock, December 17, 1981

Line

Mr. Ennis (Attorney for the ACLU):

1

Q   Dr. Gentry, didn't you get the opportunity to review the transcript of your deposition on November 24 of this year and make every correction that you considered appropriate?

A   Yes, I did.

Q   You have been a member of the Creation Research Society since the formative date of that organization, haven't you?

A   Yes.

Q   I believe you testified on direct examination that you do subscribe to the statement of beliefs of the Creation Research Society, is that correct?

A   Yes, I do.

Q   You believe, then, that all the assertions of the Bible are scientifically true in all of the original autographs. Is that correct?

A   My understanding is that all the assertions in the Bible which pertain to science would be true.

Q   Do you believe that Genesis is literally true?

A   I believe that the Bible record in Genesis is a factual account of the creation narrative.

Q   You believe that the description of creation in Genesis is

20   literally true, do you not?

A   Yes.

Q   You believe that the earth was created in six literal days, do you not?

A   Yes, I do. Twenty-four hours.

Q   Do you believe that only by scriptural chronology can we determine how long ago that six-day period was?

A   Yes, I do.

Q   Isn't one of the primary reasons that you began to rethink the entire issue of evolution and creation is because of the moral perspective of the Fourth Commandment?

A   Absolutely.

The judge asked about the question and response.

Mr. Ennis: The question was, Your Honor, was it not true that one of the major reasons that Dr. Gentry began to rethink the whole issue of evolution or creationism was because of the moral perspective of the Fourth Commandment? And that is—that is— [inaudible]

Gentry: Yes.

Q   It is fair to say, is it not, Dr. Gentry, that for several years you tried but were unable to find scientific evidence that would support the information contained in the book of Genesis?

A   Well, there were a number—there was a period of time in which after I began my research—certainly there was a period of time that I was looking for evidence for which I didn't have. That was the whole purpose, of course, in beginning the research. So with very few exceptions I was taking what other people said to be true about time. I think—if you would you repeat that statement—I think I am in essence agreeing with what you said.

Q   Well, let me ask it this way. Is it fair to say that for several years you tried but were unable to find scientific evidence that would support the information contained in the book of Genesis?

A   Well, let me give you the entire perspective so that we don't infer the statement you're giving me is true here. I became a Seventh-day Adventist in '59. I—from that time on until the time I actually began my research in 1963, I was puzzling in my mind about things concerning creation, reading books and trying to find, trying to find out from other people if indeed there was evidence—

Q   I recall that in your direct testimony, Dr. Gentry. I'm not asking you to say that that's your belief today. I'm simply asking you, is it not true that for several years you tried but were unable to find scientific evidence to support the information contained in the book of Genesis?

A   Okay.

Q   Is that true or not?

A   Okay if we use the word "tried" to define my thinking from 1959 to 1963—generally speaking in 1963 before I began this research I thought there was a paucity of evidence, and after I began this research in 1963—

Q   I am not asking you after that time, I am simply asking you if that was true up to that time—

A   From 1959 to 1963 I understand I think there was a paucity of evidence—

Q   You tried to find some and had not found it—

A   Well, I was reading what other people said—...[inaudible].

Q   You testified in your direct examination—I think you will quite candidly acknowledge—that you had what you described as a biased perspective—that you were interested in finding evidence if it exists.

A   Yes.

Q   And your question was, did my religious belief have any evidence in science, this is what I was very much interested in [inaudible].

A   Absolutely, yes.

Q   And didn't you then decide that the only way you could rationally live with yourself would be to undertake a research project to determine if there was evidence to support your belief(s) in Genesis?

A   I think this is true.

Q   So you began your research into radiohalos as a result of your inquiries into the Bible and of becoming a Seventh-day Adventist.

A   Absolutely.

Q   Is it fair to say that the last person before you to do any substantial work on radiohalos was Henderson in 1939?

A   I think so, yes.

Q   The research you do is very specialized and requires quite sophisticated equipment including ion microprobe, microprobe spectrometers, cyclotrons, and other equipment like that, does it not?

A   Well, initially all you have to have is a microscope, a razor blade, and a piece of rock. That's all it takes to find the halos. Now, to actually demonstrate the experimental support for what I've said, you do need sophisticated equipment. But Henderson identified the polonium halos basically only with the, only with using the microscope.

Q   Dr. Gentry, let me ask you this. Are you aware of any changes in the constancy of alpha decay or beta decay rates that have been identified experimentally?

A   No, I'm not. At the present time, no.

Q   You testified to some extent about singularities. You said that singularities were something that could not be explained on the basis of known physical laws.

A   This is how I formulate the hypothesis of the Big Bang versus the creative event.

Q   Given our current understanding, would it be fair to say that the singularity would have to be thought of as an extension of natural law?

A   I think that's fair. Yes.

Q   Is it your present opinion that there is no physical process short of a singularity which could cause any significant alteration of radioactive decay rates?

A   Yes, I agree with that.

Q   You believe the occurrence of a worldwide flood was the result of such a singularity or extension of natural law, do you not?

A   Yes, let me qualify and say that when I say extension of natural law, what I am basically saying is the processes in operation at that time were above and beyond what we normally consider today. Yes.

Q   And you believe that those processes were caused by the direct intervention of the Creator?

A   Yes, I do.

Q   In 1976, you published a paper suggesting that there was evidence for primordial superheavy elements—

A   Yes.

Q   And because that paper questioned more conventional understandings, it did receive wide notice, did it not?

A   Well, now the reason that the paper on primordial, primordial superheavy elements elicited a lot of interest is because people had been looking for and had spent a lot of money looking for superheavy elements for 10 years at least. And so whenever I said anything about superheavy elements, it was like ringing a bell all over the world—it wasn't necessary that it had to be primordial, although—

140

Q   Well, let me ask it this way. If the existence of primordial superheavy elements had been confirmed, that would have required drastic revisions of many existing ideas concerning nucleosynthesis and nuclear theory, would it not?

A   This was generally understood to be the case, depending on what element it was.

Q   When your data was re-examined using more sensitive techniques, it was found that superheavy elements were not present, is that correct?

A   Well, the techniques that we used to re-examine—actually the original results were made, of course, using protons, and the people who did—

Q   Dr. Gentry, I am not asking you about which techniques were used. I am simply asking you, it is true, is it not, that when your data was re-examined using more sensitive techniques that it was found that superheavy elements were not present. Is that true or false?

A   What I am trying to tell you is that in examining the inclusions again we used the same techniques we used to begin with. So

160

it wasn't that we necessarily had to use more sensitive techniques— ...[partly unclear or inaudible]. It turns out we didn't do, the people doing the [original] experiments didn't [properly] do the blank background experiments. [Note: The original experiments used protons and the later ones used x rays to fluoresce the giant-halo inclusions, but x-ray analysis was the detection technique used in both experiments.]

Q   You testified at some length about a letter from the National Science Foundation, July 11, 1977, which denied your application for a particular grant.

A   Yes.

Q   Is it not fair to say that that letter concluded that one of the reasons they denied your grant application at that time was that the panel felt that you and your colleagues were to be faulted for the techniques you used in coming to your initial conclusion that there were superheavy elements?

A   Yes, I believe it did say that.

Q   Did not that rejection letter go on to say that the panel felt that the principal investigator and his colleagues should have checked out all such possible reactions before publication because we know that that technique might produce the results you found? Is it not true?

A   I think what you are saying is generally true.

Q   Now, Dr. Gentry, I am not trying to embarrass you on this point because you yourself candidly acknowledged by your own admission—

A   There is no problem.

Q   You acknowledged by your own admission, did you not, that the evidence described in that earlier paper was not due to superheavy elements but was due to a more conventional phenomena?

A   That's right.

Q   That's not the only time you have published conclusions you later retracted, is it?

A   No, that is right.

Q   In fact, didn't you once invent new alpha activity to account for some ghost rings in radiohalos?

A   Yes, if you are referring to the slides of the Wölsendorf fluorite—the slides that I showed yesterday—yes.

Q   And did you not later acknowledge that you erred in inventing new alpha activity?

A   ...[inaudible] I surely did.

Q   So you have published conclusions in this field before which later have turned out to be wrong.

A   Which I later said that were wrong, yes.

Q   In August of this year did an attorney named Wendell Bird ask you if you would be willing to testify for the State in this case?

A   Yes, he did. We discussed that.

Q   You would concede, would you not, that a scientist can have observations in accord with a theory but that would not necessarily confirm the proof of it.

A   That is correct.

Q   Henderson's theories do explain the existence of Po-210 halos even in the absence of uranium halos in coalified wood in a conventional, natural-law way, do they not?

A   ...[inaudible] no. There are uranium halos and polonium halos in coalified wood.

Q   But not occurring exactly in the same halo rings. There is migration, is that not correct?

A   Yes, the uranium halos and the polonium-210 halos are different.

Q   Yes, that's what I am asking. You mentioned in your testimony some scientists, I believe you mentioned Wheeler—

A   Yes—

Q   And Anders. Is it not true that Wheeler and Anders and other scientists who have read your material think that a conventional natural law explanation will be found for the existence of other polonium halos in granites?

A   Yes, they do.

Q   I have no further questions.

Transcribed from audio tape recorded by P. Merkel, Official Court Reporter, U.S. District Court, Little Rock, Arkansas.

INDEX

igneous, 127
lava, 127, 129–131
lunar, 44
magma, 2, 98, 130, 185
metamorphic, 128, 133, 184
pegmatites, 131
primordial, 2, 30, 32–34, 45, 71,
129, 133, 152, 179, 185
primordial liquid, 129, 184
rhyolite, 130–131
sedimentary, 33–34, 36, 52–53,
133–134, 184
synthetic, 66
time of formation, 14
Rossbach, P., 166–167
Roth, Ariel, 148
Ruse, M., 146

Sasser, Sen. Jim, 166–169
Scanning electronic microscope, 47
x-ray fluorescence capabilities, 47
Science, 41, 43–47, 57, 59, 70, 90, 145,
152, 154–155, 165, 186, 188–194
Science 81, 91–92
Science and Creationism, NAS booklet,
8–9, 194–197
Science education. *See* Public education
Science News, 75, 84–85, 97, 159–160,
165
Scopes, John, 87–90
Scopes Trial, 1, 87–90, 92, 107, 136
Scott, E.C., 186–190, 194
Sinclair, Rolf, 84–85, 96–97, 175
Skow, J., 92
Smith, S., 101–105, 109–110, 111–122
Solar nebula, 34
Sparks, C.J., 70, 75
Spilhaus, A.F., 63–64, 66–69, 71
Stars, star formation, 9, 13, 34, 178.
See also Evolution model
Stieff, L.R., 53
Strother, D., 173
Students for Origins Research (SOR),
178
Stutzer, O., 59
Superheavy elements, 74–76, 78,
156–157, 171
SuperHeavy Element Project (SHEP), 74
Supernova(e), 14, 29–32, 34

Talbott, S.L., 47, 141, 159
Taxes, to support evolution, 6, 85,

182–183
Ten Commandments, 12, 150, 201
Thrash, C.L. and A.M., 25
Thwaites, William, 175, 178
Tiny Mystery. *See* Radioactive halos
Todd, E.P., 77–78, 83

U.S. Geological Survey, 107, 130, 175
Uniformitarian principle, 10, 15, 33,
53, 65, 81, 106, 109, 116, 134, 149,
151, 176, 179, 181, 183, 195, 199
University of California-Davis, 74–76
University of Florida, 12–13
Uranium decay chain, 18, 26, 164, 170

Vulcanism, 133, 185

Walker, Rep. Robert, 82–83
Walsh, R.E., 198
Wheeler, John, 158
Wickramasinghe, N.C., 160
Williams, David, Arkansas Deputy Attorney General, 93, 110, 124, 133,
172
cross-examination of Dalrymple,
111–122
recross-examination of Dalrymple,
134–135
Wirth, Kevin, 182, 296

X-ray fluorescence studies, 47, 157

York, Derek, 66–71, 143

Zircon. *See* Minerals

About the Author:

Robert V. Gentry (D.Sc., Hon.) is a research physicist whose area of expertise is the geophysical phenomena of radioactive halos. For thirteen years he was a visiting scientist in the Chemistry Division of the Oak Ridge National Laboratory, from Columbia Union College, Takoma Park, Maryland. After completing an M.S. in physics at the University of Florida in 1956, Gentry spent several years in the defense industry (General Dynamics-Fort Worth and Martin Marietta-Orlando) and in college and university teaching.

Dr. Gentry has authored or co-authored over twenty research papers in scientific publications. He is a member of the American Association for the Advancement of Science, the American Physical Society, the American Geophysical Union, Sigma Xi, the New York Academy of Sciences, and is listed in Who's Who in America.

The writing of *Creation's Tiny Mystery* spanned three and a half years, and upon its completion he plans to again return to full-time research. He feels quite fortunate that his wife, daughter and two sons have been very supportive of his research because his work is his only hobby.